Preparing for Biological Terrorism

An Emergency Services Planning Guide

GEORGE BUCK, Ph.D.

THOMSON LEARNING™

Australia Canada Mexico Singapore Spain United Kingdom United States

Preparing for Biological Terrorism: An Emergency Services Planning Guide

George Buck, Ph.D.

Delmar Staff

Business Unit Director:
Alar Elken

Executive Editor:
Sandy Clark

Acquisitions Editor:
Mark Huth

Editorial Assistant:
Dawn Daugherty

Executive Marketing Manager:
Maura Theriault

Channel Manager:
Mary Johnson

Marketing Coordinator:
Erin Coffin

Executive Production Manager:
Mary Ellen Black

Production Manager:
Larry Main

Production Editor:
Betsy Hough

Cover Design:
Michael Egan

Production Services:
TIPS Technical Publishing, Inc.

Printed by:
Phoenix Color

Printed in the United States of America
1 2 3 4 XXX 04 03 02 01

For more information contact Delmar at 3 Columbia Circle, PO Box 15015, Albany, NY 12212-5015.

Or find us on the World Wide Web at http://www.delmar.com or www.fire-science.com

ISBN: 1-4018-0987-1

To those serving to protect America—on the front lines, behind the scenes, overseas, and at home.

CONTENTS

Foreword xi

Preface xiii

Acknowledgments xv

Chapter 1 Overview of National and International Terrorism 1

Overview 1
Types of Terrorists 3
Terrorist Targets 5
Recognizing Terrorist Events 5
An Overview of International Terrorism 6
A Brief History of Terrorism 15
Response Overview, Strategies, and Tactics 18
Emerging Response Planning and Controversial Issues 19
Summary 20

Chapter 2 The DNA of Emergency Management Planning: The Basic Concepts of Emergency Management 21

Overview 21
Emergency Management Agencies 22
Fire Service Involvement 23
Factors that Affect Emergency Management 24
Threats Facing the United States 25
Types of Threats 26
Ranking of Threats 26
Potential Hazards 27
The Changing Context: A Window of Opportunity 28
The Emergency Manager 29
The IEMS Overview 32
The IEMS Concept 33
The IEMS Process 34
Starting the Real World Planning Process 36

Hazard (Risk) Analysis 37
The Planning Process 38
The Cascade Effect 38
Capability Assessment 40
Setting Goals 43
Getting Organized 43
Multi-hazard, All-Hazard, Functional Planning 44
History of Incident Command System 44
Summary Exercising 46
Summary 47

Chapter 3 The Basics of Emergency Management: Putting It All Together 49
Overview 49
Your Role as the Emergency Manager 50
Job Title and Organizational Chart 51
Hazard Identification and Planning 51
Staffing Issues 51
Office and Professional Duties 52
Resource Lists 52
Management 53
Information 53
Leadership 53
Professional Training 53
Non-Emergency Management-Related Duties 54
Responsibilities to Employer or Organization 54
Planning 54
Why plan? 55
The Case for Planning 55
Planning Guidelines 55
Writing the Plan 56
Planning vs. Improvisation 57
Plan Components 57
Technical Writing 57
What Your Local Emergency Operations Plan Should Include 58
Emergency Support Functions (ESFs) 59
Terrorism Consequence Management Plan 59
Summary 64

Chapter 4 System Implications **65**
Overview 65
Interoperability and Self-Sufficiency 66
Responder Casualties 67
Treatment Roles and Responsibilities 67
Summary 67

Chapter 5 Event Management **69**
Overview 69
Incident Operations 70
Emergency Operations Center Operations 72
Metropolitan Medical Response System (MMRS) 73
Integration with Federal Government 73
Integration with State and Local Government 77
Medical Management 77
Medical Facilities 79
MMRS Organizational Chart 80
Hospital Emergency Incident Command System (HEICS) 80
Summary 88

Chapter 6 Local Plans and Resources **89**
Overview 89
Local Emergency Operations Plan 90
Local Hazards Analysis 90
Vulnerability Analysis 91
Functional Annexes 91
Local Resources and Special Teams 93
Community Emergency Response Teams (CERTs) 94
Disaster Community Health Assistance Teams (DCHATs) 98

Chapter 7 Federal Medical Resources **101**
Overview 101
Federal Response Plan Assumptions 102
The Plan's Concept of Operations 104
The National Disaster Medical Service 117
Historical Development of The NDMS 121
Recent Experiences of the NDMS 124
Other Resources from the Public Health Service 129

Federal Urban Search and Rescue Resources (US&R, USAR) 129
Task Force Leader 130
Search Team 130
Rescue Team 130
Medical Team 131
Technical Team 132
Summary 133

Chapter 8 Medical Consequences 137
Overview 137
Introduction 138
B-NICE 138
Patient Care Mainstays 139
Physiological Effects of Terrorist Weapons 143
Nervous System 144
Blood, The Mediator of Inflammation 154
Mass Patient Decontamination 159

Chapter 9 General Biological Terrorism Concepts 163
Overview 163
Biological Response 164
Integrating Federal, State, County, and Local Responses 165
Concept of Operational Management Issues 170
Emergency Operations Center 170
Crisis Management 170
Situation and Threat Background 171
Incident Types: Package, Covert Release, Threat 171
Threatened Use 172
Confirmed Presence 173
Actual Release 173
The Agents 173
Planning Strategies 174
Administration 175
Plan Maintenance 175
Outline of Biological Response Operations 176
Conduct Internal Notification 176
Determine the Course of Action 176
Action Planning 176

Conduct Public Notifications 179
End Event 179
Stages of Severity 179
Strategies and Actions 180
Decision Factors 180
Specific Treatment Models 181
Biological Terrorism Operations 182
Suspicious Outbreak of Disease Notifications 182
In the Event of an Alert 183
In the Event of a Warning 183
In the Event of an Actual Incident (Response) 185
Threatened Use 185
Confirmed Presence 186
MMRS Biological Agents 186
The Historical Biological Agent Scenarios as a Baseline 186
Treatment vs. Prophylaxis 188
Discovery of Agent Prior to Release 188
Actual Release 190
Initial Response Concerns 194
Suspicious Outbreak of Disease 195
Synopsis of BW Agent Characteristics 197
BW First Responder Concerns 198
BW Agent Dissemination 198
Weather Effects 199
Decontamination (Decon) Considerations 199
Recovery Concerns 200
Site Decontamination and Restoration 201
Oversight 201
Investigation 201
Summary 201

Chapter 10 Planning and Action Guidelines 203
Overview 203
Passive Surveillance 204
Active Surveillance (Epidemiological Services) 204
Biological Terrorism Stakeholders Group 206
Detection 207
Agent Surety 208

Notification 214
Diagnosis 214
Mass Prophylaxis 216
Medical Response Expansion Program 225
Logistics for the National Pharmaceutical Stockpile Program 227
EMS Resources 235
Hospital Plan 235
Emergency Support Function #8 (ESF#8),
Health and Medical 236
Mass Fatality Management Implementation 237
Medical Examiner Expansion Program 238
Environmental Clean Up 240
Summary 241

Bioplan References 243
Appendix A Check List for Agencies 245
Appendix B Public Safety Precautions/Actions 249
Appendix C Public Health Anthrax Threat Advisory 259
Appendix D Complete Agent Descriptions 263
Appendix E Area Hospital Listings 271
Appendix F County Public Health Clinics/Centers 273
Appendix G An Anthrax Threat Field Guide 275
Appendix H Pharmaceutical Needs for the
Five Types of Biological Agents 277
Appendix I Pharmaceutical Push Package Contents 283
Appendix J Biological Agent Signs and Symptoms 285
Appendix K Israel's Fixed Hospital Decontamination System ... 287
Appendix L Sample Domestic Preparedness Training Courses:
Public Health Focus 295
Appendix M Biological Incident Preparedness Training 297
Appendix N Sample School Emergency Plan 311
Appendix O Internet Resources for Terrorism/
Disaster Planning 321

Acronyms 329

Glossary 339

Index 357

The unprecedented terrorist attack on September 11, 2001, on American citizens in New York and Washington, D.C., reinforces the need to "Plan for the Unthinkable".

No one wants to plan for a bioterrorism attack. The thought of dealing with tens of thousands of casualties is not a palatable task, but we must plan, we must prepare, we must act as a nation united to limit the effects of an intentional or naturally occurring epidemiological disaster.

This book should be a building block to develop an effective local, state, and national program to deal effectively with biological terrorism.

God Bless America.

—A.D. Vickery, Deputy Chief
Special Operations
Seattle Fire Department

PREFACE

There has always been a need for planning and preparedness for disaster situations in our country and we have been doing a good job. Large-scale disasters in other countries may have heavy casualties, while the same incident in the United States may only have a few, as in the case of the Seattle Earthquake. The many training and education programs at the federal, state, and university levels for first responder and related areas that exist now have led to an increase in the preparedness this country enjoys. However, on and after September 11, 2001, the books, the plans, and the approaches to those plans have been and are being rewritten. What once was considered a possible but remote threat of biological terrorism is now a reality in many parts of the United States. Hazardous materials teams are working overtime to respond to multiple calls for suspicious packages; envelopes with powder on or in them, and powder on cars, mail boxes, and other everyday items. Many of these are hoaxes, but unfortunately a few are not. The time is here to deal with the real incidents of biological terrorism and the associated public fear that has evolved from it.

One outcome of these past few weeks has been the unification of emergency services. Barriers often associated with planning and preparedness between various agencies have dissolved. In the basic planning steps prior to September 11, 2001, battles would, in some cases, ensue over who is in charge of an emergency scene and who should write emergency plans. Now, in meetings throughout our country, the common good is the focus and progress is occurring faster than ever before. In times of crisis, emergency services come together in a way like no other, making the public the major and foremost concern. Public health, fire, police, emergency management, hazardous materials, medical, and 911 agencies are working in virtually every city to review, revisit and revise their emergency response plans. This book will assist in those efforts. The book does not provide the answer, but is designed to stimulate the thought process to mitigate against, prepare for, respond to, and recover from a biological event.

There are many people and agencies who can benefit from this book besides first responders. School systems should look at their planning and response guidelines for biological situations, or for that matter, all their hazard emergency/disaster plans. Anthrax hoaxes today are the bomb scares of the 1990s. Are they prepared with current and correct information about risk and response? This book will help to write those plans. Businesses need information on planning for biological threats and events, as many of them have received envelopes containing powder, some of which have tested positive for anthrax. Even the children of business workers could become victims of anthrax exposures simply by visiting a parent's work site. Public utilities and other local/county agencies should look at plans for these types of events and determine, based on lessons learned so far, how to prepare and respond if needed. Anyplace where there are large gatherings of people, such as stadiums and theme parks, should be prepared for even just the mention of a biological event. A concentrated, coordinated plan will allow these areas to maintain calm and respond appropriately when needed. The public has an overwhelming need for information to calm their concerns for their safety. How are their local agencies preparing and responding? What if there is an outbreak in their town? Will CDC respond with medicine? What is the federal government doing to assist local agencies with the threat of exposure? Much of this information is available in this book. I hope that it gives the general public a sense that your national, state and local community leaders are taking the safety and welfare of you and their communities as a top priority.

This book was started before September 11 and is the second in a series of terrorism preparedness books. The World Trade Center attacks and the deaths of several personal friends created the need to complete this book. I know that emergency planners are looking closely at their plans during this time of crisis in our country and I hope that in some way this book will assist them, (you) in their efforts. This book is published in honor of the first responders and friends who died in the attacks on September 11, 2001. None of the victims will ever be forgotten. May God bless America.

I would like to extend my appreciation to a number of individuals whom I call friends, that helped make this book a possibility. I would like to thank Chief A. D. Vickery of the Seattle Fire Department, Mike Makar, Dr. Sagar Galwankan, my colleagues and friends at the University of South Florida, College of Public Health's Administration staff, and The Center for Disaster Management and Humanitarian Assistance for all their support. Dr. Tom Maston, Eric Matos, Bob Tabler, Dr. Wayne Westhoff, Erin Hughey and Brenda Gonzalez, and special thanks to Kristina Marsh and Krista Uhde of the University of South Florida's Center for Biological Defense. Thank you to the Seattle Fire Department for their assistance and support in the development of this book. I also would like thank the staff at Delmar and extend a special thank you to Mark Huth and Besty Hough for their assistance in the development of this book. Finally, thanks to the folks at TIPS Technical Publishing who worked all hours to help bring this book to you quickly, including Juanita Covert, Tracee Hackett, Darryl Hamson, Robert Kern, Maria Mauriello, Kara Minoui, and Paige White.

I would like to offer a word of appreciation to my family, my son George, daughter Taylor, my stepsons Joshua and Nathan, and my mother and father for their support and encouragement in the quick development of this book. This was made easy with the support my wife Lori and for that I thank her.

I dedicate this book to Americans serving to protect America whether on the front lines, behind the scenes, overseas, or at home.

Overview of National and International Terrorism

OVERVIEW

This chapter will identify the basic response strategies to Weapons of Mass Destruction (WMD) events that build upon existing doctrine, while addressing the unique considerations of the terrorism environment and reviewing a brief history of terrorism. You will be able to

- Define terrorism and the terrorist ideology
- Identify considerations for the selection of potential targets by terrorists
- Demonstrate an understanding of the difference between strategies and tactics

DEFINING TERRORISM

The United States Department of Justice defines terrorism as "The Unlawful Use of Force against Persons or Property to Intimidate or Coerce a Government, the Civilian Population, or Segment Thereof, in the Furtherance of Political or Social Objectives."(Source: FBI)

While this may be the official definition, the news media define terrorism in their own way. A reporter in Southern California once made the statement that someone was shooting out car windows on the freeways and called this "highway terrorism."

This suggests a broad definition of terrorism as an illegal act intended to cause a change in politics or social issues through the use of intimidation. To better understand terrorism, we must look at five commonly accepted variables:

1. *The violence need only be threatened.* Terrorist acts are designed to do one thing—instill fear. People will feel that they cannot protect themselves and that the government is unable, or unwilling, to provide adequate protection. Therefore, a terrorist act need only be threatened to instill that level of fear, provided that the threat is perceived to be genuine and valid. Consider the impact on the United States airline and travel industries if a believable threat were made to bomb twenty United States planes over the next six months. If the public believed the threat to be valid, the resulting reduction in travel could be financially devastating to the industries and would result in increased pressure on the government to respond to the terrorists' agenda. Recall, for example, the incident in the summer of 1986, when Middle Eastern terrorists threatened United States tourists in Europe with a campaign of terror and bombings. This simple threat resulted in a reduction of United States tourism to the area of almost forty percent and placed a severe economic burden on the European tourism industry.
2. *Fear is the actual agent of change.* For a terrorist to be successful, fear must be instilled. It is the population's fear that results in the pressures for the change that the terrorist desires. As fear grows, distrust in the government's ability to protect the public increases. It is this distrust that will result in either a policy change or an overthrow of the government. What would happen if a terrorist instilled the fear of becoming targets in the minds of America's emergency first responders? If responders become overly fearful and reluctant to act, public levels of fear will increase dramatically. Sometimes, however, a terrorist plan to instill fear will backfire. Sometimes a terrorist act is so reprehensible that it incites anger rather than fear. If terrorists cross this line from fear to anger, their agenda is unlikely to be furthered. Such examples are fresh in our mind. The bombing of servicemen in a West German nightclub in the 1980s resulted in a strong military response against Libya for supporting the operations. More recently, the images burned into our minds of the deaths of 168 people in Oklahoma City resulted in an effect opposite to what was desired by the perpetra-

tor, Timothy McVeigh. The American public was not frightened but appalled by the act and rallied behind the victims and Oklahoma City as if in the common defense of the country during a war. These strong and unyielding responses and sense of community will do much to prevent terrorists from carrying out such acts.

3. *Terrorists' victims are not necessarily the ultimate targets.* Generally speaking, the actual victims (whether injuries or deaths) are not the specific targets of terrorist acts. Victims are only pawns in a terrorist's attempt to instill fear in those who witness the attack. Unfortunately, the victims many times just happen to be in the right place ("right" meaning somewhere they feel comfortable and secure) at the wrong time.
4. *Those who observe the act are the intended audience.* Add to the situations discussed above the media coverage that such events generate. Extensive media coverage is a double-edged sword. On one hand, it helps to pull the country or community together, while, on the other hand, it furthers the terrorist's agenda by allowing more people to witness, almost instantly, an event that is designed to instill fear.
5. *A terrorist's desired outcome is a political or social change.* A terrorist is trying to create enough individual fear or distrust of the government to force changes in social or political situations. Whether that change is to stop abortions, change a public policy, or get the public to relinquish a certain freedom or liberty (thus giving the government more power to protect them) the people who are fearful demand the change. A terrorist sets the terms for the cessation of hostilities. If we buy into those terms, we must remember that we have allowed the terrorists to achieve their goal.

TYPES OF TERRORISTS

There are numerous ways to categorize or define terrorists—domestic or international, left or right, ideological, special interests, anarchists, neo-fascists, and so forth. We will discuss only a simple breakdown of the types of terrorists.

Domestic terrorists originate within the United States; more often than not, they hold extreme right-wing beliefs. This is not to say that domestic terrorists are never from the left-wing politically (we will discuss left and right in just a moment); but the right is by far the largest and most active group within the United States. Luckily, to this point, there is little organization among the groups and they do not currently operate in concert. However, recent meetings between some of the larger groups are raising concerns that they are becoming better organized.

The continuum developed by Vetter and Perlstein in 1991 (see Figure 1–1), demonstrates how terrorist groups range from the radical far left to the reactionary far right. The major belief structure of the far left is that of the fair and equitable distribution of power, wealth, prestige, and privilege. This belief structure is expressed by many as the "Marxist" left, inasmuch as members of the far left believe in the writings of Karl Marx and therefore follow socialist or communist agendas.

These types of groups are more likely to engage in terrorist activities designed to prompt the public to allow the government greater power. This is accomplished by instilling sufficient fear so that the public demands the government do more to protect them. This increased government involvement generally results in the reduction of individual liberties or freedoms in the interest of protection. Alternatively, the agenda could be to institute more extensive social programs and the redistribution of wealth.

On the other end of the continuum is the reactionary far right, whose values are based on order and a binding and pervasive morality. The far right may include religious, separatist, or racial supremacy groups. Essentially these groups believe in less government intervention in social issues or, in many cases, no government intervention at all.

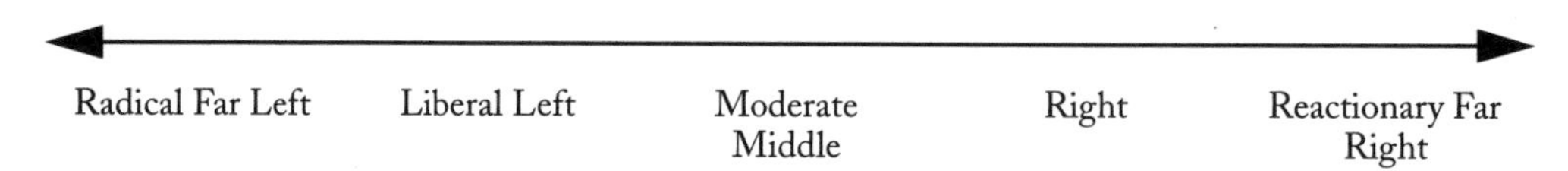

Figure 1–1 The political continuum developed by Vetter and Perlstein

This is not to say that all left- or right-wing organizations are terrorist organizations. Other factors must be present before a group resorts to terrorist activities. In order for a group to be considered terrorist, it must meet three criteria:

Extremist viewpoint—First, it must hold an extremist viewpoint. Simply put, an extremist viewpoint is the belief in the "one truth." That is to say, the group believes that there is only one answer to a particular issue, be it abortion, sexual orientation, or another religious, social, or political issue. But holding an extremist viewpoint in itself by no means makes a group terroristic, any more than it does an individual. Many of us have one point of view on a specific topic—pro-life versus pro-choice is an excellent current example. Because the vast majority of people are tolerant of other points of view, however, we do not resort to terrorist activities provided others do not force their points of view on us.

Intolerance—It is when the group becomes intolerant of other points of view that it moves closer to the terrorist belief structure and meets the second criterion, intolerance. At this point, the group becomes unable to accept differences of opinion. This results in a belief that anyone who does not believe in the one truth is the enemy. Still, the group isn't yet a terrorist group; one can be intolerant and still be a law-abiding citizen. It is the third and final step that defines a terrorist.

Vilified enemy—In the third criterion, those persons, governments, or countries with beliefs other than the terrorists' are not only considered an enemy, but are also vilified. The enemy now becomes a hindrance to accomplishing the belief or is seen as directly jeopardizing the one truth. Once this shift occurs, the enemy loses all value. The enemy is thus worthless and a direct threat to the individual's or group's belief structure. Therefore, any means necessary to defeat or overcome the enemy becomes acceptable. In other words, the end justifies the means. This can even be taken one step further to create the most dangerous of situations. If the terrorist accepts the belief that anyone who is not fighting the enemy *becomes* the enemy, then such people also become worthless and may themselves be attacked regardless of age, gender, or relationship to the primary enemy.

TERRORIST TARGETS

Remember that terrorists want to instill fear in those who witness their attack. Therefore, they will look for a target that will give them as much media coverage as possible. The target may be a government facility if the terrorist is trying to portray the government as weak or inept (right-wing terrorist), or public facilities may be selected if the terrorist is trying to show the public that they need greater protection (left-wing terrorist).

Targets can be people, places, or infrastructure. People will be targeted to build the fear that no place is safe. Places could be of historical or ideological significance or of value to the public. Infrastructure targets include those elements necessary for a community to function (e.g., roadways, bridges, water treatment plants). The more critical the infrastructure and the harder it is to restore to service, the greater its potential for targeting.

RECOGNIZING TERRORIST EVENTS

We mentioned earlier that terrorists may specifically target emergency responders. Therefore, it is essential that responders understand how to recognize a terrorist event as early as possible so that changes can be implemented immediately to ensure that we maintain the tactical advantage. As soon as it is recognized that we are dealing with a terrorist event, we must change the way we do business. We want to avoid getting "blocked in," stay away from "choke points," use different response routes, implement security measures, be on the lookout for secondary devices, and so forth. Therefore, we need some tools for recognition.

There are two stages of terrorist event recognition—pre-event recognition, which is predominantly an awareness of conditions that might increase the likelihood of terrorism; and response phase recognition. We will discuss this in the planning section of the text but simply put, we watch out for people looking at the way we do business and we communicate closely with local law enforcement officials who should be monitoring trends (intelligence) with regard to groups that might carry out such acts.

During the response phase we need to be alert for other indicators of terrorist incidents. These include:

- Occupancy and location
- Symbolism/History
- Public assembly
- Controversial facility
- Infrastructure
- Critical facility
- Vulnerable facility

TYPES OF EVENTS

- Bombings and incendiary fires
- Events involving firearms
- Nontraumatic mass casualty incidents
- Epidemiological events

CONDITIONS

- Ideal attack weather conditions (unstable atmospheres with inversions and little air movement) which can be either natural or manmade, such as in a building or a subway
- Situations that appear to deliberately place people at a tactical disadvantage (choke points, high grounds, unexpected traffic diversions)

TIMING OF THE EVENT

- Timed for maximum casualties
- Historic or significant dates

OTHER OBSERVATIONS

- Unusual casualty patterns or symptoms
- Odors
- Out-of-place containers or dissemination devices

AN OVERVIEW OF INTERNATIONAL TERRORISM

INTRODUCTION

The trend in recent years has been a decrease in international terrorism; the number of international terrorist incidents fell from a high of 665 in 1987 to 304 incidents in

1997, up by eight incidents from the previous year but nevertheless reflecting a general downward trend.

However, taken at face value, this is no indication of how serious the problem posed by political violence has become. Deaths as a result of international and domestic terrorism and insurgency appear to be on the increase, although figures for so many parts of the world are elusive and inaccurate, not least because there is a problem of definition.

There is no universally acknowledged definition of terrorism, but for the purposes of this discussion we will use the following definitions:

> **Terrorism**—"The threat or use of extraordinary violence for political ends"; or "Premeditated, politically motivated violence perpetrated against noncombatant (including military/police who are not on duty) targets by subnational groups or clandestine agents, usually intended to influence an audience."
>
> **Guerrilla warfare**—"Military operations conducted by irregular troops against conventional government forces."
>
> **Insurgency/Insurgent**—"A rising, a revolt... a rebel, a revolutionary."

Included in this publication is a profile of every significant terrorist, guerrilla, and insurgent group in the world, with analysis of tactics, assessments of numbers, leaders, and political affiliations. There is also a section detailing major incidents committed by the groups in the past two years, and details of the world's more significant modern terrorists, guerrillas, and revolutionaries, many of whom are still at large. Most of the groups in the text are still active; some have been dormant for a number of years but have not officially been disbanded and, in a few cases, are important for reference since they were involved in significant recent insurgency campaigns.

CATEGORIES OF INSURGENT AND TERRORISM GROUPS

During the 1970s and 1980s it was *de rigueur* to divide guerrilla and terrorist organizations by motivation, and such categories are still useful:

Nationalist Groups

Nationalist organizations were once regarded as the "aristocrats" of political violence and they remain a persistent and often deadly threat. Extremist Palestinian groups, such as the military wing of Hamas; the Provisional Irish Republican Army (PIRA); the Corsican separatist Front de la Liberation Nationale de la Corse (FLNC); and the Basque Euzkadi Ta Askatasuna (ETA), are clear examples of groups which have waged a sustained campaign of terror in an attempt to achieve a nationalist goal. They conduct isolated incidents of violence, often against civilian or economic targets. In Myanmar (formerly Burma), Sri Lanka, India, Bangladesh—to name just a few—the authorities have fought against insurgency campaigns which are examples of rural guerrilla warfare waged for a nationalist cause.

With the end of the Cold War and the removal of the shackles of East-West politics, new groups have emerged with their own nationalist agendas. In Chechnya, an insurgency campaign against the Russian Army humiliated Moscow and may ultimately pave the way for full independence. The newly independent republics of Georgia and Azerbaijan have also faced effective, damaging insurgencies by ethnic groups seeking to break free or join their kinsmen in another state.

In Europe, the break up of Yugoslavia not only led to the continent's first 'hot war' since 1945 but also spawned ethnic/nationalist terrorist groups in Macedonia and Kosovo.

In Africa, where tribal and ethnic divisions have frequently threatened the concept of statehood, fighting has occurred between rival ethnic groups in Rwanda and Burundi over their respective nationalist agendas.

"Politicos"

Left-wing terrorist groups, of which the axis of Euro-terrorists—Action Direct, Brigate Rosse, Rote Armee Fraktion and the Cellules Communistes Combatants—were the most notorious examples, have largely declined in importance. Their aims were perhaps overambitious and unlike nationalist groups, there was not a ready pool of new recruits. In the 1990s, such organizations seem like an irrelevance of the radical 1960s to the next generation, and with the end of the Cold War, Marxism has been discredited.

Nevertheless, left-wing ideology does still have an important role to play among guerrilla groups such as Sendero Luminoso in Peru and Fuerzas Armadas Revolutionaries de Colombia (FARC) in Colombia.

There has also been a decline in the number and influence of right-wing organizations, although they were never as numerous as were those on the left.

Religiously Motivated Groups

Perhaps the most significant change in patterns of insurgency and political violence in the last twenty years has come in the increase in the number of groups which are religiously motivated. Where secular ideologies have failed, a spectrum of radical religious organizations has sprung up. As we approached the end of the millennium it was perhaps inevitable that cults like the Branch Davidians and the Aum Shinrikyo should attract members—such organizations are often categorized as propagating "messianic terrorism." However, increasing numbers of political and nationalist movements are also expressing a religious identity.

That ethnic or national divisions sometimes coincide with indigenous religious beliefs is nothing new: Muslim Kashmiris and Sikh Punjabis in Hindu India mainly, and Hindu Tamils (Muslim Tamils have tended not to rebel) in Buddhist Sri Lanka predominantly, for example, have rebelled. But religious radicalism—be it Christian, Islamic, Jewish, etc.—is increasingly setting the agenda in a number of arenas of conflict and providing rebels with a new impetus.

Religious extremist groups, sometimes referred to as subconflict organizations, aim to provoke full-scale insurrection.

For Islamic groups, the experience that outsiders gained fighting in the war against the Soviet Union in Afghanistan has proved invigorating; indeed, the emergence of "Afghanis" (foreigners, often Arabs or Pakistanis who fought alongside the various Mojahedin groups between 1980 and 1989) is one of the most important trends in international terrorism and insurgency violence. An estimated 10,000 Islamic mercenaries fought in Afghanistan and the irony is that they were indirectly equipped and trained by the CIA, via Pakistani Intelligence, and funded by Saudi Arabia. Conservative Arab governments encouraged young men to defend the honor of Islam.

In the great valleys of Afghanistan, these young men were taught a firebrand message and given first-rate military training, including training in the use of sophisticated weapons and explosives. Once the Soviet Union retreated, the Afghanistan conflict became just another civil war and thousands of these foreign recruits returned home to face unemployment. In North Africa in particular, their return proved disastrous for their governments; in Kashmir, moderate insurgent groups like the Jammu and Kashmir Liberation Front (JKLF) have found themselves contending with Islamists determined to turn their campaign from a struggle for self-determination into a jihad (holy war).

Other Afghanis have used their experience when fighting in insurgency campaigns as far afield as the Middle East, India, China, Tajikistan, Bosnia, and the Philippines, or have sought renewed training in Sudan, Somalia, Lebanon, Iran, and Yemen. They are not mercenaries in the proper sense: they will not perpetrate acts of political violence on behalf of anyone and their only cause is the Islamic revolution.

Christian extremist groups have made a particular impact in the United States. Christian fundamentalist, white supremacist, and extreme right-wing politics have become synonymous, largely through the activity of antigovernment militia groups and organizations like the Ku Klux Klan. The United States has the dubious accolade of having suffered one of the worst domestic terrorist outrages in modern times, the Oklahoma City bombing, the work of at least one paranoid, anti-government former soldier who had links to militia groups.

Such organizations and individuals are not in the habit of compromise; they are unlike politically mature organizations such as the Provisional Irish Republican Army (PIRA), which will usually steer clear of wholesale slaughter because it is invariably counterproductive. (PIRA learned this after the bombing of Christmas shoppers in Harrods, the Warrington bombing, and similar attacks lost the group sympathy in key constituencies, particularly in the United States.)

Religious extremist groups, sometimes referred to as subconflict organizations, aim to provoke full-scale insurrection.

Single-Issue Groups

Single-issue violence, sometimes referred to as "consumer terror," has also gained in prominence with the growth of violent groups campaigning over issues such as animal rights, the environment, and abortion.

In the United States, federal agencies also talk of the "narcissistic terrorist," the loner whose deep sense of alienation pushes him to harbor a grudge and wage war on society. The "Unabomber," Theodore Kaczynski, was one such example.

Surrogate Terrorism

Finally, there are groups and individuals dealt with in this report who are basically mercenaries, "guns for hire."

The most significant examples are the Abu Nidal Organization, the Japanese Red Army—which for the time being at least has abandoned its domestic motivations in favor of training other insurgency groups—and Carlos, who is now serving time in a French prison convicted of committing acts of terrorism. Groups not normally associated with this category, but which arguably belong here as comfortably as they do in "politicos" and nationalists, are those organizations whose prime motivation has shifted to the pursuit of wealth. In Colombia, for example, groups such as FARC have long been linked with drug cartels, and the divisions between political violence and narcoterrorism has become ever more blurred.

In light of the events in the world, Central Asia has become a focal point for terrorism today. Hence, let us not forget the middle east and homegrown terrorism. It is important to be briefed on the events going on in the Central Asian region.

The most populous traditional and economically strong country in Central Asian Region, Uzbekistan is facing important threats to her national sovereignty and independence for the last five years. The rise of the Islamic fundamentalism; neo-imperialist ambitions of Russia, China, and Iran in the region; and increasing drug trafficking from Afghanistan to Europe via Russia are the basic causes of turbulence in the region. This short analysis is an effort to explain dynamics of the regional security policies.

Russia is very uncomfortable with the unipolar world system and the United States' supreme position in the post-Cold-War era. To balance the United States' position, Russia tries to establish a series of "strategic partnerships" with a series of countries, beginning with China. After the establishment of the Shanghai Five for security concerns in Central Asia, Russia increased its strategic relations with Iran and India. Russia simply wants to create a new block in Asia, first to stop the spread of Islamic fundamentalism in Central Asia and the Caucasus, and second, to counterbalance the United States' position as a super power. On this parallel, Russia has been using the "Afghan ghost" quite effectively to influence Central Asian leaderships. There have even been claims that it was Russian military presence that enabled the Islamic Movement of Uzbekisan (IMU) terrorists, under the leadership of Cuma Kodjayev (Namangani) and Tahir Yuldash, to operate in Ferghana Valley and to pass freely through the border between Afghanistan and Tajikistan.

Uzbekistan, the most populous and most powerful state in the region in many senses, has been claiming the leadership position in Central Asia for the last ten years. It is the only country in the region that has an operational armed force, and it has the biggest economy in the region. Additionally, Uzbeks historically have always been the most dynamic people of the region. Although all of Central Asia's records of human rights is quite poor, it should not be forgotten by analysts that both capitalism and democracy entered this part of the world after 1991. It is possible to compare the current state of affairs in this region with the fragile politics of the young Turkish Republic during the 1920s and 1930s.

According to many observers, the "hot" spot of the world is becoming the Central Asian region, given the large quantities of opium produced in Afghanistan, which finances international terrorism in the middle-run. It is a well-known fact that in the summer of 2000, the IMU militants combined their forces with those of Usama Bin-Laden to achieve the Islamic Caliphate in the region. The Russians, at this point, are the best experts to play with the religious, clan, and ethnic differences in this part of the world. Accordingly, they seem to be the main enemy of Islamic fundamentalism in the region in order to increase their military presence, just like in the case of Tajikistan. Although Kazakhstan and Kyrgyzstan (now Kirgizia, the Russian form of the name, after the decision by the Kyrgyz Parliament in 2000) stick together with Russia in the former Shanghai Five (now the Shangay Cooperation Organization) and the Eurasian Economic Union, it is simply because of their inability to deal with even internal crime and corruption. Uzbekistan has long been the only country to resist Russian military presence and political hegemony. Russia wants to bypass Uzbek resistance to military presence by threatening to bomb Chechen bases in northern Afghanistan, which would automatically trigger direct Taliban reaction to both Uzbekistan and Tajikistan. Both countries would have little chance against the well-known, organized, and Pakistani-backed Taliban military threat.

Although the Uzbek administration has launched a series of reforms in the army and a new armament program, it needs foreign support, especially in terms of know-how, to continue its struggle against both Islamic fundamentalism and the Russian hegemony in the region. Both Uzbek and Kyrgyz officials have managed to learn lessons from the 1999 Batken invasion of the IMU militants. However, military reform and new tactics to stop the spread of fundamentalism are urgently needed. Not developing the state of human rights and political freedoms on the same parallel with fighting against terrorism creates a vicious circle, as in the case of Turkey for the last twenty years. But at least the Uzbek administration has plans to follow both policies together, with international help, of course.

It is the vital interest of the United States not to isolate Uzbekistan and leave the country open for Russian, Chinese, and Iranian influence. Hosting one of the world's oldest Jewish communities in Bukhara, the secular Uzbek state would be the sole guarantor of the survival of this culture as well. One of the main arguments of both IMU and Hizb-e Tahrir in Uzbekistan is to "clear" this world heritage in Bukhara.

It seems that it will would be very difficult to recreate regional security in Central Asia without creating a buffer zone in northern Afghanistan, which would separate the Taliban from Central Asia. On the other hand, Russia's plans for sending 50,000 troops to Central Asia by the end of the year 2003 would directly mean the loss of national sovereignties of Central Asian states. Russia has been politically forcing Uzbekistan for the last two years to reenter the CIS Common Security Treaty and open its borders for Russian troops. In fact, it was only a month after Uzbekistan's declaration of retreating from the treaty that car bombs aiming to kill the Uzbek president exploded in Tashkent. The following summer, the IMU suddenly acquired the military power to launch attacks to Uzbek and Kyrgyz territories.

Russian propaganda in the region is spreading the news of Bin Laden being, in fact, an American puppet whose brother lives freely in Cambridge, Boston and whose financial operations have never been seriously subject to U.S. restrictions. The same propaganda also tells people that it might be U.S. oil companies who do not want Central Asian natural resources to have access into the world market soon, helping the spread of Islamic fundamentalism in Central Asia and the Caucasus. The isolation of the Uzbek administration because of the human-rights problems in the country just helps Russian propaganda become more and more effective in the region's paranoid atmosphere.

The Uzbek administration has always supported the idea of a common economic space in Turkestan, as well as a common defense identity for the region. This means a Russia- and China-free Central Asia and more integration with the Western economies, as well as democratic values.

The important point here is not to alienate the Uzbek administration, and, if possible, officially invite Uzbek leadership to the United States soon for detailed negotiations on the reform program for human rights and political freedoms, as well as to discuss the potential for cooperation in Uzbekistan's fight against terrorism. Increasing their numbers in Afghanistan with the joining of some Afghan Uzbeks into their lines, the IMU might become powerful enough to capture at least some part of the Ferghana Valley in the fall of 2001. Divided into three parts among Uzbekistan, Kyrgyzstan, and Tajikistan, the valley is the most dangerous spot of the region, which may cause a series of social and political explosions soon.

Uzbekistan is strategically located at the very heartland of Asia, and there are Uzbek minorities in Tajikistan, Turkmenistan, Kazakhstan, Kyrgyzstan, and Afghanistan. A radical Islamic epidemic would directly influence all the people living in the neighboring countries. The stability of Uzbekistan means the stability of all of Central Asia. If Uzbekistan can balance the aggressive Russian claims in Central Asia, it would have the possibility to fight against terrorism by itself. This is only possible through a solid U.S. response to the Uzbek leadership's urgent need of security and international support for its independence. Uzbek leadership is more than ready to work closely with the United States toward reforming the political system for more freedoms and rights.

Simply because United States and the West left Uzbekistan alone, the country became a member of the Shanghai Cooperation Organization and have no choice in the long run but to accept Russian military presence. The Putin administration's neo-imperial policy would soon make Uzbekistan a part of the anti-American alliance of Russia, China, and Iran and possibly India (to counter Pakistani support for the spread of Talibani influence in Central Asia). Another consequence would be the total control of all Central Asian wealth by Russia. Without Uzbekistan's leadership, other Central Asian states would be too weak against Russia, even for negotiating.

Taking into account that Uzbek leadership would resist this Russian dominance till the "end," the other scenario would be the capture of the Ferghana Valley by the IMU forces and then the spread of Islamic fundamentalism to all of Central Asia. This might have a great impact on the traditionally secular but thirsty-for-religious-freedom masses of Central Asian states. If this scenario or part of it becomes true, Central Asia might very well experience a similar situation to that of Afghanistan. More importantly, the region may become a core for international drug traffic as well as terrorism.

As stated before, Uzbek leadership has no preconditions with any country to communicate and cooperate as long as Uzbekistan's independence and sovereignty are respected. Uzbeks now name Russia and the IMU as their two worst national security threats. The existence of American military experts and advisers in Central Asia has been used by Russian propaganda often as Central Asian borders have been opening for an American military presence, which would also take an important amount of reaction from the masses too. Uzbek leadership, eager to strengthen its relations with the United States and the West, has became the target of both Russian and IMU propaganda in this respect. However, for the survival of Uzbek independence and sovereignty, Uzbek leadership will continue its policy to increase the number of "channels" with the United States, not just for acquiring arms and military expertise to fight against terrorism, but also for finding common grounds to work on human-rights issues and democratic values.

The survival of secular, market-oriented, moderate democratic regimes in Central Asia is vital for Turkish, Israeli, and U.S. interests. The existence of the international network of radical Islamic groups associates the IMU and Bin Laden with all other Islamic groups in Turkey and Israel conducting terrorist activities against both the countries. This causes a direct threat to U.S. interests, both in the region and even domestically, simply because the United States has been seen as a "greatest devil" by these Islamic fundamentalist groups. A well-planned cooperation among the United States, Israel, Turkey and Uzbekistan would create a common front against both the spread of Islamic fundamentalist terror and neo-imperialist ambitions of China, Russia, and Iran in the region.

CATEGORIES OF RISK

The following definitions categorize the levels of risk that terrorist, guerrilla, and insurgency groups pose to their governments in each of the states discussed in this document. The levels are not an indication of the number of deaths or casualties that occur in each state or conflict, but rather are a guide to foreign interests and state security. Neither are the levels of risk for an overall security rating for the country, but they relate specifically to the groups in this document.

Very High Risk

- There is a state of civil war or widespread political or religious violence that undermines the effective running of the state and directly targets foreign interests or personnel.
- The government's authority does not extend across large parts of the country, and the violence is the dominating factor in the operations of the state. The violence also directly threatens to bring down the government.

High Risk

- There is an ongoing terrorist, guerrilla, or insurgency campaign in which foreign interests or personnel have been directly and indirectly targeted and where the authorities have no control in some significant area(s). However, it is possible to conduct normal daily life in less affected areas. The violence could indirectly bring down the government.
- Although the violence is a significant consideration in the government's planning and activities, the state and administration are not under constant siege.
- The state is recovering from a full-scale guerrilla war or insurgency campaign.

Medium Risk

- There is an ongoing terrorist, guerrilla, or insurgency campaign, but it is one that is confined and does not threaten to undermine the state or its institutions.
- Whereas foreign interests have been caught up in the violence, they are not necessarily being directly targeted and an attack against them is rare.
- While the insurgency concerns a regional issue, acts of violence may be conducted nationwide.
- The state is recovering from a high-risk insurgency campaign.

Low Risk

- The government is fighting a terrorist, guerrilla, or insurgency campaign, but it is confined to remote geographical areas that can be easily avoided. The campaign does not pose a significant threat to foreign or domestic interests, although security forces will be involved in counterinsurgency activities.
- The insurgency concerns a regional issue and acts of violence are, on the whole, confined to that region.

Minimal Risk

- A state has experienced past problems with terrorist groups but the threat has been effectively crushed. Although the state has other security concerns, it is not a major target of terrorist groups.

A BRIEF HISTORY OF TERRORISM

Terrorism has a much longer history than bombings, since its use predates the discovery of gunpowder. As far back as the sixth century Before the Common Era (BCE), various peoples poisoned wells and used powerful herbs and infected corpses as tools of biological warfare. The English infected blankets with smallpox and distributed them to American Indians early in colonial American history. While a student of history will find plenty of terrorism as early as colonial times, we will look at some examples of various kinds of terrorism from the more recent past, with a special emphasis on America.

Eighteenth-century examples of terrorism

- Infected corpses were used by the Russians against areas held by Sweden.
- The Puritans performed many acts of violence against other religious groups, especially the Catholics and Quakers before and after the Revolutionary War.
- Organized violence against government taxation included Shays' Rebellion (1786) and the Whiskey Rebellion (1791).
- British officials provided blankets from smallpox patients to Native Americans.

Nineteenth-century examples of terrorism

- Planned assassination of Tsar Alexander II.
- Twelve additional assassinations of public office holders in the United States took place following that of President Lincoln.
- The Ku Klux Klan begins acts of terrorism.
- In 1886, a peaceful labor rally at the Haymarket Square in Chicago was disrupted when an unknown person threw a bomb, killing seven people.
- Catholic churches were burned in Boston, Philadelphia, and other cities from the 1830s to the 1850s.
- Industrial Workers of the World (IWW) activists used 3,000 pounds of dynamite to blow up the Hill and Sullivan Company mine at Wadner, Idaho, along with a boarding house and bunkhouse.

Twentieth-century examples of terrorism

- In 1950, Puerto Rican nationalists attempted to assassinate President Truman. One District of Columbia police officer was killed during a gun battle outside the Blair House, where Truman was living at the time.
- In 1954, five members of Congress were wounded by gunfire during an attack by Puerto Rican nationalists.
- One of forty-nine bombings attributed to the Puerto Rican group FALN between 1974 and 1977 was the Fraunces Tavern bombing in New York City. Four people died in the event.
- In 1975, a bombing by Croatian nationalists killed eleven people and injured seventy-five others at La Guardia Airport in New York City.
- In 1976, Orlando Letelier, a former Chilean ambassador to the United States, was killed along with one of his associates in a car bombing in Washington, DC.

- In 1981, a man was killed in a men's bathroom at Kennedy International Airport in New York City by a bomb planted by a group calling itself the Puerto Rican Armed Resistance.
- In 1983, a bomb detonated in the cloak room next to the United States Senate chamber in the Capitol Building. Two left-wing radicals pled guilty to the attack.
- During the latter half of the twentieth century there were numerous airline hijackings and bombings.
- The 1993 World Trade Center bombing.
- The 1995 Oklahoma City bombing.
- The 1994 Matsumoto and 1995 Tokyo chemical attacks.
- Multiple bombings in Atlanta in 1996 and 1997.

Twenty-first century examples of terrorism

- The September 11, 2001 terrorist attacks on the World Trade Center and Pentagon.
- The spread of anthrax utilizing the United States Postal Service in the later part of 2001.

Extremists have used property destruction and violence to generate fear and compel societal change throughout history. Prior to modern times, terrorists usually granted certain categories of people (e.g., women, children, the clergy, the elderly, the infirm) immunity from attack. Like other warriors, terrorists recognized innocents—people not involved in conflict. For example, in late nineteenth-century Russia, radicals planning the assassination of Tsar Alexander II aborted several planned attacks because they risked harming innocent people.

Historically, terrorism was direct; it intended to produce a political effect through the injury or death of the victim.

The development of bureaucratic states resulted in a philosophical change among terrorist groups. Since modern governments are designed to be more dependent upon processes and structure than unique individuals, the death of a single individual, even a president or prime minister, will not necessarily produce the major disruption terrorists desire.

Terrorists have reacted by turning from targeting prominent individuals to attacks aimed at a wider range of targets, including those historically considered immune.

Events such as the World Trade Center and Oklahoma City bombings, and various clinic bombings, are designed to create a public atmosphere of anxiety and undermined confidence in government. The unpredictability and apparent randomness of these events make it virtually impossible for governments to protect all potential victims.

Modern terrorism offers its practitioners many advantages. First, by not recognizing innocents, terrorists have an infinite number of targets. They select their target and determine when, where, and how to attack.

The range of choices gives terrorists a high probability of success with minimum risk. If the attack goes wrong or fails to produce the intended results, the terrorists can deny responsibility.

RESPONSE OVERVIEW, STRATEGIES, AND TACTICS

You will be presented with various strategies and tactics for an Emergency Medical Services (EMS) response to terrorist events. It is important to understand both the difference between strategies and tactics and the role of EMS responders in the overall response plan.

Strategies are broad statements of overall desired outcomes. The incident commander develops the necessary strategies for the management of the event. During the early stages of the event, the incident commander may be the paramedic in charge or the first arriving fire-company officer or law-enforcement officer. As the event progresses, command of the incident is passed to higher levels and the strategies initiated by earlier commanders are reviewed and revised as deemed necessary. The earlier officer then acts under the strategic direction of the new command officer.

The major strategies with regard to EMS response to terrorist events are

- Recognition and identification
- Safety
- Security
- Assessment of incident extent
- Acquisition and management of resources
- Patient care
- Protection of evidence
- Termination and documentation

In contrast to strategies, tactics are specific actions. Tactical objectives are the necessary actions that must be taken to achieve the successful completion of a strategy. If we were to apply this concept to a routine multicasualty incident, one likely strategy would be "patient care"; others might include safety and extrication. The tactical objectives necessary to provide patient care at such an event would include actions such as triage, treatment, and transport. As we progress you will see the interrelationship of EMS strategies and their associated tactics.

This book focuses on the tactical considerations of emergency response to terrorism. However, as a person who will be implementing these tactics, it is important that you understand that you are working within an incident command system that is identifying strategic goals that must be accomplished.

The incident commander will identify broad overall goals that must be accomplished to bring the event to a successful conclusion. At the incident commander's level, the attention to actual tactics are employed to achieve the strategic goal would be overwhelming; therefore, the incident command structure is developed to manage the actual operations so that strategic goals can be completed in a timely, safe, and efficient manner.

Strategic goals are broad statements of desirable outcomes and tactical options are more refined definitions of specific measurable actions that must be carried out to achieve the strategic goal.

Emergency services throughout the country currently operate under a variety of strategic goals systems. These systems are designed to provide organized direction to the management of a particular type of event. Whether in response to a mass-casualty event, structure-fire event, hazardous-materials event, or technical-rescue event, strategic goals systems are a part of our lives on a day-to-day basis. An attempt to redesign a strategic goal system for the management of a terrorist event in the midst of responding to it, or to create a totally new strategic goal system under those circumstances, would be impractical and confusing.

EMERGING RESPONSE PLANNING AND CONTROVERSIAL ISSUES

The planning of emergency response procedures to terrorist events in the United States is a young and constantly changing set of ideas and information. As these events occur both domestically and internationally, we are constantly learning new and improved ways of managing the events and their consequences from other response organizations and military agencies. Therefore, the information in this book must be understood to be dynamic in nature. The standard of care today may become outdated in weeks or months to follow. With this in mind, it is important that we constantly monitor changing information and the standard of care just as we would monitor the changes in the other medical procedures that we perform on a more routine basis. Please refer to the Appendix O for numerous sources considered by the National Fire Academy to be authoritative. Most of the resources are U.S. government and military sources.

SUMMARY

It is important that we understand the terrorist's motivations and identify potential targets of terrorist attacks in order to recognize terrorist events quickly. If we fail to have pre-event or event phase recognition in place, we are likely to find ourselves in over our heads.

Know your community, the political and social conditions around you, likely targets, and how to recognize terrorist acts. The number of casualties we suffer is up to you, our leadership.

POINTS TO PONDER

Was George Washington a freedom fighter or a terrorist? It depends on how one views him: from the American side, or the English side. When preparing for terrorist events and when considering various groups of people, use an open mind and try to view all sides. Two examples of this approach are the January 22, 2001 television show, "60 Minutes," which featured two interviews on the Israeli-Palestinian conflict; and the History Channel's "100 Years of Terror: the War Against Colonialism."

What is a terrorist if his cause fails? A criminal.

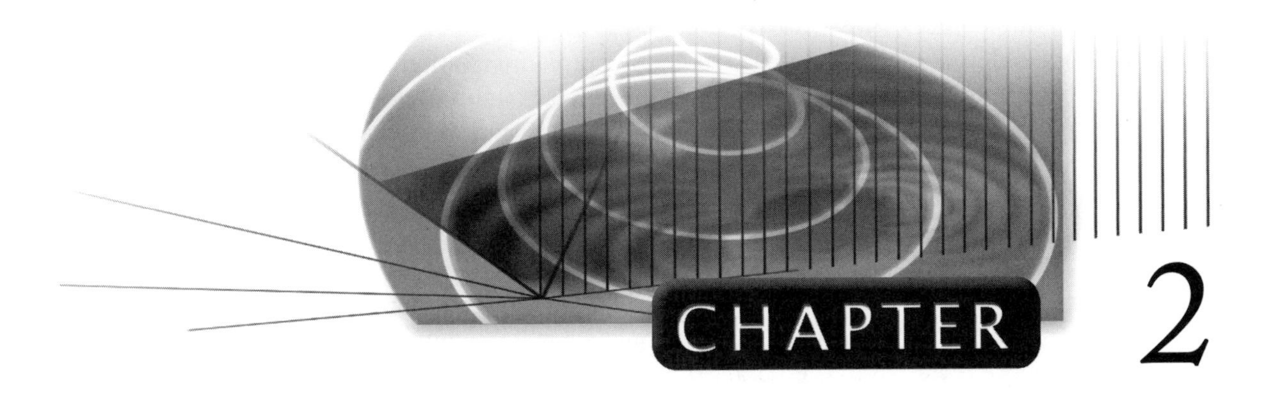

The DNA of Emergency Management Planning: The Basic Concepts of Emergency Management

OVERVIEW

In this chapter, you will be shown the building blocks of an emergency management system. Topics included here are

- Comprehensive Emergency Management System
- Integrated Emergency Management System
- An overview of the Incident Command System
- An overview of emergency operational planning

One of the most common issues heard from emergency responders is the lack of training. Police officers, fire fighters medical personnel, health department employees, and everyone involved in emergency planning offer similar concerns. Who offers training, where is the training, where is training money, what happens when only one person can get trained, what else can we do to get training? Most people want more information as it relates to their job and the events that may unfold around them. Many responders want to know a little about everything while some want to specialize in a particular area of emergency response. Those who are assigned to a specific team or function will get that specific training. But what about the average emergency responder? How do they get training? The first and foremost way is to "read." Read all the professional magazines your department subscribes to. Responders can go to the library or contact the National Emergency Training Center's Learning Resource Center (LRC) at 1-301-447-1000 and ask for specific information on terrorism or specific agents, and obtain copies of publications related to terrorism preparedness and response. You may ask for a general publication search for related information. The LRC will process your request and send you copies of the information you have requested. They receive all of the major magazines and periodicals for emergency responders and emergency management. Read current books on the market about the subject and share the books and materials with your department. Search the web for: CDC, FEMA, FBI. Discuss with your department for the possible development of a department library on WND. Watch the news and keep current with technology and new governmental missions. There are also self-study courses available at www.FEMA.gov.

The following chapter covers basic information and will assist you in understanding the role emergency management plays in bio-terrorism.

EMERGENCY MANAGEMENT AGENCIES

Emergency management agencies over the years have taken on numerous roles and responsibilities for their communities. These have included emergency medical services, hazardous materials containment, and now overall emergency management.[1] The fire service's role in emergency management is now a focal point in some agencies; it has become a necessity, not a luxury. The fire service role is to review, focus, and plan for an all-hazard emergency situation. This is not a new concept for fire services. They have often used many components of emergency management, such as command and control, incident command systems, and integrated emergency management systems in responding to emergencies and disasters.

---

1. Emergency management is organized analysis, planning, decision making, and assignment of available resources to mitigate (lessen the effect of or prevent), prepare for, respond to, and recover from the effects of all hazards. The goal of emergency management is to save lives, prevent injuries, and protect property and the environment when an emergency or disaster occurs.

Emergency management focuses on two basic tasks: preparing the organization to manage emergencies or disasters today, and preparing for such situations tomorrow. The fire department leader who promotes knowledge of emergency management helps meet the department's responsibilities by improving the department's and the community's capability to prepare for disasters and to mitigate their effects. Taking a leadership role and making improvements in emergency management not only develops the department's ability to cope with tomorrow's disasters, but also improves the management of day-to-day operations. The fire department will manage emergencies and disasters as part of an all-hazard system. This, in turn, will enhance the protection of its response personnel, increase its own professional reputation, and further protect the citizens it serves. Additional responsibilities may help to increase budgets and personnel. Additionally, it will be cost effective for many departments to inherit this activity instead of establishing new services. Naturally, a decision of this nature must be based on the jurisdiction's needs and its ability to provide and fund such services.

FIRE SERVICE INVOLVEMENT

In any emergency or disaster, small or large, involving fire department response, the goal is to mitigate the effects of the incident. A comprehensive emergency management system must be developed to assess potential situations and available resources, determine an appropriate proactive action plan, monitor the plan's effectiveness, and continually modify the plan to meet the realities of the situation.

If emergency service personnel are not functioning as part of an integrated emergency management system, their response, as well as that of fire, law enforcement, and public works services, is effectively reduced; as is the potential for communication and coordination with other agencies that may respond to the situation.

An effective emergency management system will benefit the community it serves by

- Saving lives and reducing injuries
- Preventing or reducing property damage
- Reducing economic losses
- Minimizing social dislocation and stress
- Minimizing agricultural losses
- Maintaining critical facilities in functioning order
- Protecting infrastructure from damage
- Protecting mental health
- Lessening legal liability of government and public officials
- Providing positive political consequences for government action

MANAGEMENT AT THE OKLAHOMA BOMBING

Planning is a must for the management of mutual aid from local, state and federal agencies. Whether you request it or not, you will receive assistance during any major disaster. Soon after the explosion, 300 to 400 firefighters arrived on the scene to assist. Although the Oklahoma City Fire Department (OCFD) never requested mutual aid fire resources, when personnel and equipment from surrounding cities and counties arrived, they were immediately utilized. Using their advanced planning system, the OCFD designated a department command vehicle and gave its crew the responsibility of organizing the responders. The responders were then assembled within a parking structure and requested through the mutual-aid coordinator as needed.

Thus, unrequested resources were utilized with maximum results by the OCFD because of preplanning.

Emergency management systems are useful at all locations, for all types of situations, and for all types of fire organizations. To be effective, emergency management systems must be suitable for use regardless of the type of jurisdiction or agency involvement. This may include single jurisdiction/single agency, single jurisdiction/multi-agency, and multi-jurisdiction/multi-agency involvement. The organizational structure must be adaptable to any disaster, applicable and acceptable to users throughout a community or region, readily adaptable to a new technology, and capable of logical expansion from the initial response to the complexities of a major emergency.

Common elements in emergency organizational management, terminology, and procedures are necessary for maximum application of a system and use of existing qualifications and standards. In addition, it is necessary to quickly and effectively move resources committed to the disaster with the least disruption to existing systems.

FACTORS THAT AFFECT EMERGENCY MANAGEMENT

The most distinguishing characteristic of emergency management is the element of danger to lives and property. For example, over each of the past ten years, an average of 130 firefighter fatalities and 100,000 injuries (accounting for the loss of at least one workday for each injured) have occurred. Each year an average of 5,000 civilians died in fire incidents and 95,000 were injured. Property loss from 1989 to 1999, according to the Federal Emergency Management Agency (FEMA), totaled $64 billion in direct costs.

Untold numbers of people have been exposed to toxic materials, resulting in immeasurable numbers of injuries and causing enduring pain and suffering. Among the health risks for firefighters and other rescue workers are hepatitis B, AIDS, and other infectious diseases. Cancer-producing chemicals are also a danger, both for firefighting forces and for civilians.

Emergency management is carried out in a constantly changing environment; although the situation may get better or worse, it seldom stays the same. The dynamics of such an environment present additional challenges to the Incident Commander (IC). The effectiveness of the incident action plan depends on a building's construction and contents, factors that may be difficult to determine. Danger increases due to flashover, backdraft, or the presence of hazardous contents. The dynamics of the incident may create difficulty in gathering accurate and current information, especially because of the limited time available at an incident scene. Additionally, emergency personnel reporting to the IC may not be able to adequately comprehend the overall situation.

A dynamic situation may require frequent shifts from an offensive to a defensive mode. Offensive modes include an aggressive interior attack and a direct attack on wildland fires. Defensive modes include an indirect attack on wildland fires, exposure protection, resource gathering, and the transition from offensive to defensive operations. Also, changes in priority may occur with regard to life safety, incident stabilization, and property conservation.

In later reports of an incident, any compromise of firefighter and rescue worker safety, poor management of resources, or an inability to enlarge the command organization to meet the demands of the situation may have a negative impact on public perception of the emergency management department. Departments should be ready for any type of incident. Since there is no guarantee that adequate resources will be available for every incident, departments must be prepared to handle every incident, regardless of its size or complexity, with whatever resources are available.

THREATS FACING THE UNITED STATES

Every day, the population of the United States is at risk from a broad spectrum of threats. These threats range from an ordinary house fire, to a hazardous materials accident on a major interstate highway, to a natural disaster, such as a major earthquake affecting many thousands of people. They also include the social threats posed by various forms of terrorism and civil disturbances.

The possible range of threats was brought home vividly during 1992, which until 2001 had ranked as one of the most devastating years in U.S. history. Among the most notable incidents in that year were the Chicago Tunnel Flood (a public works disaster), the series of earthquakes that occurred along California's San Andreas Fault, and the Los Angeles riots. Perhaps the most significant threats in 1992, though, were posed by the hurricanes that battered states on both the Atlantic and Pacific coasts. Hurricane Andrew, which tore across Florida and Louisiana, resulted in fifteen deaths and billions of dollars in damages (an estimated $27 billion in Florida alone). Hurricane Andrew was probably the most costly natural disaster in U.S. history to date. It was followed by typhoons Omar and Brian in Guam, and Hurricane Iniki in Hawaii which caused over a billion dollars in damages. The events in 1992 demonstrate the need to prepare for all potential hazards, both common and newly recognized. The cost of of the attacks of September 11, 2001 is still being counted.

TYPES OF THREATS

For the purposes of this book, there are two basic categories of threats:

Natural threats—The largest single category of repetitive threats to communities come from weather, geological, seismic, or oceanic events. They can pose a threat to any part of the country, and their impact can be localized or widespread, predictable or unpredictable. The damage resulting from natural disasters can range from minor to major, depending on whether they strike small or large population centers. Extremely severe natural disasters can have a long-term effect on the infrastructure of their location. Natural threats include avalanches, dam failures, droughts, earthquakes, floods, tropical storms or hurricanes, landslides, thunderstorms, tornados, tsunamis, volcanos, winter storms, and wildfires.

Man-made threats—These represent a category of events that has expanded dramatically throughout this century with advancements in technology. Like natural threats, they can affect localized or widespread areas, are frequently unpredictable, can cause substantial loss of life (besides the potential for damage to property), and can pose a significant threat to the infrastructure of their area. In this category are social threats that primarily come from actions by external, hostile forces against the property, population, or infrastructure of the government in the form of domestic civil disturbances. Other man-made threats include hazardous materials accidents at fixed facilities or in transport, power failures, structural fires, wildfires, telecommunications failures, transportation accidents, and terrorism.

RANKING OF THREATS

It is important to note that any ranking of a threat to communities and emergency services is potentially misleading because of

- The wide variations that can occur with the application of different criteria to the same threat
- The significant differences that can occur in the impact of a particular threat on a region and the individual states within the region
- The fact that threats in one region are not necessarily applicable to another region

- Variances in the types of data collected on each threat.
- The lack in some cases of available data with which to develop a reasonable ranking

POTENTIAL HAZARDS

(These were identified by local emergency managers,[1] and ranked by number of responses.)

- Hazardous materials incident—highway
- Power failure
- Winter storm
- Flood
- Tornado
- Drought
- Radiological incident—transportation
- Hazardous materials incident—fixed facility
- Urban fires
- Hazardous materials incident—rail
- Wildfire
- Hazardous materials incident—pipeline
- Civil disorder
- Earthquake
- Air transport incident
- Dam failure
- Hazardous materials incident—river
- Rail transportation incident
- Hurricane/tropical storm
- Subsidence
- Radiological incident—fixed facility
- Nuclear attack
- Landslide
- Avalanche

1. *FEMA Capability Assessment and Hazard Identification Program for Local Governments.* 1992.

- Volcano
- Tsunami

THE CHANGING CONTEXT: A WINDOW OF OPPORTUNITY

One of the most important structural changes for emergency management came with the end of the Cold War and the dissolution of the Soviet Union. While many uncertainties remain about the disposition of thermonuclear weapons formerly under the control of the U.S.S.R., public perceptions of the threat they pose have been significantly reduced, substantially altering the context of emergency management. For example, the term "civil defense" is properly applied to some programs that are relevant to all hazards. However, it has become so identified with preparedness for attack, that a program with such a label is now much more difficult to justify in terms of size and resources.

While the perceived threat of nuclear war has diminished, low profile threats posed by the proliferation of nuclear, chemical, and biological weapons are increasing. The chances of such deadly items falling into the hands of unstable or fanatical leaders has increased manyfold, but absent a galvanizing event such as the September 11, 2001. The events of that day will have a long term impact on the rewriting of disaster/terrorism preparedness programs. The national security emergency preparedness programs that have underpinned emergency management are very difficult to justify when perceptions of threats have diminished, available revenues have declined, and demands for attention to domestic problems have increased instead.

One of the most dramatic changes for emergency management is the greater intrusiveness and influence of the news media. Disaster and emergencies provide dramatic news, and the appetites of the media, particularly television, are insatiable. This means that emergency management agencies will have to perform under media scrutiny. It also means that few emergencies and disasters will remain local—most will now be nationalized and politicized as a result of media coverage. This presents particular problems for maintaining the tradition that local and state governments take primary responsibility for emergencies, while the federal government merely supplements their efforts. The media will pressure reluctant local and state leaders to "ask for federal help", will call on presidents to dispatch such help, and will urge representatives and senators to demand it on behalf of their constituents. This "CNN Syndrome" or "camcorder policy process" disrupts and distorts normal procedures and response patterns. The best laid plans and procedures are now vulnerable to disruption, indeed destruction, by one dramatic sound bite that the media can turn into a political shock wave.

The public expects more from all levels of government today, and especially from the federal government. The reasons for this are not clear. They may stem from the general nationalization of the political system that has come with population mobility and the consolidation of the news media. It may also be that the general erosion of community, mutual aid, and self-help is resulting in people turning directly to the government for assistance in emergencies An increase in expectations may fall upon the president as the only official elected by a national constituency and as the chief executive and commander-in-chief. Or it may simply be that the president is the most visible symbol of our government; and for people whose lives have been disrupted or who are in shock, symbols of competentcy and caring on the part of government are extremely important.

These changes in the context of emergency management present unprecedented challenges and opportunities. With memories of several disasters still fresh, a change of administration, and renewed attention on the nation's domestic problems, government and private industry have the greatest opportunity in more than a decade to address and ameliorate the enduring problems of the fire services and other emergency management agencies.

THE EMERGENCY MANAGER

When you take or are given the job of emergency manager, you become a leader, looking at all the components of an all-hazard comprehensive emergency system. You also have a job that is specifically defined by law. You have been appointed and can be terminated by an elected mayor, a city manager, or a county executive. How does your legally defined position help determine your priorities? You also must consider your moral responsibility to save lives and reduce property damage, not only from the threat of fire, but now from all hazards facing your community. You may be well prepared to handle a fire emergency or disaster, but where do you start with the planning process for a comprehensive, integrated emergency management system? Where do the next higher levels of government come in? What can you as the emergency manager expect from county, state, and federal governments? Finally, one of the biggest questions right now may involve your role as emergency manager.

EMERGENCY MANAGER: ROLES AND RESPONSIBILITIES

An emergency manager has the responsibility of coordinating all the components of the emergency management system in the jurisdiction. The components of an effective emergency management system include the fire service, law enforcement, elected officials, public works, parks department, emergency medical services, volunteers, other government agencies, the private sector, etc.

Emergency management involves knowing the possible threats to the community, planning for emergencies or disasters, operating effectively in a disaster, and conducting recovery operations after a disaster. Emergency management is the vital ingredient in the development of an effective emergency program. You will become the key leader in planning, the coordinator of operations, the chief of staff to the city executive during an emergency response, the community liaison in building the emergency program, and the supporter of mitigation efforts.

Coordination of emergency services, public and private, is a matter of personal style. Frequent contacts with colleagues, sharing advice with other personnel, and combined training among agencies are all ways to make coordination easier. Most important of all, however, is knowing the boundaries of coordination. For example, coordination means that police and firefighters cooperate in setting up the security of a crowd-control line. The emergency manager should make certain that responsibility is assigned and action taken without conflict or controversy. The emergency manager should definitely not, for example, tell a police chief how or where to set up security forces. The emergency manager assists in policy development for a time of disaster, not the development of standard operating procedures during a disaster.

In summary, the emergency manager serves the jurisdiction as the cement that holds together all the various components of a mitigation, preparedness, response, and recovery program. The emergency manager's job is to draw together the other emergency response managers into an effective, coordinated response team. In addition, the emergency manager keeps a constant eye out for opportunities to avoid disasters through hazard mitigation efforts. In short, the emergency manager draws on a wide body of resources to produce the most effective emergency management system.

In the late 1970s the Defense Civil Preparedness Agency, which later became a division of the Federal Emergency Management Agency (FEMA), supported a study of emergency management at state and local levels. The study was conducted by the National Governors Association, working closely with the federal reorganization project that created FEMA in 1979. The study called for a comprehensive emergency management policy and organizational structure at the federal level; and articulated the concept of Comprehensive Emergency Management (CEM), including the description and fuller development of principles concerning the nature and relationship of the four phases of disaster management.

These events have greatly influenced the nation's emergency management environment. CEM is now widely accepted as a useful emergency management framework. Former FEMA Director James L. Witt, who was appointed by President Clinton, had developed many proactive planning and management strategies, including CEM and the Integrated Emergency Management System (IEMS).

Comprehensive Emergency Management and the Integrated Emergency Management System are the corner stones of all emergency management programs.

COMPREHENSIVE EMERGENCY MANAGEMENT

CEM views emergency activities occurring in four separate but related phases: mitigation, preparedness, response, and recovery. These phases are visualized as having a circular relationship to each other. Each phase results from the previous one and establishes the requirements of the next one. The phases are related to the disaster by time and function and utilize different personnel skills and management orientation. Activities in one phase may overlap those in the adjoining phases. For example, preparedness moves swiftly into response when disaster strikes. Response yields to recovery at different times, depending on the extent and kind of damage. Similarly, recovery should trigger mitigation, motivating attempts to prevent or reduce the potential of a future disaster.

- **Mitigation**—Mitigation refers to activities which reduce or eliminate the chance of a disaster's occurrence or the effects of a disaster. Recent research has shown that, while natural occurrences cannot be avoided, much can be done either to prevent major emergencies or disasters from happening or at least to reduce their damaging impact. For example, requiring protective construction to reinforce a roof will reduce damage from the high winds of a hurricane. Preventing construction in hazardous areas like flood plains can reduce the chance of flooded homes.
- **Preparedness**—Preparedness means planning how to respond when an emergency or disaster occurs and working to increase the resources available to respond effectively. Preparedness activities help save lives and minimize damage by preparing people to respond appropriately when an emergency is imminent. To respond properly, a jurisdiction must have a plan, trained personnel, and the necessary resources. The objectives for local emergency management will describe the importance of a preparedness plan for every community and the value of human and material resources.
- **Response**—Response activities are designed to provide emergency assistance to victims of the event and to reduce the likelihood of secondary damage. The emergency services first responders and other support services are primary responders. Building and maintaining the capability to respond will be described in several sections of this book.
- **Recovery (Public)**—Recovery is the final phase of the emergency management cycle. Recovery continues until all systems return to normal or near normal. Short-term recovery returns vital life support systems to minimum operating standards. Long-term recovery from a disaster may go on for years until the entire disaster area is completely redeveloped, either as it was in the past, or for entirely new purposes that are less disaster-prone. For example, portions of a town can be relocated and the area turned into open space or parkland, thus providing the opportunity to mitigate future disasters. Recovery planning should include a review of ways to avoid future emergencies.
- **Recovery (Private)**—One area that is often overlooked by communities is the recovery of the private sector. If local government does not assist in the

recovery of businesses, the community loses tax revenue. Later chapters will lay out a working knowledge of a business continuation plan and will assist local government officials in the development of a public/private partnership.

The common factors among all types of natural and technological disasters, including terrorism attacks, indicate that many of the same disaster management strategies can apply to all kinds of hazards. These common management approaches are a principal component of CEM.

The burden of disaster management and the resources for it require a close working partnership among all levels of government (local, state, and federal) and the private sector (business and the public). This final part of CEM calls for a conscientious effort to draw on the widest possible range of emergency management resources.

THE INTEGRATED EMERGENCY MANAGEMENT SYSTEM (IEMS)

IEMS is a management tool that reduces property damage, prevents injuries, and saves lives. The system has been implemented and tested time after time with great success. FEMA developed the system in 1983 to improve the ability to respond to major emergencies and disasters. Past successful implementations of the system are documented throughout the United States. IEMS has been used on a barge collision and fire in Tampa Bay, Florida[1], a train derailment in Louisiana, and most recently, the Oklahoma City bombing.

Agencies using IEMS as an effective emergency management tool realize that no single agency can respond adequately during a major disaster. When public and private groups agree that emergency preparedness and efficiency in time of emergency are of vital importance, development of IEMS has begun. IEMS is the first step towards working together to protect emergency responders and the community they serve.

THE IEMS OVERVIEW

Emergency management[2] and planning are essential for effective disaster response. IEMS emphasizes teamwork in preparing for and responding to disasters. As described here it offers a conceptual framework for organizing and managing emergency protection efforts.

An integrated approach to emergency management incorporates all available resources for responding to man-made and natural emergencies or disasters, as well as the full range of issues relating to the four phases of a CEM system. Once the system is in place, it provides a means of efficiently incorporating resources from the private sector and other levels of government.

1. International Associations of Fire Chiefs *IEMS News*. November 1993.
2. The term "emergency management" as used in the United States equals the terms "disaster management" and "civil defense" as defined by the United Nations Department of Humanitarian Affairs.

Every local area has distinct groups with differing capabilities to perform specific functions in an emergency. These include program management, super and tactical Emergency Operations Centers (EOCs), law enforcement, emergency medical services and fire departments, voluntary organizations such as the Red Cross or Salvation Army, and many other resources. When these resources are linked through planning, direction, coordination, and clearly defined roles and functions, they are components of an integrated emergency management system.

Carry that integrated approach into the larger universe of regional, state, and federal resources and support relationships, and an IEMS is established. This can be achieved through mutual support with other jurisdictions, good lines of communication with other governmental levels, and dual use of emergency management resources.

Outside factors that will affect the system include the hazard or emergency to be faced and associated political, social, and economic issues.

THE IEMS CONCEPT

The goal of the IEMS is to develop and maintain a credible emergency management capability throughout the United States by integrating activities along functional lines at all levels of government and, to the fullest extent possible, across all hazards.

Agencies can begin to achieve this goal by

- Establishing an emergency management program tailored to the specific requirements of each community
- Developing plans which consider functions common to all types of emergencies, as well as activities unique to specific types of emergencies
- Incorporating existing personnel assignments, operating procedures, and facilities used for day-to-day operations into the disaster response plan
- Integrating emergency management planning into overall national and local government policy-making and operations
- Encouraging the operational use of an Incident Command System (ICS) to incorporate government functions (including law enforcement, public works, environmental protection, fire protection) and elected officials into the emergency management system
- Incorporating all areas of government jurisdictions (i.e., surrounding counties and other cities) into the emergency management system
- Establishing close working relationships with private-sector organizations such as the Red Cross, construction contractors, associations, and private search and rescue teams, etc.

- addressing all phases of a CEM system: mitigation, preparedness, response, and recovery (short- and long-term, public and industry)

THE IEMS PROCESS

The IEMS has been introduced into the United States network of emergency management organizations representing thousands of jurisdictions, not all confronted by the same hazards, and not all having or requiring the same capabilities. Implementing the IEMS process, therefore, will require different levels of effort by each jurisdiction and will result in the identification of different functional areas requiring attention. The process, however, is logical and applicable to all jurisdictions regardless of their size, level of sophistication, potential hazards, or current capabilities.

In order to provide a complete description of the IEMS process, each step is described below as it would apply to a jurisdiction that has done little toward developing the capability required, given its potential hazards.

Although IEMS emphasizes capability development, the process recognizes that current operations must be conducted according to existing plans and with existing resources, and that these operations can contribute to the developmental effort. The process, therefore, includes two paths: one focusing on current capabilities and activities (Steps 1-7), and the other emphasizing capability improvement (Steps 8-13).

Step 1. *Hazards Analysis.* Knowing what could happen and the likelihood of it happening, and having some idea of the magnitude of the problems that could arise, are essential ingredients for emergency planning. The first step, then, is for the jurisdiction to identify the potential hazards and to determine the probable impact each of those hazards could have on people, property, and the environment. This task need not be complicated or highly sophisticated to provide useful results. What is important is that any hazard that poses a potential threat to the jurisdiction is identified and addressed in the jurisdiction's emergency response planning and mitigation efforts.

Step 2. *Capability Assessment.* The next step for the jurisdiction is to assess its current capability for dealing with the hazards that have been identified in Step 1. Current capability is determined by standards and criteria FEMA has established as necessary to perform basic emergency management functions, e.g., alerting and warning, evacuation, and emergency communications. The resulting information provides a summary of the capabilities that exist and upon which current plans should be prepared (Step 3), and leads to the identification of the jurisdiction's weaknesses (Step 8).

Step 3. *Emergency Operations Plan.* A plan should be developed with functional annexes common to the hazards identified in Step 1. Those activities unique to specific hazards should be described separately. This approach is a departure from previous guidance which stressed development of hazard-specific plans.

Existing plans should be reviewed and modified as necessary to ensure their applicability to all hazards that pose a potential threat to the jurisdiction. The exact format of the plan is less important than the assurance that the planning process considers each function from a multi-hazard perspective.

Step 4. *Capability Maintenance.* Once developed, the ability to take appropriate and effective action against any hazard must be continually monitored or it will diminish significantly over time. Plans must be updated, equipment must be serviced and tested, personnel must be trained, and procedures and systems must be exercised. This is particularly important for jurisdictions that do not experience frequent large-scale emergencies.

Step 5. *Mitigation Efforts.* Mitigating the potential effects of hazards should be given high priority. Resources utilized to limit the effects of a hazard or reduce or eliminate the hazard can minimize loss and suffering in the future. For example, proper land-use management and stringent building and safety codes can lessen the effects of future disasters. Significant mitigation efforts can also reduce the level of capability needed to conduct recovery operations, thereby reducing the capability shortfall that may exist. The results of these efforts will be reflected in future hazards analyses (Step 1) and capability assessments (Step 2).

Step 6. *Emergency Operations.* The need to conduct emergency operations may arise at any time The operations must be carried out under current adopted plans and with current resources, despite the existence of plans for making improvements in the future. Such operations, however, can provide an opportunity to test existing capabilities under real conditions.

Step 7. *Evaluation.* The outcome of the emergency operations (Step 6) should be analyzed and assessed in terms of actual *vs.* required capabilities and considered in subsequent updates of Steps 2 and 8. Identifying the need for future mitigation efforts should be an important part of each evaluation. Tests and exercises should be undertaken for the purpose of evaluation, especially where disasters occur infrequently.

Step 8. *Capability Shortfall.* The difference between current capability (Step 2) and the optimum capability reflected in the standards and criteria established by policy represents the capability shortfall. The areas not currently meeting the assessment criteria should receive primary consideration when preparing the jurisdiction's multi-year development plan (Step 9).

Step 9. *Multi-Year Development Plan.* Based on the capability shortfall identified in Step 8, the jurisdiction should prepare a multi-year development plan tailored to meet its unique situation and requirements. The plan should outline what needs to be done to reach the desired level of capability. Ideally, this plan should cover a five-year period so that long-term development projects can be properly scheduled and adequately funded. The plan should include all emergency management projects and activities to be undertaken by the jurisdiction regardless of the funding source.

Step 10. *Annual Development Increment.* With the multi-year development plan serving as a framework for improving capability over time, the next step is to determine in detail what is going to be done next year. Situations change each year and what was accomplished the year before may have been either more or less than what had been planned. These factors should be reflected in modifications to the multi-year development plan and in determining next year's annual increment. Through this process, emergency managers can provide their local officials and national counterparts with detailed descriptions of what they plan to accomplish in the coming year as well as their requirements for financial and technical assistance in support of these efforts.

Step 11. *State and Local Resources.*[1] State and local governments are expected to continue their own capability development and maintenance efforts as they have done in the past. Some activities identified in the annual increment may be accomplished solely with local resources, while others may require state and/or federal support. Whatever the source of funding and other support, each project and activity should represent a necessary building block in the jurisdiction's overall capability development program.

Step 12. *Federal Resources.* The federal government continues to provide policy and procedural guidance, financial and technical support, and staff resources to assist state and local governments in developing and maintaining capability. FEMA's Comprehensive Cooperative Agreement with states will remain the vehicle for funding FEMA-approved projects and activities on an annual basis though out the United States and its territories.

Step 13. *Annual Work Increment.* As capability development projects and activities are completed, the jurisdiction's capability assessment shortfall (Steps 2 and 8) changes as the results of the process are reviewed each year. Emergency operations plans should then be revised to incorporate these improvements. Multi-year development plans also should be modified in view of these changes and in view of the experience gained through exercises and the conduct of actual emergency operations. Each state should provide a method for recording and consolidating local annual work increments.

STARTING THE REAL WORLD PLANNING PROCESS

The ability of a community to respond and mitigate a large-scale emergency or disaster depends on its planning process. The CEM and IEMS steps overlap and are repeated many times. Emergency managers are constantly reexamining their

1. The three levels of government in the United States are: local (city, township, village, county etc.), state (State of New York, State of Washington, State of Hawaii, etc.). and federal (Washington D.C.).

needs, reassessing their goals, planning actions, and evaluating results. The disaster plan, therefore, is a living, functional document and must be maintained throughout the years.

The first step in the planning process is to identify the problems and needs faced by the community. Every community is different, both in terms of the danger it faces and the resources it has available. As an emergency manager, one of your first tasks is to determine the potential for emergency situations to occur in your community and to evaluate your ability to respond. This is accomplished by utilizing two IEMS steps: Step 1, hazard analysis, and Step 2, capability assessment.

HAZARD (RISK) ANALYSIS

A good way to determine what may happen in the future is to review the past. Therefore, the first step in conducting a hazard analysis is to review your community's history of disasters and major emergencies and to determine which of them were caused by hazards still present in your community. Talk to citizens who know local history, and teachers in the local high schools, community colleges, or universities. Find out if any major emergencies or disasters occurred before the records of your department were initiated.

The information which is collected and analyzed helps the emergency manager ascertain how vulnerable the community is to different kinds of disasters. Such information establishes part of the basis on which resource allocation decisions can be made. The hazards analysis system is essentially a process for using a common set of criteria to determine and compare the risks that the community faces from a variety of threats. Hazards are rated and scored in a way that allows for easy comparison with each other.

The rating and scoring system involves the following four criteria, each rated as *low, medium,* or *high* for each hazard identified:

- History: the record of previous disasters in the community
- Vulnerability: the number of persons who might be killed or injured, and the value of property that might be destroyed or damaged
- Maximum threat: the "worst case" scenario of a hazard, that is, that set of circumstances in which the emergency will have greatest impact
- Probability: the likelihood that the disaster will occur

To perform a hazard analysis we must first review the elements of an emergency and then consider the cascade effect (see below). The elements are

- Probability
- Vulnerability (effect on the community)

- Predictability (likelihood of the emergency's occurrence)
- Frequency
- Controllability
- Speed of onset
- Duration
- Protection action options
- Scope/intensity
- Sources of assistance

This analysis is necessary so that we may identify the resources needed to manage these hazards. It also allows us to plot our hazards on a community map, to set planning priorities, and to develop response models.

THE PLANNING PROCESS

- Identify problems and needs
- Set goals
- Determine objectives
- Set priorities
- Design action programs
- Evaluate results

The objective is to systematically identify and analyze the natural and technological hazards that threaten the jurisdiction, including the new threat of domestic terrorism, and use the results as a basis for multi-year program planning. Emergency planning should be based on those hazards that pose potential threats and significant consequences to the local jurisdiction. It is vital to understand the nature and implications of the hazards to which the population is, or may become, vulnerable.

A vital first step in this process is for the emergency manager to develop a comprehensive hazards analysis. An effective analysis must address all hazards to which a jurisdiction might be susceptible and the relative risk involved in each. The completion of a hazard analysis should result in the development of an agenda of hazard mitigation efforts and preparedness activities. In summary, use the completed hazard analysis as a factor in formulating the multi-year development plan.

THE CASCADE EFFECT

Part two of a hazard analysis is to develop a cascade effect model. The cascade effect is what happens when one emergency triggers others. For instance, what other sorts of hazards might be triggered by a truck bomb? By recognizing the cascade effect and relating this phenomenon to community hazards, the emergency manager can identify several primary hazards capable of triggering a wide range of other hazards (see Figure 2–1).

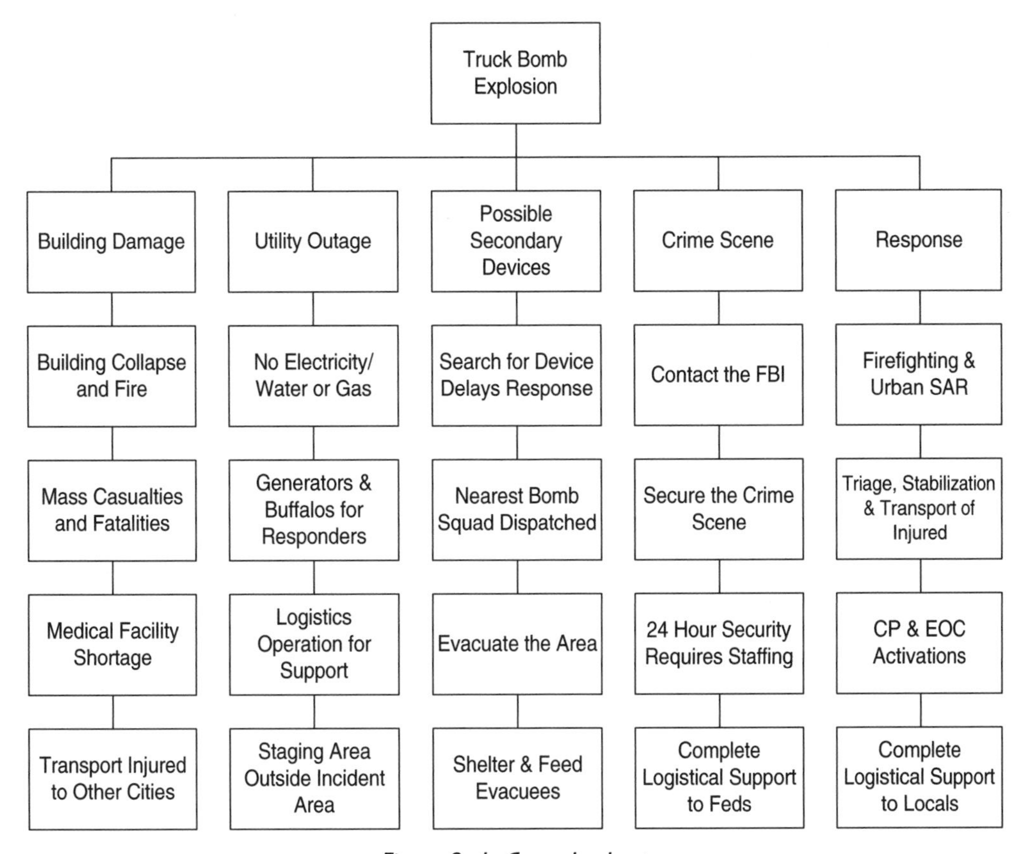

Figure 2–1 Cascade chart

KEY TERMS IN HAZARD ANALYSIS TOOLS

Elements at risk

- The population
- Structures
- Public services
- Continuity of government
- Economic
- Political activities

Emergency/Disaster

- Hazard/hazard agent
- Incidence
- Local conditions
- Physical properties
- Probability
- Risk
- Risk reduction measures
- Vulnerability

CAPABILITY ASSESSMENT

Hazards analysis describes the dangers that a community may face. A capability assessment describes the likelihood that a community will deal well with those dangers. A list follows of the functional areas, and the major elements within each area, for which assessment criteria have been developed. A review of this list will provide a snapshot of the activities and resources that an effective emergency management organization should be able to provide.

EMERGENCY MANAGEMENT ORGANIZATION

- Legal authority
- Budget development
- Selection and training of the coordinator and staff
- Hazards analysis and multi-year development
- Written agreements for aid and resources
- Private-sector support

EMERGENCY OPERATIONS PLANNING

- Responsibilities
- Plan components
- Plan content
- Approval and promulgation
- Plan distribution
- Plan maintenance

RESOURCE MANAGEMENT

- Evaluation of needs
- Planning and preparedness
- Timely and effective utilization

DIRECTION AND CONTROL

- State and local facilities
- Assignment of responsibilities
- Protection
- Emergency operations
- Life support
- Damage assessment
- Maintenance

EMERGENCY COMMUNICATIONS

- Primary emergency communications
- Backup emergency communications

STAFFING AND TRAINING

- Protection
- Emergency power
- Standard operating procedures
- Emergency Broadcast System
- Maintenance

ALERTING AND WARNING SYSTEMS

- Point of contact for official information
- Warning points
- Staffing and training
- Emergency power
- Standard operating procedures
- Special locations and arrangements
- Maintenance

EMERGENCY PUBLIC INFORMATION

- Point of contact for official information
- Coordination and authoritative spokesperson
- Development and distribution of Emergency Public Information (EPI) materials
- Rumor control
- Media
- Reentry
- Emergency Broadcast System

CONTINUITY OF GOVERNMENT

- Lines of succession
- Safeguarding essential records
- Predelegation of authority
- Alternate headquarters
- Protection of government resources

SHELTER PROTECTION

- Congregate lodging facility
- Standard public shelters
- Shelter upgrading
- Expedient shelters
- Stocking
- Marking
- Reception and care
- Shelter managers

EVACUATION

- Preparation
- Movement

PROTECTIVE MEASURES

- Prevention of individual exposure to hazards
- Mitigation of impending or actual exposure
- Classification system and action level system
- Supply and maintenance of protective devices
- Procedures for recovery and reentry

EMERGENCY SUPPORT SERVICES

- Preparedness
- Law enforcement
- Fire and rescue
- Public health
- Transportation
- Medical services
- Public works
- Utilities

EMERGENCY REPORTING

- Monitoring
- Collection
- Processing
- Analysis
- Ddissemination

SETTING GOALS

Determining objectives is the first step in developing an Emergency Operations Plan (EOP). In this plan you begin to decide what you want to accomplish in terms of the kind and extent of emergency management capabilities, and in what period of time you want to accomplish it.

The objective of all-emergency management functions is to develop and maintain a comprehensive EOP based on hazard analysis, existing resources, and current operational capabilities in order to deal effectively with any kind of emergency or disaster.

GETTING ORGANIZED

To a great extent the abilities of you, your department, and your community to respond quickly and effectively to a large-scale emergency depend upon the quality and scope of the planning process.

In some ways the phrase "planning process" is misleading because it may imply that planning is a one time effort performed in advance of a disaster operation. People sometimes think that once a document titled "Disaster Plan" has been produced, the process is over. On the contrary, the plan itself may be less important than the process that produced it.

Plans and standard operating procedures grow old and can quickly become outdated. The planning process, on the other hand, is ongoing. Planning is as much an attitude involving management orientation as it is group of procedures that produce a document. Thus, in a real sense, planning is never finished.

An EOP is a document that contains information about actions that may be taken by a governmental jurisdiction designated by the emergency manager to protect people and property in the time of a disaster or the threat of a disaster. It describes actions that may be required for any emergency, natural or man-made. It details the tasks that are to be carried out by specified organizational elements, public and private, during a disaster, based on established objectives and assumptions, and on a realistic assessment of capabilities.

MULTI-HAZARD, ALL-HAZARD, FUNCTIONAL PLANNING

There are numerous emergency management requirements that are common to any disaster situation regardless of the cause. Experience has shown that plans developed for one type of hazard can be very useful in coping with other emergency situations. EM capabilities can be developed by building a foundation of broadly applicable functional capabilities in such areas as command and control, warning, communications, evacuation, and sheltering. Multi-hazard functional EOPs, therefore, begin by providing for basic capabilities without reference to any particular hazard. Hazard-specific planning, within the multi-hazard planning process, is to focus on those requirements that are truly unique and are not properly covered by the planned generic capability.

HISTORY OF INCIDENT COMMAND SYSTEM

The identification of a need for an Incident Command System (ICS) stems from a series of fires and multi-agency responses in the 1970s. In Southern California, a series of wildland fires over a thirteen-day period burned more than 500,000 acres in seven counties, destroying 800 structures and killing sixteen people. The economic loss approached $233 million. An analysis of the emergency response to these fires indicated that there was no management mechanism or resource allocation process in place which could coordinate the federal, state, and local resources needed to respond effectively to future wildland fire emergencies.

In 1972, Firefighting Resources of Southern California Organized for Potential Emergencies (FIRESCOPE) was established to assist the fire service in Southern California. FIRESCOPE's charge was to assist in the coordination of multi-agency emergencies or disasters which could not be contained by one jurisdiction's resources. Congress provided the U. S. Forest Service Fire Research Laboratory with funds to assist in the development of this coordinated response plan.

FIRESCOPE developed what is known today as the Incident Command System, or ICS. It is designed to be used as a multi-agency, multi-jurisdiction, all-hazard system of command and control. Although originally designed for fire service responses to wildland fires in Southern California, the ICS is now being utilized throughout the state of California by most of the various emergency response agencies in almost all disaster situations. The ICS is designed to include all agencies which may be involved in emergencies or disasters, such as fire service, emergency medical service, law enforcement, emergency management, public works, etc.

During the early 1980s, fire departments across the country saw the success of the ICS during emergencies and disasters, and identified the need for some type of command and control system for everyday fireground use. Most fire departments used the ICS as a model, and began developing their own incident management systems for day-to-day emergency incidents, rather than for multi-agency, multi-jurisdiction use. The most notable of these systems was the Fireground Command System, developed by the Phoenix Fire Department. A very popular training package for the Fireground Command System was developed and marketed to fire departments nationwide by the National Fire Protection Association (NFPA).

In 1987, shortly after the passage of NFPA 1500, the standard on the Fire Department Occupational Safety and Health Program, the NFPA Technical Committee on Fire Service Occupational Safety and Health initiated a project to look at the different types of incident command and incident management systems being utilized across the country. The intent of the project was to identify the main components of successful systems, and to develop a generic standard on these systems to ensure that such systems would have certain common characteristics. That way, regardless of the system used by a department, there would be some level of compatibility among systems that met this standard.

Key players from FIRESCOPE, the Phoenix Fire Department, and many other organizations, agencies, and departments met several times during the development of the standard. The group found more commonalities than differences in the systems; however, the procedures within the systems were by nature different for a "routine" emergency than for a large, disaster-type incident. The group agreed that each of these systems was designed to help manage an incident, and that they should generically be called "incident management systems". This would also eliminate confusion between FIRESCOPE's ICS and a department's incident management system used for routine emergencies.

In 1991, following the adoption of the NFPA 1561 Standard on Fire Department Incident Management Systems, many of the groups involved in the development of the standard recognized the need to continue their efforts, working toward a single, national incident management system. Representatives of these groups formed the National Fire Services Incident Management System Consortium, which now includes representatives of some forty fire service agencies across the country.

One of the initial efforts of the consortium was to integrate the strategic and tactical procedures of the Fireground Command System with the overall structure of ICS. They have developed the *National Fire Service Incident Management System Model Procedures Guide for Structural Firefighting*. This Model Procedures Guide, published by Fire Protection Publications of Oklahoma State University, is intended for use by the fire service for incidents in which fewer than twenty-five units have been committed to the scene. The consortium recognized that ICS is the foundation for any model incident management procedures, and that beyond twenty-five units, an incident has escalated into an area best managed by FIRESCOPE's ICS.

The consortium, including representatives from FIRESCOPE and the Phoenix Fire Department, has requested that FEMA's United States Fire Administration include the Model Procedures Guide in the instructional curriculum of USFA's National Fire Academy (NFA) for structural firefighting operations. FEMA selected ICS as a model system in the early 1980s; ICS currently serves as the standard instructional underpinning of the curriculum offered by the NFA. The Model Procedures Guide would be a natural extension of NFA's program, and would enhance their curriculum. The consortium has agreed to turn over its model procedures to NFA for instructional purposes.

ICS can be used as a framework for managing hazardous materials emergencies, emergency medical service and mass casualty incidents, and other all-hazard situations. More and more organizations, including the Integrated Emergency Management System National Advisory Committee, have endorsed the ICS concept and the use of the consortium's model procedures.

SUMMARY EXERCISING

After the planning process has been completed, the testing operations must begin. Exercising provides an organized approach and a controlled environment for plan evaluation.

Why exercise? Because of their structured and controlled approach, exercises can provide better evaluation of the EOP than can actual events where results cannot be attributed directly to the emergency operational plan, but may be the product of chance or luck. Also, under the Superfund Amendments and Reauthorization Act (SARA)/Title III, exercises are mandated for the testing of hazardous materials contingency planning. FEMA Region VI has one of the best exercising programs in the nation.

The variety of needs for exercising has produced a selection of exercises, each designed to suit specific circumstances. The five types of exercise activities include

- Orientation seminars
- Drills
- Tabletop exercises
- Functional exercises
- Full-scale exercises

The ongoing process for an effective emergency management system is a plan maintenance program. The program defines the circumstances that warrant upgrading of the plan and the appropriate strategies for maintaining a plan, including

- Post-emergency response evaluation
- Exercise evaluation
- Resource base changes

- New hazards
- Identifying appropriate strategies for maintaining the EOP

Maintaining the EOP is part is of the cyclical planning process. The maintenance of emergency management systems is never ending.

SUMMARY

CEM and the IEMS steps described in this chapter are intended to serve management at each level of government by providing basic information upon which reasonable and justifiable plans can be made and effective action can be taken to increase emergency management capabilities. Government agencies will realize benefits from the process almost immediately. It will take time, however, to achieve total integration of emergency management activities and to develop the capabilities required to perform the functions necessary to deal effectively with all hazards. It will also take time and practical experience to refine the process and to develop the best guidance to assist in its implementation. Utilizing cooperation and constructive criticism from emergency managers and professionals at all levels, public and private, you will continue to make progress toward your goals.

Just as there is no typical emergency or disaster, so there is no single emergency management organizational structure that is ideal for every community. The CEM and IEMS concepts, however, do recognize that disasters have common phases and that a systematic emergency management approach is best for dealing with these phases.

Although communities and their emergency management organizations differ, three positions are key to effective local emergency management. In some places each position is filled by a different individual; in others the same individual fills more than one position. These positions and their functions are:

Director of emergency management—This person is responsible to the governmental entity for emergency management and provides overall policy and direction.

Emergency management coordinator or planner—This person generally is responsible for overall community disaster planning and coordination.

Incident commander—This person actually directs the disaster operation in the field.

The Basics of Emergency Management: Putting It All Together

OVERVIEW

In this chapter you will be shown how to put all the basic building blocks of an emergency management system together. Topics included here are

- Your role as the emergency manager
- Job title and organizational chart
- Hazard identification and planning
- Staffing issues
- Office and professional duties
- Management, leadership, and training

In assisting other jurisdictions with bio-terrorism planning, several issues seem to always be apparent. The most common issue is that of turf and unfamiliarly of the other agency. It has been interesting to hear the opinions one agency may have of another—yet they have rarely interacted with each other. The opinions seem to be coming to an agency second hand through other agencies that deal with them, yet the reality never seems to match the opinions formed. When multiple agencies get together, the lack of interagency understanding can lead to turf battles. One great way to start the planning process is with a brief overview from each agency. This could include the agency mission statement, roles and responsibility, specialty teams, and available resources. Then all agencies will have an equal understanding of each other. If you are in the middle of the planning process and are not functioning well then going back to the beginning may help immensely. In some cases a particular agency may overestimate it's own role in an emergency. Departments need to look at their responsibilities and learn to give up some of that control as needed during emergency situations. During large events state and federal resources may overwhelm local personnel. The overall goal is to learn to work together at a local, state and federal biological terrorist event. The following material covers coordination of multiple agencies.

One goal of emergency management is to coordinate a unified plan for, and response to, a crisis; using the CEM and IEMS ideas. The terrorism/emergency manager and planner is not expected to be an expert in every area, but will be expected to know where to get expert help. The emergency management office is a broker for resources. The emergency management office should have a resource list to help locate experts, equipment, and supplies for disaster emergencies.

Many emergency management offices are "out of sight, out of mind". Local people and agencies have little idea of what the emergency program does day-to-day. This chapter outlines some of the daily and long range activities emergency management offices perform. The topics below expand upon the issues discussed in the readings.

YOUR ROLE AS THE EMERGENCY MANAGER

Emergency management is not unlike any other job: staff are often overworked, underpaid, and receive little recognition. Emergency management responsibilities themselves are often secondary to the immediate needs of the day, making the completion of long term projects difficult. In this profession, the saying "disasters ensure job security" holds some truth. Communities don't think of emergency management until a disaster strikes.

The FEMA standard for number of personnel assigned to the emergency management function is one full-time staff person for every population of 25,000. However, this standard is rarely met, and it is not unusual to see a 1:200,000 ratio. One reason for this may be budget constraints, and another may be the perceived lack of need for program managers.

Daily chores for emergency managers are many and diversified. As the previous chapter indicated, there is a need to submit budgets, make purchases, administer programs, write plans, attend meetings, and train and educate the public. Resource listings and phone numbers are continually changing and need revision. Emergency managers must allocate their time to allow some attention to each issue. Priorities change based on the hazard environment (e.g., hurricane season) as well as political changes (e. g., elections and public demands). Consequently, an emergency manager could easily become reactive rather than proactive.

One important daily responsibility of emergency managers is to monitor world, regional, and local events. The term "situation awareness" best describes this. Using television, newspapers, and other forms of news media, you should continually seek to understand current events and the implications they may have for your jurisdiction. An event occurring thousands of miles away may ultimately impact your community as a hazard or a demand for resources.

JOB TITLE AND ORGANIZATIONAL CHART

Where the emergency management office is located on the organizational chart is indicative of how important is it perceived to be by the local jurisdiction. Many offices are located in National Guard departments, sheriff's departments, or fire departments. A few are stand-alone and report directly to the jurisdiction's highest officer, such as the county administrator or city manager. And some are part of other, unlikely departments such as Public Works or Roads and Sanitation.

There are many different titles for emergency managers. The job title may include the word coordinator, director, planner, manager, operations officer, etc. It may also indicate the level of influence the position commands. It is important to have an appropriate title for each staff position.

HAZARD IDENTIFICATION AND PLANNING

Hazards change; old ones disappear and new ones emerge. The emergency manager must continually be aware of this process. We are all rewriting and/or reexamining our books and plans after the terrorist attacks of September 11, 2001. The hazard vulnerability analysis is the bedrock document on which to build your program. Keeping this document updated is very important, since many other agencies rely on it to develop their own plans.

STAFFING ISSUES

Whether there is full time or part time staff, the emergency manager must continually seek out people whose services may be needed during a disaster. To do this, the emergency manager must maintain an "emergency partnership" with government agencies and must coordinate plans and activities with local, state, federal, and private organizations. Many volunteer agencies provide specialized services. Some will provide labor pools.

For example, consider the Explorer Cadet Corps. During an emergency, they can provide "runner" services and perform simple tasks, as long as they are never put in harm's way. Their preparedness and response training makes this Corps familiar with emergency management operations. The long term benefits for the community are obvious as these young men and women grow up and enter the workforce.

OFFICE AND PROFESSIONAL DUTIES

The emergency program manager develops and implements plans to establish and maintain the facilities and equipment. The EOC is the most important facility for emergency management. Without the proper equipment and configuration, the community may not function efficiently during a disaster.

Office administration duties include ensuring that enough consumable supplies are available. Remember that when the EOC is activated, the demand for pens and paper will escalate.

We live in a changing environment in which new technology simplifies many tasks. The emergency manager should continually search for new and simpler methods to automate emergency program tasks. Flexibility is the key to a successful emergency operation. When the EOC in New York City was destroyed in the terrorist attack of September 11, 2001, city officials needed to develop an alternate facility immediately. That they did so is a tribute to the men and women of that city. Emergency personnel around the nation will need to be able to match the level of professionalism of those in New York.

RESOURCE LISTS

The emergency program manager must keep an accessible resource database. The most comprehensive listing is the telephone company Yellow Pages. Beyond that, the database should be filtered and refined into an updated, comprehensive listing for immediate use. This listing should also include any state and federal agencies that could assist during a disaster.

An outdated database is as bad as no database. After agencies and businesses that can provide services are identified, they should be contacted at least annually to update names, phone and pager numbers, and addresses. The database should include a short summary of the type of resources each agency can provide. Reimbursement procedures and contracts should also be summarized.

Developing and maintaining a comprehensive listing could easily be a full-time job. Emergency managers may wish to develop an abbreviated listing that prioritizes the most likely or the most difficult services needed.

MANAGEMENT

The primary functions of an emergency manager are

- To activate the EOC
- To implement emergency plans
- To serve as the director of emergency activities
- To take charge of damage assessment and recovery
- To support the Chief Executive Officer (CEO)

INFORMATION

Disaster situations require many types of information. Technical data, such as the chemical makeup of a hazardous material or structural information about a building or a bridge, may be needed. Historical data may be needed to answer weather and flood questions. Fortunately, much of the needed information is becoming readily available on the Internet. Researching websites and developing a professional library is crucial to having necessary information available for decision making.

LEADERSHIP

The emergency manager is viewed as a leader during disasters. If an emergency manager fails to exercise the appropriate leadership, someone else will, and regaining control may be difficult. Sometimes the emergency manager may be called a "coordinator" rather than a "director". As a coordinator, it may be difficult to tell the police or fire chief what their departments should be doing, yet you must gain their cooperation and support. Learning the techniques to become a leader and to overcome the organizational obstacles is so important in emergency management that entire courses are devoted solely to that.

PROFESSIONAL TRAINING

It is imperative that emergency managers seek continuing training and knowledge. A great deal of training and information is readily available by computer. Many colleges now offer courses in emergency management. FEMA offers an in-residence course and home-study courses similar to the extra credit Emergency Program Manager course. Additionally, state emergency management offices also offer training at regional locations. Some of the courses offered include EOC management, radiological training, management and leadership, and hazard-specific courses. Some emergency management training is specialized enough that a local organization may be the only one which offers it.

NON-EMERGENCY MANAGEMENT-RELATED DUTIES

Sometimes emergency management offices are perceived as doing little except during a disaster, a perception that can lead to the assignment of non-emergency management-related tasks. This can hinder accomplishing the daily responsibilities each office may have. It can also lead to "mission creep", when the emergency program manager is assigned additional responsibilities.

For small agencies, one task likely to be added is the operation of the 911 or dispatch communications center. This is a logical and many times beneficial responsibility, since the EOC is an extension of the communications center. Still, the emergency manager must ensure that mission creep does not consume valuable staff time and leave true emergency management tasks undone.

RESPONSIBILITIES TO EMPLOYER OR ORGANIZATION

Emergency management offices coordinate with, and many times are assigned within, other organizations (police, fire, paramedical, etc.) that are military in style. Personnel wear uniforms, have a visible rank structure, and have high expectations that their members will perform their duties as needed. Many of these people are at risk themselves from the hazards inherent in their jobs. Such organizations cannot function efficiently without rules. For example, though some offices enforce strict time standards, many are flexible enough to allow emergency program managers to move about the community at will. Although meetings and training events will fill your calendar, and many of those events will depend on your participation, still, learning good work habits is essential in gaining the respect of your peers. Punctuality is one trait which can gain you a good reputation very quickly. As the saying goes, "If you can't be on time, be early!"

PLANNING

There are two approaches to disaster planning: the military approach and the IEMS approach. Though the military is relatively rigid in its planning, it also has the resources to train and exercise as needed to develop the desired capabilities. In civilian organizations, the level of interorganization cooperation is lower because of the much greater number of organizations. Lack of resources to train and exercise contribute to lower levels of readiness. However, planners who adopt the principles of planning discussed here can significantly increase their chances of success.

Using IEMS contributes to the standardized planning concept for which each organization strives. The military uses a "command and control" model, whereas civilian organizations use a "cooperation, coordination, and communication" model. These concepts are not mutually exclusive. Both military and civilian organizations use the two concepts and with much success. (Remember these terms for discussion in later lessons.) With the IEMS concept comes the idea of all-hazard planning vs. disaster-specific planning. IEMS extracts the common disaster management functions and can create a "one plan fits all" document. Combining functions reduces the size and improves the functionality of the plan. It must, however, still contain disaster-specific information not covered in the common functions. This information should be in a disaster-specific addition to the plan. Some hazards, such as hazardous materials and nuclear power plants, may be discussed in stand-alone documents, either by choice or by law.

WHY PLAN?

The following quotations help illustrate why planning is both important but many times inadequate:

> "It would not surprise me to find that actual Soviet war plans are completely inappropriate. War plans usually are so, because they are drawn up in peace rather than in war. There once was a saying among staff planners that the purpose of war plans is not to give the initial tactical operations but to give people a chance to exercise their skills in peacetime and to jog their memories on their responsibilities. Thus the first thing one does when a war starts is to throw away the war plans and write something sensible. But there is no longer time to do that. If the war plans are not sensible, the war will not be sensible." – Herman Kahn, *On Thermonuclear War*, (1977) p. 165.

> "I cannot imagine any condition which would cause a ship to founder. I cannot conceive of any vital disaster happening to this vessel. Modern shipbuilding has gone too far for that to happen." – Spoken about the *Adriatic* by Captain Edward Smith, 1906. Captain Smith was lost when the *Titanic* sank in 1912. The belief that the ship was unsinkable, coupled with the shortage of lifeboats, caused the loss of over 1,500 lives. Technology, no matter how advanced, is still vulnerable to the forces of nature, human failure, technological failure, and terrorist attack.

THE CASE FOR PLANNING

- Planning makes an important difference.
- Planning reduces unknowns.
- Planning should be based on research and knowledge.
- Planning should evoke the appropriate actions.

PLANNING GUIDELINES

- A plan must be based on facts or valid assumptions.
- Planners should use the best available knowledge base (behavioral and organizational).
- Do not limit planning experience to one or two disasters; use a larger sample size. Do not rely on past personal experience.

- Use scientific research. Researchers and practitioners should work together to create a rational emergency plan. A good base of scientific research is available now at many universities around the United States. For additional information on this subject, see FEMA's EMI Higher Education project (www.fema.gov, or contact the University of South Florida in Tampa).
- Disaster planning is not disaster preparedness. A completed plan gives a false sense of security. A plan must remain a living, fluid docment, always changing and being updated. Without training and exercising, a plan will probably not work during an actual disaster.
- The best plans are the shortest. Many times, plans are large because of institutional (corporate or legal) requirements. The state of Florida's Fire-Rescue Disaster Response Plan, managed by Florida's Fire Chiefs Association for the deployment of overhead teams, is less than fifty pages and is designed very well.
- A planner must accept the research-based general principles of disaster planning. A comparison can be made with a physician. A physician (planner) who treats (develops a plan for) a patient (community) using false or misleading assumptions, no scientific research, and little specialized training is creating a recipe for failure (a dysfunctional plan).
- Planners should try to anticipate problems and possible solutions.
- Planners should try to evoke the appropriate actions rather than simply improve response speed.
- Planners should be based on what is likely to happen: on what people will do, not what you want them to do.
- Planners should focus on general principles and not specific details.
- Planners should focus on training and education.
- The planning document is not the final solution.

WRITING THE PLAN

To write a plan requires a strategy. Who is going to contribute to and coordinate the plan? What areas and entities should the plan encompass? For instance, it should include response and support agencies; anyone who may be actively involved should be considered. The plan does not negate an individual agency's plan but the two should be coordinated.

Use the Hazard Vulnerability Analysis developed for the community and include the threat of terrorism. The analysis explains what hazards should be addressed and who and what is vulnerable. The IEMS concept introduced in Chapter 2 is nationally recognized, so it makes good sense to create the plan using this concept. Planning integrates the community's resources and trains agencies and individuals. After training, an exercise tests the plan to further refine the planning process.

PLANNING VS. IMPROVISATION

No disaster is managed without both planning and improvisation, but they must be balanced. A overly detailed plan leaves the user little room to make decisions on the run. Allowing improvisation adds flexibility for unforeseen events.

PLAN COMPONENTS

There are various ways to organize the plan. Many plans follow the FEMA Community Preparedness State and Local Government (SLG) 101, formerly called Civil Preparedness Guide (CPG) 1-8. Its outline contains headings for:

- Basic plan
- Annexes
- Appendices
- Standard operating procedures

TECHNICAL WRITING

Technical writing conveys specific information about a technical subject to a specific audience. Technical writing in our context is the science of communicating the plan to the user. The user may be a state governor or a volunteer relief worker. Research who your audience is and make the plan readable for them.

Creating a plan requires technical writing skills. To illustrate how technical writing can convey information quickly, consider the following experiment. During a class, a United States Air Force technical manual over 500 pages long was given to a student and he was asked a technical question. Within three minutes he had the answer. With a good table of contents, a comprehensive index, and a logical subject progression, the answer to any question can be found with little difficulty.

Technical writing should be characterized by:

- Clarity: single meaning
- Accuracy: no mistakes
- Comprehensiveness: providing all needed information
- Accessibility: ease of locating information
- Conciseness: the shorter the better

WHAT YOUR LOCAL EMERGENCY OPERATIONS PLAN SHOULD INCLUDE

WHAT MAKES A GOOD PLAN

- Addresses all functions
- Contains a basic plan, functional annexes, and hazard-specific appendices (for requirements unique to certain hazards)
- Is updated regularly
- Is based on facts or valid assumptions
- Is based on community resources inventory
- Provides organizational structure
- Uses simple language
- Has coordinated elements
- Has EOP characteristics
- Is a "Living Document"
- Is the first thing you reach for when an incident occurs
- Keeps concepts simple with few details
- Assigns responsibility to those who know the mission best

SIGNS OF A BAD PLAN

- Is used as door stop
- Weighs over three pounds
- Has dust on it
- Cannot be found during a disaster
- Agency has only one copy
- Contains outdated phone numbers and names

TESTING THE EOP TYPES OF EXERCISES

- Orientation
- Table top
- Functional
- Full-scale
- Actual event (can count as exercise)

EMERGENCY SUPPORT FUNCTIONS (ESFs)

The Federal Response Plan adopted (www.FEMA.gov, on FEMA's webpage type in search box "FRP") a new format utilizing ESFs. These are the common functional areas similar to the functions (direction and control, medical, transportation, warning) outlined in CPG 1-8 or SLG-101. The ESFs are:

1. Transportation
2. Communications
3. Public works
4. Firefighting
5. Information and planning
6. Mass care
7. Resource support
8. Medical and public health
9. Urban search and rescue
10. Hazardous materials
11. Food and water
12. Energy

States may include these additional ESFs:

13. Military support (non-federal)
14. Public information
15. Volunteers and donations
16. Law enforcement (non-federal)
17. Animal issues

Others in some communities and states include business and industry, tourism, etc.

See also FEMA's website (www.fema.gov).

TERRORISM CONSEQUENCE MANAGEMENT PLAN

COMPONENTS OF A BIOLOGICAL TERRORISM PLAN

General Biological Terrorism Concepts

Overview

Biological Emergencies

Biological Response

Integraton of Federal, State, County, and Local Responses

Metropolitan Medical Strike Team (MMST)

Command and Control

Intelligence and Early Prevention

Ongoing Preparedness Efforts

Concept of Operational Management Issues

Emergency Operations Center (EOC)

Crisis Management

Situation and Threat Background

Incident Types: Package, Covert Release, Threat

Threatened Use

Confirmed Presence

Actual Release

The Agents

Planning Strategies

Threat Response Management

Recognition and Evaluation Strategy

Administration

Training and Exercising

Plan Maintenance

Your State's Mutual Aid Pact

Biological Response Operations Outline

Detect

Event

- Unannounced
- Announced

Assess Event

Conduct Internal Notification

Determine the Course of Action

Respond to Emergency

Conduct Public Notifications

End Event

Stages of Severity

Strategies and Actions

Decision Factors

Specific Treatment Models

Biological Cross-Function Action Chart

Biological Response Operations

Suspicious Outbreak of Disease

Notifications

In Case of an Alert

In Case of a Warning

Threatened Use

Confirmed Presence

MMST

MMRS Biological Agents

MMRS Biological Agent Scenarios

Treatment vs. Prophylaxis

Evaluate the Device/Release Potential

Consider Incident Consequences

Develop a Course of Action

Evaluate and Implement Protective Actions

Address Potential for Secondary Attack

Develop a Site Security/Force Protection Plan

Actual Release

Make Essential Notifications

Define Incident Objectives

Activate the MMST

Evaluate Device/Release Potential

Define the Disease/Agent

Consider Incident Consequences

Develop a Course of Action

Request Specialized Resources

Evaluate and Implement Protective Actions
Address Potential for Secondary Attack
Develop an Ongoing Intelligence Collection Plan
Real-Time Intelligence
Initial Response Concerns
Downwind Potentials
Traffic Restrictions and Congestion
Self-Transport to Medical Providers (Convergent Casualties)
Panic Victims
Scarce Supplies
Suspicious Outbreak of Disease
Consider Different Criteria
Community Awareness and Public Information Concerns
Synopsis of Biological Warfare (BW) Agent Characteristics
BW First Responder Concerns
BW Agent Dissemination
Weather Effects
Decontamination (Decon) Considerations
Recovery Concerns
Managing Mass Fatalities/Deceased Disposition
Recovery, Site Decontamination, and Restoration
Investigation

Attachments

Passive Surveillance
Active Surveillance (Epidemiological Services)
Biological Terrorism Stakeholders Group
Detection
Agent Surety
Notification
Diagnosis
Mass Prophylaxis

Medical Response Expansion Program

Logistics, National Pharmaceutical Stockpile Program

EMS Resources

Hospital Plan

ESF#8

Mass Fatality Management Implementation

Medical Examiner Expansion Program

Environmental Cleanup

Biological Agents

Pharmaceutical Needs for the Five Types of Biological Agents

Domestic Preparedness Training Courses

Appendices

Sample Check List for Agencies

Public Safety Precautions/Actions

Public Health Anthrax Threat Advisory

Complete Agent Description

Area Hospital Listings

Your County's Public Health Clinic Listings

Anthrax Threat Field Guide

Pharmaceutical Needs for the Five Types of Biological Agents

Pharmaceutical Push Package Contents

Table of Agent Types

Sample Chem/Bio Casualty Cards

Your County's Hospital Decontamination Preparedness

MMST Equipment List

- Personal Protective Equipment
- Detection Equipment
- Decontamination Equipment
- Communications Equipment
- Ancillary Equipment

Federal Response Plan, Terrorism Annex

Your State's Terrorism Annex

SUMMARY

When developing and reviewing your plans, remember to keep them short and clear and stick to the basics of a comprehensive and integrated emergency management system.

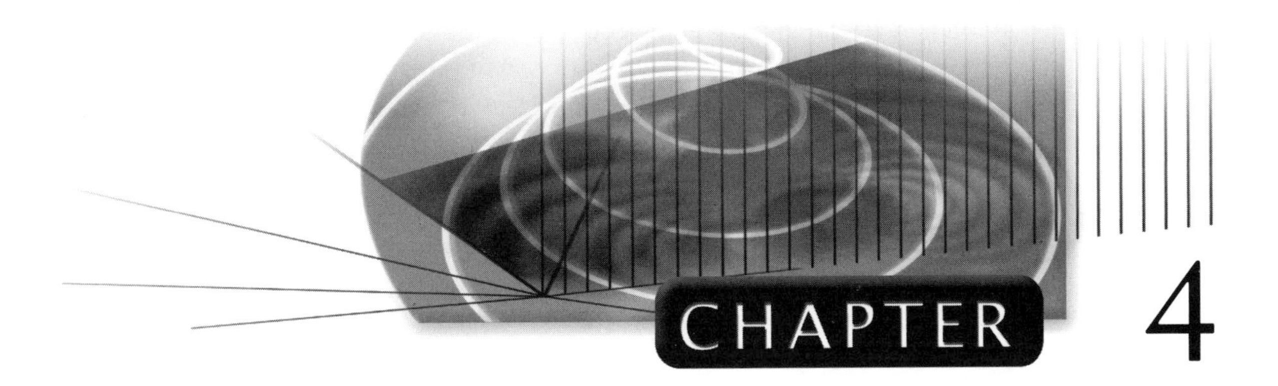

System Implications

OVERVIEW

Given a hypothetical event, provide a critical analysis of your medical systems' abilities to deal with various terrorist event impacts. This chapter covers how to

- Recognize response systems that can be targeted
- Describe system self-sufficiency and limitations to capabilities
- Secure standing agreements with other regional medical facilities
- Identify impacts of long transport times upon field providers
- Recognize possibility of responder casualties if not properly managed

MEDICAL SYSTEM VULNERABILITY CONCERNS

Medical facilities are a critical part of a community's ability to respond to incidences of terrorism. Careful examination of response system vulnerabilities must include an evaluation of the medical community and its ability to protect itself from being compromised during a terrorist event.

There are no rules of conduct; no "Switzerlands" in this game. Everyone is vulnerable. Responders and any part of the response system can become targets. A common objective of the terrorist is to create mass casualties to get recognition. Other objectives might include demonstrating that the local system is incapable of protecting its citizens. For example, a primary event may be staged to overwhelm the medical community. Then when people converge on a medical facility, the medical facility is hit. First responders, traditionally thought to be the "good guys," now find themselves being targets. The first responders on the scene of the Sept. 11, 2001 attack on the World Trade Center are a tragic example. Consequently, first responders need to be trained to take measures to protect themselves from secondary devices and direct attacks. Paranoia should not reign in our thoughts. Anxiety and apprehension among the public can be dispelled through frequent, straightforward communication from community leaders.

MEDICAL INFRASTRUCTURE SECURITY

A primary consideration for any facility must be the security and integrity of the medical complex. The same vigor of protecting the facility against environmental hazards, such as high winds or earthquakes, should be applied to minimize the medical complex's susceptibility to sabotage by a terrorist bent on instilling fear in, and crippling, the community.

Pre-emptive, protective actions can include 24-hour surveillance by cameras and sensors, or a conventional security force that monitors the complex, paying close attention to suspicious activities, persons, or packages.

Sometimes our own lack of attention to processes can render us incapable of delivering services, just as if the terrorist had physically compromised the medical facility. Consider the following: if a biological or chemical agent is used to inflict the injury, what measures does the medical system have in place to detect the hazard, isolate the contaminated patients, and protect its employees (and current facility census) from also becoming contaminated? A facility must have a plan that addresses these threats.

INTEROPERABILITY AND SELF-SUFFICIENCY

How long can a facility operate if it must, for security reasons, be isolated from the community? Are there redundant systems in place to provide continuous services? These may include duplicate power feeds that are not on the same power grid, an emergency generator for electrical power, or back-up computer processing systems in the event that a computer virus is introduced into the system. How long can a facility sustain patient treatment processes before it needs additional staffing or supplies? Where is the closest relief? Is it from a "sister" corporate facility across town or in another community, or are there alliances that have been formed to share supplies and staff? Does the facility rely on state activation of Disaster Medical Assistance Teams (DMATs) or other NDMS resources to provide the relief?

How long should a facility be expected to operate without outside support? Is it reasonable to expect that municipal power, water, and waste services will continue uninterrupted? What if the terrorist picks infrastructure as one of the first targets in order to cripple a community and compound the effect of the primary event? What caches are available throughout the community to offset or extend the operating capability of medical facilities from hours to days or weeks?

RESPONDER CASUALTIES

A community must be prepared to triage victims, providing preferential treatment to responders. Preferential treatment is not really foreign to us. Firefighters will cease offensive operations at a structure fire if the structure collapses and immediately conduct a role call to determine who is missing and what part of the structure needs to be searched. In the heat of a gunfight, law enforcement will evaluate the feasibility of attempting to rescue an injured officer.

A community depends upon its front-line service providers for safety, security, and well-being. In this context, hospital staff are considered a front-line force that can be compromised if protective measures are not available to prevent contamination of the facility and staff. When a community's first responders are injured, where do replacements come from?

TREATMENT ROLES AND RESPONSIBILITIES

The medical resources available within a community to meet the needs imposed by a terrorist event might include the local health department, nursing homes, EMS, public hospitals, private hospitals, VA hospitals (NDMS member hospitals), as well as nontraditional medical facilities such as home health nursing services and walk-in emergent medical practices. What agreements can be fashioned to create a synergistic solution to the medical demands created by a terrorist event?

SUMMARY

Terrorists want to instill fear in the public by making it seem that their local state and federal governments and protective systems cannot keep them safe. This is the basic premise of terrorism. If the group or individual can create this overwhelming fear, then the public will pressure the government to bring about the political or social change the terrorist desires.

The medical establishment is part of the protective system. Therefore, instilling in the public the belief that the medical system is no longer effective may become one of the terrorist's targets. If the system can be paralyzed with fear, paranoia, overloading or destruction, then the terrorists are closer to the achievement of their agenda.

It is interesting that the public will assume risks that they believe to be acceptable. Aircraft and public transportation have been targets of terrorist acts in the past, yet the public still rides subways and flies on planes daily. After grounding all American aircrafts on September 11, 2001 due to the terrorist attacks on the World Trade Center and the Pentagon, the government immediately responded by providing a higher level of safety to the traveling public. As we resumed our normal daily activities, or as close to normal as we could make our lives after September 11th, once again many of us deemed the risk to be acceptable.

Event Management

OVERVIEW

Covered in this chapter are the basic concepts of the Incident Command System (ICS) and the application or integration of that system to the hospital setting. Identifying and defining roles provides the structure by which the local community can operate to achieve the most favorable event outcome. These steps include

- Defining crisis and consequence management and identifying the federal agencies responsible for these management roles
- Identifying the five roles of the Incident Commander (IC)
- Identifying the importance of a unified command during the Weapons of Mass Destruction (WMD) event
- Describing the role of the local EOC and the interface between the field IC and the EOC
- Describing the organizational concepts of the Metropolitan Medical Response System (MMRS) and the Hospital Emergency Incedent Command System (HEICS)

CRISIS AND CONSEQUENCE MANAGEMENT

According to Presidential Decision Directive 39 (PDD-39), the Federal Bureau of Investigation (FBI) is the federal agency responsible for the management of any crisis created by an act of terrorism. Initially, the local law enforcement agency will assume this responsibility, but it will pass authority to the FBI when they arrive on the scene. Management of the consequences of the act (e.g., building collapse, search and rescue, medical treatment) begins with the local responders and is supplemented by FEMA. FEMA is able to provide many resources that may be beyond the reach of the local community or even the state.

Local first responders are going to be saddled with the initial brunt of the event. Just as a building needs a foundation that is engineered to hold up the building, so must the local community provide the foundation for the multitude of resources that will soon arrive. The sooner a management system is developed, not only at the site of the incident but also throughout the entire community, the better the event will be handled. Having an incident management system in place is not an option.

There are different versions of incident management systems throughout the country, within the military, and in local, state, and federal government. The concepts presented here are therefore somewhat generic. The goal of this section is to show how these different incident management systems can work together using their common elements to fit them into one operating process.

The agencies that are coming to assist will be expecting some sort of management system to be in place and operating when they arrive. Failure to provide the framework for the outside resources to fit into will jeopardize responder safety, result in unproductive actions, and may be measured in lives lost or increased injuries.

The intent of this discussion is to present examples of management systems and specific organizational structures so that the common elements can be observed and utilized to ensure the best response and recovery operations possible.

INCIDENT OPERATIONS

The most common form of an Incident Management System (IMS) or Incident Command System (ICS) found in this country consists of five sections: Command, Operations, Planning, Logistics, and Administration or Finance. Each of these sections is responsible for certain aspects of the incident. With a unified command directing regularly scheduled planning and briefing cycles, many agencies can work cooperatively to bring the incident to a safe conclusion.

A unified command structure should be implemented whenever multiple agencies have jurisdiction in an event, when the event crosses jurisdictional boundaries (e.g. floods covering numerous communities), or when several agencies have legal responsibility in an event (e.g. hazardous materials events with involvement of environmental regulators, Emergency Medical Services (EMS), and fire services). Unified command principles ensure that objectives are developed that meet the needs of all the participating agencies. Planning meetings with all agencies represented will allow a comprehensive Incident Action Plan (IAP) to be developed for each operational period. This IAP, used in shift briefings, helps ensure consistency of efforts and therefore results in safer operations.

Local responders should practice the principles of an IMS on every call. Determining the responsibilities of, and interactions among, fire services, law enforcement services, and EMS on a scene should not be an issue. What will matter now are the cooperative efforts of each agency that will allow each of them to complete their responsibilities during the event.

A typical organizational chart, built by local responders for a reinforced and extended operation, might look like Figure 5–1. In the case of a WMD event, a

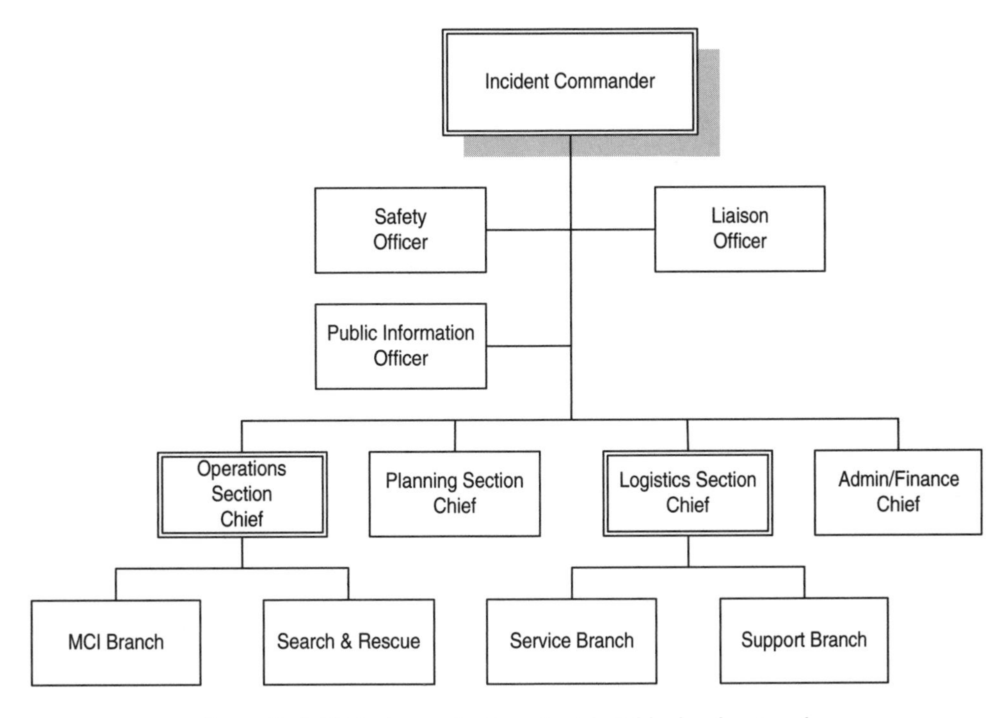

Figure 5–1 Typical organization chart buitd by local responders

Unified Command should be implemented. A typical organizational chart for Unified Command might look like Figure 5-2.

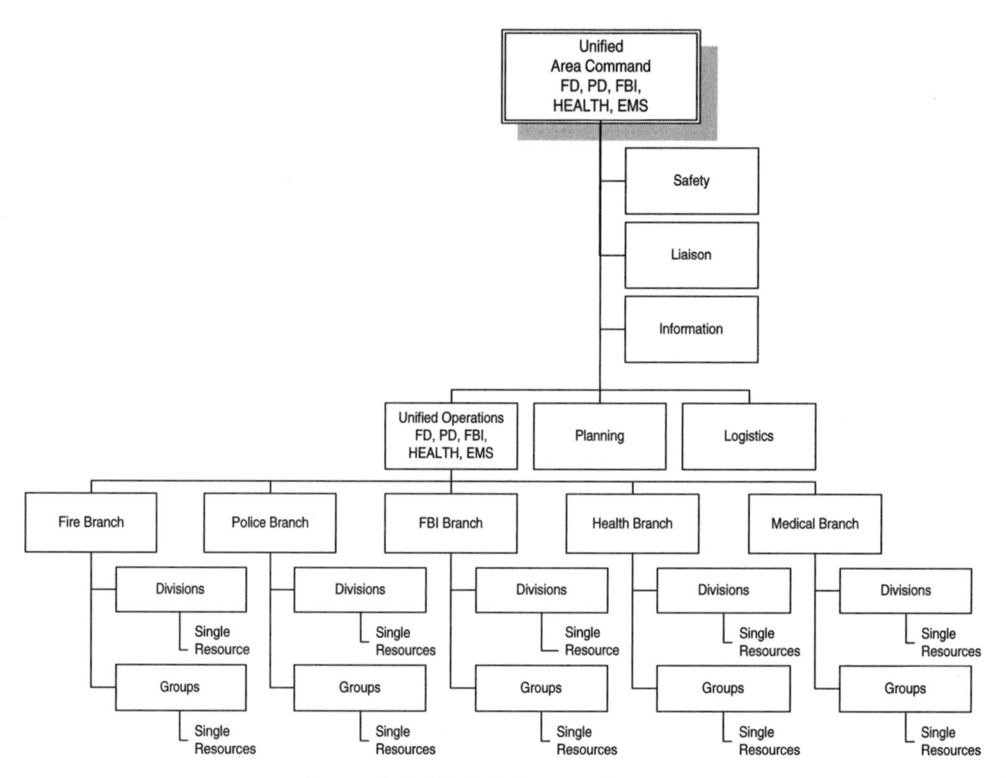

Figure 5–2 Unified Command structure

EMERGENCY OPERATIONS CENTER OPERATIONS

The next level of management and coordination within a community will most likely be the EOC. Administrative functions are usually given to the EOC staff, while the IC retains on-scene authority: the EOC manages the entire community during the incident, and the IC manages the actual incident. These areas will sometimes overlap and conflicts may develop. In such situations, the EOC, which has responsibility for the entire community, will prevail. The EOC will not dictate actions at the scene, but may influence changes in the action plan or goals by advising the IC on resource availability. While the IC is making plans and directing operations at the event site, the IC's resource requests will most likely be directed to the EOC as soon as it is activated. The EOC will make requests for state resources.

Most state EOCs have people responsible for Emergency Support Functions (ESFs). Federal resources are managed through the ESF categories and federal coordinating centers are organized around the ESF concept. Without careful planning and exercising, confusion will reign as local responders try to communicate through their EOC with resources that understand the ESF language. The meshing of IMS and ESFs usually takes place at the local EOC. This is referred to as the ICS/EOC interface.

As examples of ESF positions, ESF#4 is Firefighting and ESF#8 is Medical. On a federal level, the United States Public Health Service (USPHS) is the lead agency for ESF#8. An example of a federal resource that can come through ESF#8 is an MMST, now called a Metropolitan Medical Response System (MMRS).

If local governments expect to be able to fully utilize the state and federal resources available to them, they need to make the ICS/EOC interface a smooth transition from scene actions to support. Again, outside resources are looking for a place to fit in to what has already been started and to be assured of the safety of their personnel.

METROPOLITAN MEDICAL RESPONSE SYSTEM (MMRS)

The United States Public Health Service Office of Emergency Preparedness (USPHS/OEP) has supported the development of a MMRS as a locally available, Nuclear-, Biological- or Chemical-trained (NBC-trainded) incident response team and a component of ESF#8 of the Federal Response Plan (FRP). Within this framework, tactical medical response strike teams will be mobilized for large-scale NBC terrorist events on a local, state, and national basis.

The MMRS is composed of the following: a Task Force Leader (TFL), an assistant TFL, a medical operations physician, a safety officer, a two-person communications team, and an administrative officer. There are five sectors:

1. The Medical Information/Research Sector includes a leader, a public health specialist, and a toxicologist who is responsible for toxicology, public health issues, education, training, and a technical information repository.
2. The Field Medical Operations Sector includes a leader, thirteen EMTs/paramedics assigned to the medical ops team and ten HAZMAT technicians assigned to the HAZMAT ops team who are responsible for patient decontamination.
3. The Hospital Operations Sector, includes an emergency department RN and a doctor; and is responsible for hospital/field coordination, medical information sharing, patient dispensation and tracking, and hospital cache coordination.
4. The Law Enforcement Sector includes five sworn officers who are responsible for intelligence, team security, scene security, and evidence control.
5. The Logistics Sector includes two logisticsspecialists who are responsible for team logistics, equipment/cache, medical resupply and pharmacy.

INTEGRATION WITH FEDERAL GOVERNMENT

Local ICs and the MMRS leadership must recognize that the FBI has full authority (PDD-39) over NBC terrorist events, but will operate in a unified command structure with the IC during the response and rescue phase. Once all viable victims have

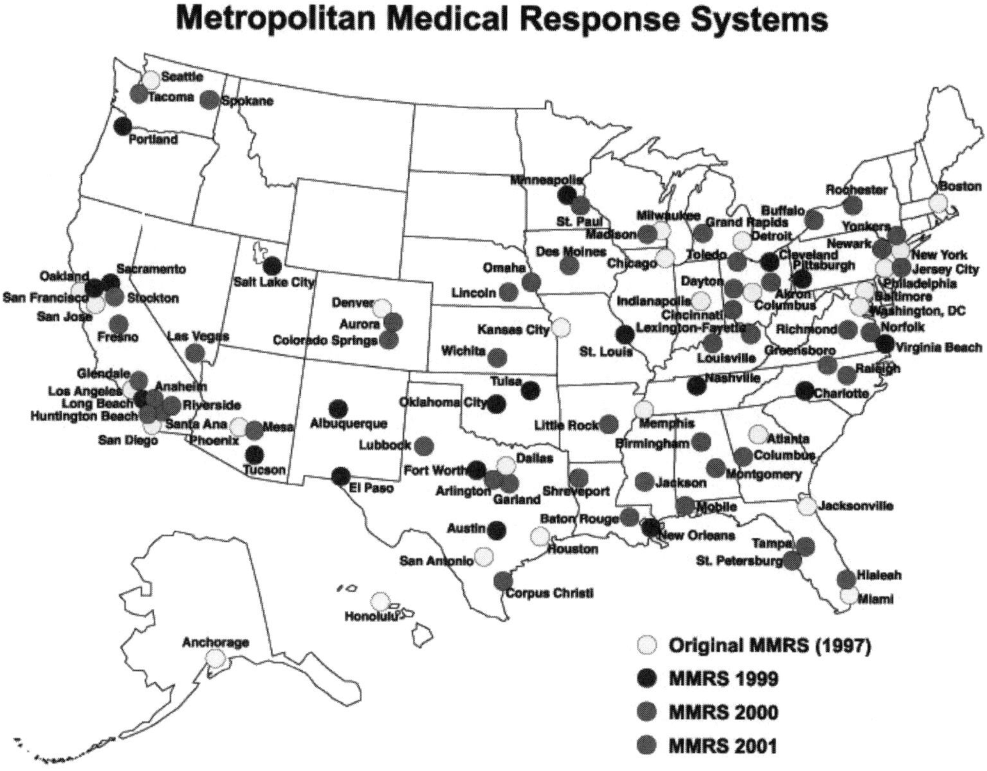

Figure 5–3 The Metropolitan Medical Response Systems (MMRS)

been removed, primary control will shift to the FBI Special Agent in Charge (SAIC) and the local responders will operate in support of the FBI.

The Federal Response Plan states that USPHS has federal responsibility for providing coordinated assistance to supplement state and local resources in response to public health and medical care needs following a significant natural disaster or man-made event when it is requested. This assistance is categorized into the following areas:

- Assessment of health/medical needs
- Health surveillance
- Medical care personnel
- Health/medical equipment and supplies
- Patient evacuation
- In-hospital care
- Food/drug/medical device safety

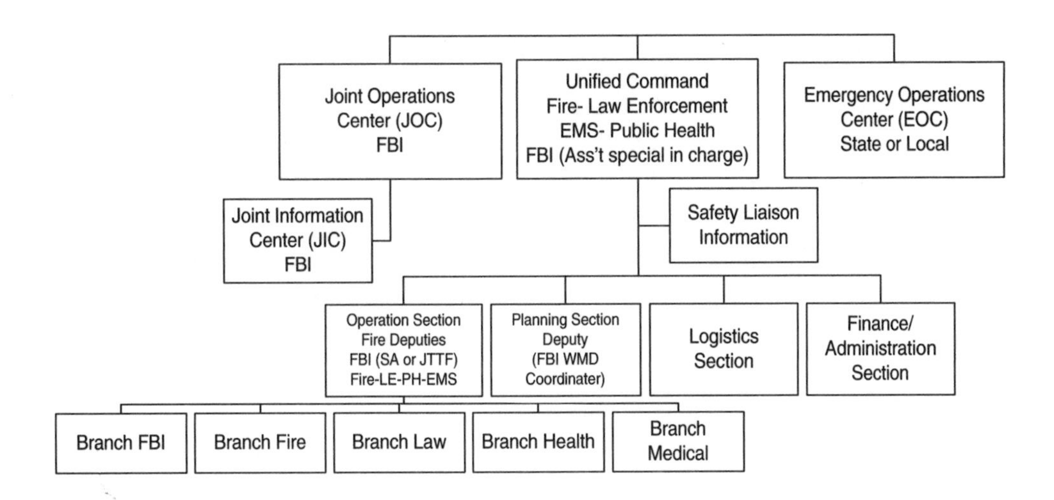

Figure 5–4 FBI interface

- Worker health/safety
- Radiological hazards
- Biological hazards
- Chemical hazards
- Mental health
- Public health information
- Vector control
- Potable water/wastewater and solid waste disposal
- Victim identification/mortuary services

Resources are available from the National Disaster Medical System (NDMS). This includes DMATs which could be mobilized for NBC events.

The Department of Defense's (DoD) role is articulated primarily through the FRP in ESF#3, "Corps of Engineers", and ESF#9, "Urban Search and Rescue"; however, there are specialized agencies within the DoD that could be of use in NBC events. These include the Technical Escort Unit (TEU) in Aberdeen, MD, the Chemical/ Biological Incident Response Force (CBIRF) in Camp LeJuene, NC, and the anti-terrorism team attached to the Edgewood Research, Development, and Engineering Center (ERDEC) located in Edgewood, MD. These teams can be activated only by another federal agency, such as the FBI or the USPHS. Because the teams' response time will likely exceed two or three hours, the determination to access them must be made early in the event.

FEMA will assume federal consequence management and public safety responsibility for NBC events once the Attorney General (AG) has determined that the priority law enforcement goals and objectives have been set or are outweighed by the consequence management concerns. As the primary agency for ESF#5, "Information and Planning", FEMA will coordinate the acquisition of federal resources for incident mitigation and activate urban search and rescue when indicated.

The Department of Energy (DOE) can play a critical role in providing specialized technical support in a nuclear terrorist event. This support may be more appropriate in a long-term scenario for agent/material removal and disposal.

The Centers for Disease Control and Prevention (CDC) are an immediate resource that should be notified as early in the incident as possible; however, it is not likely that CDC personnel can be transported to the site in a timely fashion. Therefore, a reliable communications link should be established for the rapid exchange of information and medical consultation. CDC can provide consultation on chemical antidotes, chemical decontamination practices, and medical intervention (both long- and short-term) for chemical and biological poisonings.

The Environmental Protection Agency (EPA) is the primary response agency for ESF#10, "Hazardous Materials". Its role is to provide a coordinated federal response to an actual or potential release of hazardous materials. In an NBC scenario, its role would involve the long-term remediation and decontamination of the incident site in coordination with other federal and state agencies.

The United States Secret Service (USSS), as a law enforcement agency with responsibility for protecting the United States government's leadership as well as visiting heads of state and other dignitaries, would have little role in an NBC event unless it jeopardized the safety and well-being of these officials. At that time, the USSS would focus its efforts on personnel removal and protection; very little effort would be expended in incident mitigation.

In any scenario, it must be recognized that federal personnel and resources will not likely be mobilized before the critical elements of an NBC event are addressed by local responders. Only when the event length exceeds a 24- to 36-hour timeframe will most federal personnel arrive at the Incident Command Post (ICP) to provide support. Most federal assets can be accessed through the SAIC or USPHS on-site representative.

The governor of a state can activate the Army National Guard (ANG) to support a local response. However, this requires that a state of emergency be declared. The ANG is most useful in providing security and manpower.

INTEGRATION WITH STATE AND LOCAL GOVERNMENT

Once the initial response to an NBC incident has occurred and the local responders are on-scene and have requested the MMRS, state emergency services agencies will be notified through the local 9-1-1 center or local emergency management agency. EOCs may decide to provide optimal support and coordination for the locality affected by the incident. The state hazardous materials officer or other appropriate state representative should report to the ICP to coordinate state activities in concert with the TFL. Similarly, the MMRS Law Enforcement Sector should coordinate with on-site law enforcement to accomplish the tasks assigned by the TFL and identified in other sections of this plan. Finally, the MMRS Medical Operations Sector should coordinate with state and local EMS and public health office and medical community representatives.

MEDICAL MANAGEMENT

The task force is organized, staffed, and equipped to provide the best possible pre-hospital and emergency medical care throughout the course of an incident and especially on-scene.

Task force personnel are responsible for minimizing health risks and the incidence of Critical Incident Stress (CIS) Syndrome.

Medical personnel are responsible for providing the earliest possible medical intervention for first responders and civilian victims of NBC incidents, through early identification of the agent type and proper administration of the appropriate antidote(s) and other pharmaceuticals as necessary.

Personnel must be cautious about utilizing persons offering to assist in medical management who claim to be physicians, nurses, or other medical practitioners, but who cannot substantiate their claims or provide adequate credentials.

Practitioners who provide credentials indicating that they have a medical background should be assigned responsibilities only in the cold (support) zone commensurate with their area of medical expertise and only in partnership with a known team member.

TREATMENT PRIORITIES

The treatment priorities for medical personnel (including EMS) are:

First: strike-team personnel and support staff

Second: local response personnel who become ill or injured

Third: victims directly encountered by the strike team

Fourth: other injured/affected persons as practical

It is not the intent of the Field Medical Operations Sector and Hospital Operations Sector to be freestanding medical resources at incident scenes. However, they are part of the first line of intervention in a chain of care that extends to the local hospital medical system.

TRIAGE

Triage is the process of doing the most good for the most victims by assigning them degrees of urgency and treating them accordingly. In NBC incidents, depending upon the purity of the agent, there may be few viable victims within the Hot Zone and increasing numbers of viable victims near the outer perimeter where the agent is less concentrated.

Victims should be triaged using the strike team triage protocol as follows:

Patients will be classified as:

Priority one:	exposed but not symptomatic
Priority two:	exposed, symptomatic, but salvageable if medical care is rendered within fifteen minutes
Priority three:	exposed, symptomatic, and requiring extensive medical intervention to save

The initial triage priority will be indicated by marking the number indicating the priority on the patient's forehead with a felt pen. A triage tag indicating the patient's priority is to be used as soon as possible.

DECONTAMINATION

It is extremely important that victims of NBC incidents be, at minimum, grossly decontaminated prior to being transported to medical facilities. Additional decontamination will be conducted as time and resources allow.

The degree of medical decontamination that has been performed and the solution used must be noted on patient care forms (triage tags), which are attached to the victims prior to transport.

Definitive decontamination (a more intensive scrubbing/cleansing of patients) may have to be completed at the receiving medical facilities; this should also be noted on the triage tags.

Emergency decontamination procedures will be initiated immediately when MMRS personnel are injured and/or have a Personal Protective Equipment (PPE) failure.

Appropriate basic medical care will be initiated during decontamination and continued in the treatment sector.

A rapid assessment will be initiated by a decontamination team member and the findings immediately reported to the HAZMAT Officer and Field Medical Operations Officer.

TREATMENT

Rapid Basic Trauma Life Support (BTLS) assessment will be conducted on all victims. Medical intervention will be initiated following the MMRS protocols.

Particular medical attention should be paid to airway/respiratory support and cardiovascular support.

Medical care will address the supportive needs of each patient and the specific treatment will be initiated when the agent is identified.

On-site treatment may include care for injuries sustained as a result of explosions and/or falls.

Consideration must be given to the medical and logistical implications of multiple doses of an antidote (e. g., atropine) being given to a single victim, thereby reducing the total number of patients that can be treated effectively.

The Poison Control Center should be used as a resource for product identification and determining treatment regimens not covered by MMRS protocols, and it should be notified immediately of patient problems.

TRANSPORTATION

Transportation of victims who are unconscious must be by an emergency vehicle capable of continuous treatment.

Public transportation vehicles staffed with an appropriate number of local EMS personnel can transport large numbers of victims who are minimally affected or who are suspected of having psychosomatic symptoms. These victims would normally be triaged as being among the last victims to be transported.

MMRS personnel arranging for transportation must keep in mind that vehicles used for transport may become contaminated. Vehicles and personnel used for transport that become contaminated must be decontaminated before returning them to service.

MEDICAL FACILITIES

CASUALTY COLLECTION POINT (CCP)

If a delay is encountered in transporting decontaminated victims, consideration will be given to establishing a CCP to continue medical treatment until transportation for victims is completed. This area should be located as far away from the main operating area of the incident as is possible and adjacent to the transportation sector. Staffing will be the shared responsibility of the MMRS and the local jurisdiction. Only medical care necessary to stabilize the patient's condition pending transportation will be initiated. Equipment for the CCP will come primarily from the local jurisdiction. MMRS equipment will be used only if absolutely necessary. Deceased individuals will be separated from the living.

HOSPITAL SUPPORT

It is likely that in an NBC incident of significant proportions, the local hospital system will be overwhelmed with casualties.

It may be necessary for the Hospital Operations Sector to contact hospitals by phone/fax to give advice on decontamination procedures for victims who self-refer, agent treatment protocols, and other information as requested by the medical facilities.

If the TFL determines that there is benefit in providing MMRS personnel to local hospitals to assist in the management of patients, consideration should be given to the recall of local off-duty EMTs and paramedics as well as MMRS members.

MMRS members can provide assistance to medical facilities in the following areas:

- Patient tracking
- Decontamination procedures for self-referrals
- Monitoring of vital signs
- Triage medical management
- Communication and coordination between the incident site and the hospital

Because the pharmaceuticals used for NBC incidents are not normally stocked in adequate quantities, it may be necessary to acquire additional supplies from manufacturers, local Veteran's Administration (VA) hospitals, or the USPHS.

Once the pharmaceuticals have been obtained, redistribution can be coordinated at the incident site by the Hospital Operations Sector, assisted by the Logistics Sector.

Redistribution of pharmaceuticals to medical facilities is to be accomplished by the use of fire services, law enforcement officers, or hospital personnel.

MMRS ORGANIZATIONAL CHART

An MMRS organizational chart is presented in Figure 5–5.

HOSPITAL EMERGENCY INCIDENT COMMAND SYSTEM (HEICS)

The need for clear management objectives, directed activity and precise communication is never more evident than in a time of emergency. To date, some hospitals have achieved this to a greater or lesser degree, while others have not.

For many years, fire departments have employed the ICS management program. This proven system has a chain of command that will adapt to emergency events, both large and small. It provides mission-oriented management checklists and, per-

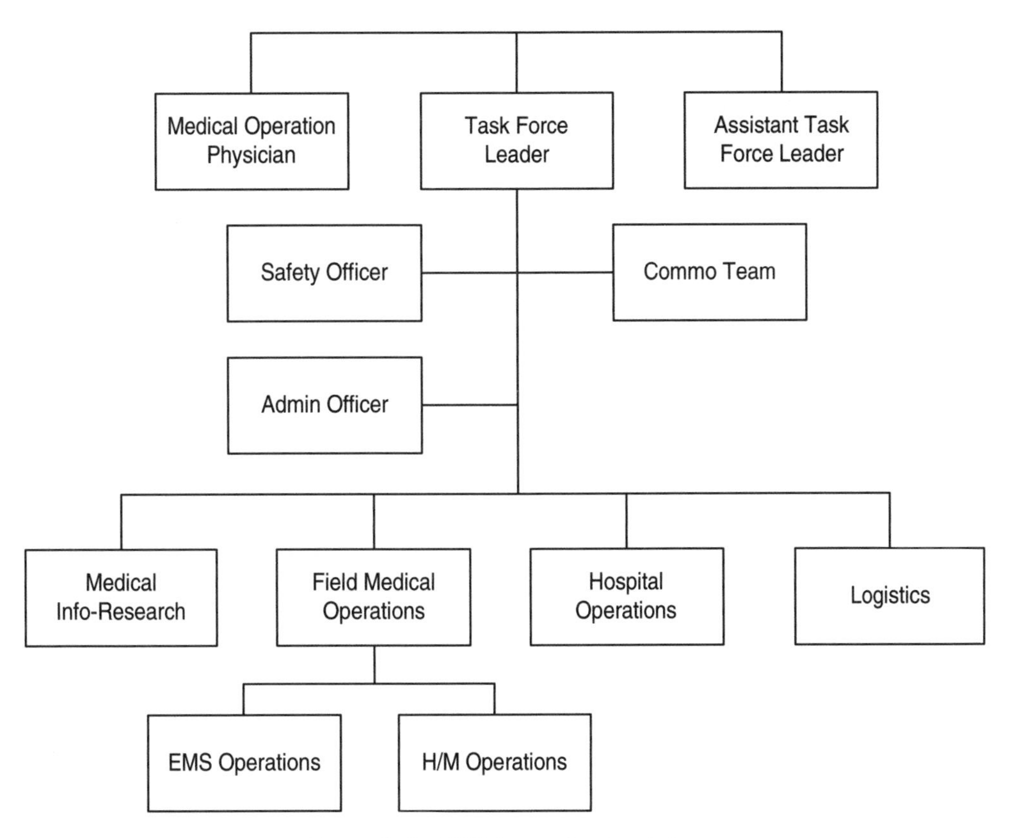

Figure 5–5 MMRS ogranizational chart

haps most importantly, provides for a common management terminology during times of stress. The system has met with such a high degree of success that it has been adopted by police departments, municipal emergency planning officials, and private industry.

The HEICS is a management system modeled after the fire service plan. The core of the HEICS is comprised of two main elements: (1) an organizational chart with a clearly delineated chain of command and with position/function titles which indicate scope of responsibility (see Figure 5–6), and (2) a prioritized Job Action Sheet (job description) which assists the individual in focusing upon his or her assignment.

The benefits of a medical facility using HEICS will be seen not only in a more organized response, but also in the ability of that institution to relate to other health-care entities and public or private organizations in the event of an emergency. The value of the common communication language in HEICS will become apparent when mutual aid is requested of, or for, that facility.

The result of the implementation of HEICS will be a medical facility with personnel who know what they should do, when they should do it, and who to report to during a time of emergency.

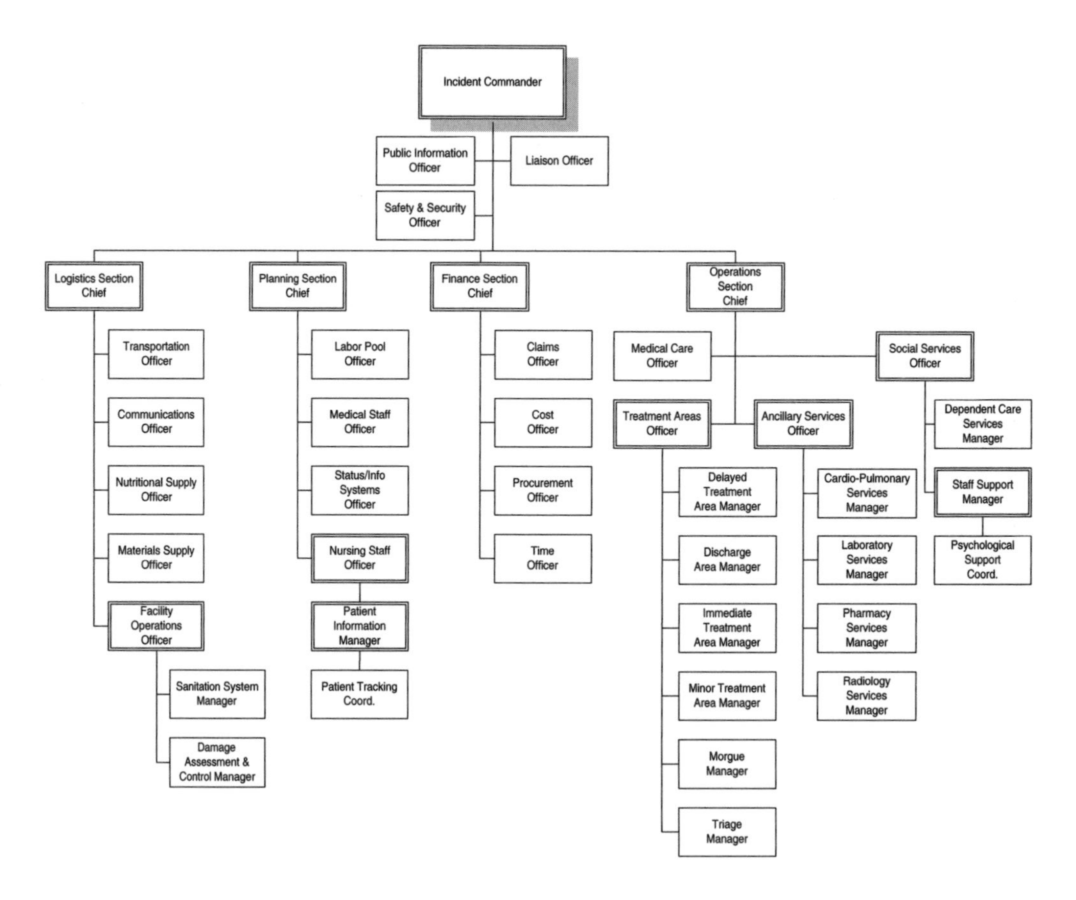

Figure 5–6 HEICS implementation

ADAPTATION OF HEICS

The Job Action Sheets are designed to meet a critical need during a time of hospital crisis. An effort has been made to write individual job descriptions compatible with the basic duties of key hospital individuals. However, there are some missions that are not entirely parallel with a given individual's day-to-day duties, although associations can be inferred.

Adapting HEICS for use within the hospital's emergency plan will be relatively easy in most situations. Most facilities will have a need to add supporting documents to address policies and procedures dealing with plan elements, such as activation, authority, and personnel recall, to name just a few. Where it is necessary to revise particular descriptions, caution should be used to retain the intent of the Job Action Sheet's mission. Changing the Job Action Sheet title or position title is not recommended, as there will be a loss of common nomenclature, which will only inhibit communication when dealing with outside agencies. Similarly, moving positions to other sections or locations will defeat the commonality of structure. Standardization of language and meaning is essential to promote communication and mission achievement, both inside the hospital and with outside entities.

Exercise of the HEICS plan can take place within the individual hospital; however, maximum insight and value will be gained when the hospital and other service providers drill together, as a system.

It is important to remember that the HEICS is a tool to facilitate accomplishing a task. The HEICS identifies critical management functions (the tools) in order for the hospital to develop and implement an emergency action plan (the task). It is the individual emergency action plan that describes how the specific emergency incident is to be addressed.

A summary of the HEICS positions and missions is presented below.

EMERGENCY INCIDENT COMMANDER

Mission: Organize and direct EOC. Give overall direction for hospital operations and, if needed, authorize evacuation.

PUBLIC INFORMATION OFFICER

Reports to emergency IC.
Mission: Provide information to the news media.

LIAISON OFFICER

Reports to emergency IC.
Mission: Function as incident contact person for representatives from other agencies.

SAFETY AND SECURITY OFFICER

Reports to emergency IC.
Mission: Monitor and have authority over the safety of rescue operations and hazardous conditions. Organize and enforce scene/facility protection and traffic security.

LOGISTICS SECTION CHIEF

Reports to emergency IC.
Mission: Organize and direct those operations associated with maintenance of the physical environment and of adequate levels of food, shelter, and supplies to support the medical objectives.

FACILITY OPERATIONS OFFICER

Reports to Logistics Section chief.
Mission: Maintain the integrity of the physical facility at the best level. Provide adequate environmental controls to perform the medical mission.

DAMAGE ASSESSMENT AND CONTROL MANAGER

Reports to facility operations officer.
Mission: Provide sufficient information regarding the operational status of the facility and information to decide whether full or partial evacuation is needed. Identify safe areas where patients and staff can be moved if needed.

SANITATION SYSTEMS MANAGER

Reports to facility operations officer.
Mission: Evaluate and monitor the patency of existing sewage and sanitation systems. Enact pre-established alternate methods of waste disposal if necessary.

COMMUNICATIONS OFFICER

Reports to Logistics Section chief.
Mission: Organize, coordinate, and act as custodian of all logged internal and external communications.

TRANSPORTATION OFFICER

Reports to Logistics Section chief.
Mission: Organize and coordinate the transportation of all casualties, both ambulatory and non-ambulatory. Arrange for the transportation of human and material resources to and from the facility.

MATERIALS SUPPLY OFFICER

Reports to Logistics Section chief.
Mission: Organize and supply medical and nonmedical care equipment and supplies.

NUTRITIONAL SUPPLY OFFICER

Reports to Logistics Section chief.
Mission: Organize food and water stores for preparation and rationing during periods of anticipated or actual shortage.

PLANNING SECTION CHIEF

Reports to emergency IC.
Mission: Organize and direct all aspects of Planning Section operations. Relay information to, and carry out directives of, the emergency IC. Anticipate future conditions and needs based on current information and conditions.

STATUS/INFORMATION SYSTEMS OFFICER

Reports to Planning Section chief.

Mission: Maintain current status information for all chiefs and officers. Ensure a written record of the hospital's emergency planning and response. Develop the hospital's internal information system. Monitor the maintenance and preservation of the computer system.

LABOR POOL OFFICER

Reports to Planning Section chief.
Mission: Gather and inventory available staff and volunteers at a central point. Receive requests and assign available staff as needed. Maintain adequate numbers of both medical and nonmedical personnel. Assist in the maintenance of staff morale.

MEDICAL STAFF OFFICER

Reports to Planning Section chief.
Mission: Gather available physicians and other medical staff at a central point. Credential volunteer medical staff as necessary. Assist in the assignment of available medical staff as needed.

NURSING SERVICE OFFICER

Reports to Planning Section chief.
Mission: Organize and coordinate nursing and direct patient care services.

PATIENT INFORMATION MANAGER

Reports to nursing service officer.
Mission: Provide information to visitors and families regarding status and location of patients. Collect information necessary to complete the Disaster Welfare Inquiry process in cooperation with the American Red Cross.

PATIENT TRACKING COORDINATOR

Reports to patient information manager.
Mission: Track the location of all patients at all times.

FINANCE SECTION CHIEF

Reports to emergency IC.
Mission: Monitor the utilization of financial assets. Oversee the acquisition of supplies and services necessary to carry out the hospital's medical mission. Supervise the documentation of expenditures relevant to the emergency incident.

TIME OFFICER

Reports to Finance Section chief.
Mission: Responsible for the documentation of daily personnel time records, including the monitoring and reporting of regular and overtime hours worked.

PROCUREMENT OFFICER

Reports to Finance Section chief.

Mission: Responsible for administering accounts receivable and payable to contract and noncontract vendors.

CLAIMS OFFICER

Reports to Finance Section chief.
Mission: Responsible for receiving, investigating, and documenting all claims reported to the hospital during the emergency incident and alleged to be the result of an accident or action on hospital property.

COST OFFICER

Reports to Finance Section chief.
Mission: Responsible for providing cost analysis data for declared emergency incident. Maintain of accurate records of incident cost.

OPERATION SECTION CHIEF

Reports to emergency IC.
Mission: Organize and direct aspects relating to the Operations Section. Carry out directives of the emergency incident commander. Coordinate and supervise the medical services branch, ancillary services branch, and human services branch of the Operations Section.

MEDICAL CARE OFFICER

Reports to Operations Section chief.
Mission: Organize and manage the delivery of medical care in all areas of the hospital. Recommend assignment of physicians within the patient care areas.

TREATMENT AREAS OFFICER

Reports to Operations Section chief.
Mission: Initiate and supervise the patient triage process. Assure treatment of casualties according to triage categories and manage the treatment area(s). Provide for a controlled patient discharge. Supervise morgue service.

TRIAGE MANAGER

Reports to treatment areas officer.
Mission: Sort casualties according to priority of injuries, and assure their transfer to the proper treatment area.

IMMEDIATE TREATMENT AREA MANAGER

Reports to treatment areas officer.
Mission: Coordinate the care given to patients received from the triage area. Assure adequate staffing and supplies; and facilitate the treatment and transfer of patients in the immediate treatment area.

DELAYED TREATMENT AREA MANAGER

Reports to treatment areas officer.
Mission: Coordinate the care given to patients received from the triage area. Assure adequate staffing and supplies; and facilitate the treatment and transfer of patients in the delayed treatment area.

MINOR TREATMENT AREA MANAGER

Reports to treatment areas officer.
Mission: Coordinate the minor care of patients received from the triage area and other areas of the hospital. Assure adequate staffing and supplies in the minor treatment area. Facilitate the minor treatment of patients and their transfer.

DISCHARGE AREA MANAGER

Reports to treatment areas officer.
Mission: Coordinate the controlled discharge (observation and possible discharge) of patients received from all areas of the hospital. Facilitate the process of final patient transfer by assuring adequate staff and supplies in the discharge area.

MORGUE MANAGER

Reports to treatment areas officer.
Mission: Gather, identify, and protect deceased patients. Assist discharge area manager in appropriate patient discharge.

ANCILLARY SERVICES OFFICER

Reports to Operations Section chief.
Mission: Organize and manage ancillary medical services. Assist in providing for the optimal functioning of these services. Monitor the use and conservation of resources.

LABORATORY SERVICES MANAGER

Reports to ancillary services officer.
Mission: Maintain laboratory services, blood, and blood products at appropriate levels. Prioritize and manage the activities of the laboratory staff.

RADIOLOGY SERVICES MANAGER

Reports to ancillary services officer.
Mission: Maintain radiology and other diagnostic imaging services at appropriate levels. Ensure the highest quality of service under current conditions.

PHARMACY SERVICES MANAGER

Reports to ancillary services officer.
Mission: Ensure the availability of emergency, incident-specific pharmaceuticals and pharmacy services.

CARDIOPULMONARY SERVICES MANAGER

Reports to ancillary services officer.
Mission: Provide the highest level of cardiopulmonary services at levels sufficient to meet the emergency incident needs.

SOCIAL SERVICES OFFICER

Reports to Operations Section chief.
Mission: Organize, direct, and supervise those services associated with the social and psychological needs of the patients, staff, and their respective families. Assist with discharge planning.

DEPENDENT CARE SERVICES MANAGER

Reports to social services officer.
Mission: Direct sheltering and feeding of staff and volunteer dependents.

STAFF SUPPORT MANAGER

Reports to social services officer.
Mission: Assure the provision of logistical and psychological support to the hospital staff.

PSYCHOLOGICAL SUPPORT COORDINATOR

Reports to staff support manager.
Mission: Assure the provision of psychological, spiritual, and emotional support to the hospital staff, patients, dependents, and guests. Initiate and organize the Critical Stress Debriefing process.

SUMMARY

Many resources will be involved in the response to a terrorist event. Many of these resources will operate at the scene while others will be activated within the community, such as hospitals and other medical care facilities. Resources from outside the local jurisdiction will come to assist and fulfill their responsibilities. The local community must provide the management structure in which these resources can operate to achieve the most favorable event outcome.

Incident management will begin with the arrival of the first responders, who will establish an incident management system for the scene. Many local resources will activate their EOPs. The community's EOC will be activated and ESFs staffed; as will the state's EOC and ESFs. Federal resources will come to the scene, to the state's EOC and to the local EOC.

It is imperative that the management structure that is developed is capable of plugging all resources into the operation.

For additional information on the MMRS and the latest updates and events, go to: *http://www.mmrs.hhs.gov/*

Local Plans and Resources

OVERVIEW

In this chapter we will evaluate local resources and planning efforts as they pertain to events involving weapons of mass destruction. The material covered in this chapter will help you to

- Describe the role of the local EOC during major emergencies and disasters
- Identify the importance of advanced planning
- Describe the concepts of hazard/risk assessment and the identification of vulnerable populations
- Assess a community's EOP for asset voids that may need to be filled during a WMD event

LOCAL PLANS AND RESOURCES

As we will see in future units, communities have access to a wide array of state and federal resources and mechanisms to respond to a mass destruction event. However, most responses are handled on a day-to-day basis by local resources. In fact, there is no other level of government that can provide for effective response during the first hours and days of a major emergency or special operations event. In addition, the medical community must look both within and outside of its organization for resources to help manage these unique and overwhelming operations.

Due to the massive amount of resources that are required during such events, it is imperative that these resources be assembled in a coordinated fashion. This assembly of resources is best facilitated by two distinct processes or plans: (1) the local EOP that identifies and coordinates the acquisition and deployment of resources, and (2) the use of an incident command system to effectively manage these resources once they have been applied to any event. The incident command system will be discussed later. For now we will discuss the local EOP which identifies and musters these resources.

LOCAL EMERGENCY OPERATIONS PLAN

The local EOP, commonly referred to as a "comprehensive emergency management plan," is the local community's method of identifying resources and organizational responsibilities for the acquisition and deployment of those resources.

Most commonly, the local EOP is designed to closely resemble state EOPs and the FRP. The local EOP contains an analysis of the hazards within the community, identifies vulnerable populations and facilities, provides a basic overview of plan operational concepts and specifically outlines the organizational responsibilities and resources that the community has available.

LOCAL HAZARDS ANALYSIS

The local hazards analysis lists hazards that are of concern to the local community. A review of laws, previous plans, and hazard identification processes by local officials is generally utilized to identify the potential hazards. From this list of hazards a comprehensive analysis of the populations and facilities that are at risk can be initiated. Some of the hazards that might be identified in your plan may include:

- Hurricanes, earthquakes, floods, tornadoes, mudslides, tsunami, hazardous materials, and confined space emergencies
- Civil disturbances, power failures, train derailments
- Plane crashes, urban conflagrations, wildland fires
- Epidemics, terrorism, cave-ins, dam failures

It is important to understand that local EMS organizations should know the potential hazards that may affect their community. In addition, EMS must have a means of providing input to this important planning function. Your EMS organization's input is essential to ensure that your identified roles are realistic regarding your agency's capabilities, and to provide you with a better opportunity to identify other resources within the jurisdiction that may be available in times of need.

VULNERABILITY ANALYSIS

Also included within the local EOP is the vulnerability analysis. This analysis identifies all of the vulnerable populations within the local community. From an EMS perspective, certain "special needs" populations need to be identified. These populations may include elderly, handicapped, non-English speaking and ethnic groups that present unique medical considerations. In addition, vulnerable facilities such as nursing homes, hospitals, adult congregate living units, high-density housing, major transportation routes and infrastucture facilities such as communication centers like 911, fire departments, sheriff's department, water facilities, and critical government buildings are also identified and cataloged. This catalog of "critical facilities" will help you identify your priorities should your system become taxed by a major disaster.

Once a vulnerability analysis has been completed, the EMS system management team can determine the vast or unique resources that may be required. Additionally, we can also begin to realize that not only will a disaster have an impact upon EMS's ability to provide emergency care, but also the ability of the public to obtain routine or special needs care.

FUNCTIONAL ANNEXES

The next major component of an EOP is functional annexes. These annexes are plans organized around the performances of tasks or goals. Each annex focuses on one of the critical emergency functions that the jurisdiction will perform in response to an emergency. The number and type of annexes included in the EOP may vary from one jurisdiction to another, depending upon needs, capabilities and jurisdictional organization. Since functional annexes are orientated towards operations, the primary audience of each annex consists of those who perform the tasks.

Although many communities have developed functional annexes which are unique to the locality, a general starting point for identifying what annexes will be required can be obtained by looking at state and federal response plans. Generally speaking, there are twelve primary annexes that are identified when discussing the FRP (see Figure 6–1).

In addition to these primary ESFs, additional annexes may be required, depending upon your specific jurisdiction. Such additional annexes may include law enforce-

Annexes (Support Functions)
Transportation
Communications
Public Works and Engineering
Firefighting
Information and Planning
Mass Care (Sheltering)
Resource Support
Health and Medical Services
Urban Search and Rescue
Hazardous Materials
Food
Energy

Figure 6–1 Primary annexes

ment and security, volunteers, donations, public information/warning, and military support. (see Figure 6–2)

The ESFs are responsible for coordinating the acquisition and deployment of resources as needs are identified.

Each ESF has one primary agency and is supported by additional agencies. For example, "transportation" may be served by the county's transportation deptartment and supported by the local school district's transportation department, the utilities department, and the road and bridge department. This group of agencies then coordinates the acquisition of resources as requested by field units or the overall commander (who is normally the county manager, an elected official or his or her designee).

In order to provide greater coordination of the ESFs they may be further subdivided along incident command functions. In our model community, the ESFs are divided along ICS lines as depicted in Figure 5–1

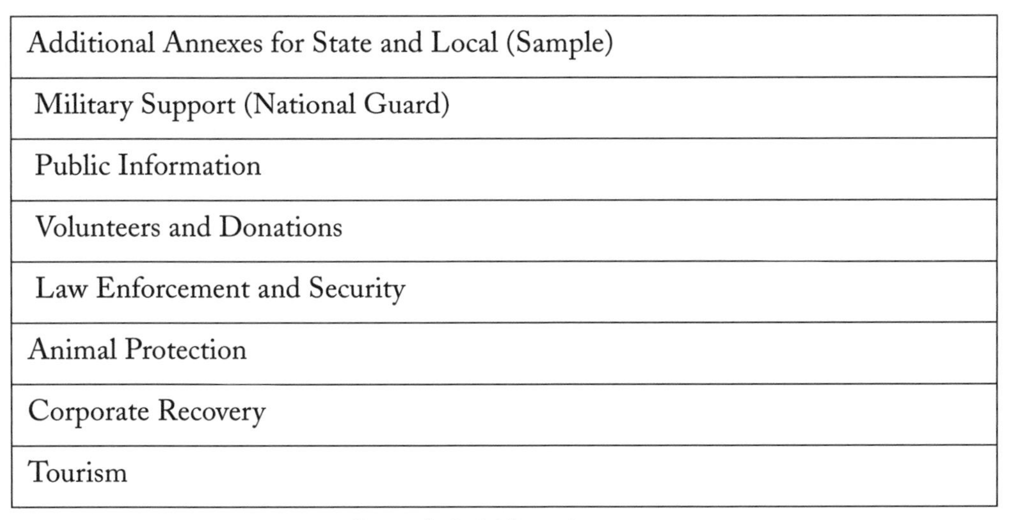

Additional Annexes for State and Local (Sample)
Military Support (National Guard)
Public Information
Volunteers and Donations
Law Enforcement and Security
Animal Protection
Corporate Recovery
Tourism

Figure 6–2 Additional annexes

By implementing your local EOP, the resources of your EMS agency and all governmental agencies within your community can be accessed to assist in mass gathering events, major emergencies or disasters. Furthermore, once the incident exceeds the local community's capabilities, a state of emergency can be declared and the state's EOP is activated, providing access to the resources and agreements that were discussed in earlier units. Finally, once the state resources have been exceeded, the governor of the state can request a federal declaration or finding of a major emergency from the president of the United States, which activates the FRP.

The enormous amount of resources which can be developed under the local EOP make it an invaluable tool to the EMS system. However, in order to ensure the effective and efficient operations of the plan, it must be utilized and tested on a regular basis. Therefore, any opportunity to implement the system should be seized rather than waiting for a disaster or major emergency. Many communities implement the plan regularly during special events and mass gatherings. Such drills can help to identify weaknesses, build upon strong points and provide an excellent training opportunity for the support function personnel.

LOCAL RESOURCES AND SPECIAL TEAMS

Once the community has completed a hazards and vulnerability analysis, EMS can begin to identify the strengths (assets) and weaknesses (voids) of its human and physical resources. The overall objective is to identify what resources could be required given a particular hazard (generally based upon the development of a worst case scenario for each hazard) and identifying weaknesses with regard to resources needed to effectively manage the scenario.

One resource that requires particular attention is that of EMS human resources. During a disaster or long, protracted event, both the demand for human resources and the demands placed upon those resources will be high. Far too often in past incidents, systems have failed to recognize the importance of ensuring that the needs of their responders are met. Only after ensuring that our responders' families are safe can we expect them to provide service at their highest possible level. Establish relief crews in advance and make sure that responders can communicate with their families so that they are aware of their family's safety and vice versa.

Remember, the major role of EMS is to provide emergency care for potentially life-threatening injuries. If we do not complement our human resources during major events, our system can quickly become taxed and our ability to respond to routine emergency calls can be compromised. Therefore, prior to major events or disasters, we need to establish mechanisms by which we can supplement and reinforce our resources. We will discuss some innovative resources that have been developed by communities in the pages that follow.

COMMUNITY EMERGENCY RESPONSE TEAMS (CERTs)

Following a major disaster, first responders who provide fire-fighting and medical services will not be able to meet the demand for these services. Factors such as the number of victims, communications failures, and road blockages will prevent people from accessing emergency services they have come to expect at a moment's notice through 911. People will have to rely on each other for help in order to meet their immediate life-saving and life-sustaining needs. In addition, any resource that can help reduce burdens on emergency services during a disaster will greatly enhance the local EMS system's ability to meet the true emergency response needs demanded by the event.

One also expects that under these kinds of conditions, family members, fellow employees, and neighbors will spontaneously try to help each other. This was the case following the Mexico City earthquake in which untrained, spontaneous volunteers saved 800 people. However, 100 people lost their lives while attempting to save others. This is a high price to pay and can be prevented through training. In the September 11th attack, emergency response personnel from surrounding areas responded to the scene to assist in the aftermath of the event. Volunteer metal workers came to assist with cutting the heavy metal apart to get to victims. These volunteers were well qualified in their areas of expertise, which reduced the risk of further injuries.

If we can predict that emergency services will not meet immediate needs following a major disaster, especially when there is no warning as in an earthquake, and that people will spontaneously volunteer, what can the government do to prepare citizens for such events?

First, give citizens the facts about what to expect following a major disaster in terms of immediate services. Second, relay the message about their responsibility for mitigation and preparedness. Third, train them in needed life-saving skills with emphasis on decision making, rescuer safety, and doing the greatest good for the greatest number of people—that's what CERTs are about. Fourth, organize teams so that they are an extension of first responder services offering immediate help to victims until fire and medical services arrive.

The CERT concept was developed and implemented by the Los Angeles City Fire Department (LAFD) in 1985. The Whittier Narrows earthquake in 1987 underscored the area-wide threat of a major disaster in California. Further, it confirmed the need for training civilians to provide emergency services. As a result, the LAFD created its Disaster Preparedness Division with the purpose of training citizens and private and government employees. As of 1993, more than 10,000 people and over 267 teams have been trained. Today, CERTs and variations of the original CERT concept operate and function on an ongoing basis.

The training program that LAFD initiated makes good sense and furthers the process of citizens understanding their responsibility in preparing for disaster. It also increases their ability to safely help themselves, their families, and their neighbors. FEMA recognizes the importance of preparing citizens. The Emergency Management Institute (EMI) and the National Fire Academy adopted and expanded CERT materials, believing them applicable to all hazardous events.

The CERT course will benefit any citizen who takes it. CERT participants will be better prepared to respond to and cope with the aftermath of a disaster. Additionally, if a community wants to supplement its response capability after a disaster, civilians can be recruited and trained as neighborhood, business, and government teams that will act as auxiliary responders. These groups can provide immediate assistance to victims in their area, organize spontaneous volunteers who have not had the training, and collect disaster intelligence that will assist professional responders with prioritization and allocation of resources. Since 1993 when this training was made available nationally by FEMA, seven communities in the states of California, Washington, Oregon, Florida, Missouri, and Kentucky have conducted the training.

Recommended steps to start a CERT:

- Identify the program goals that CERT will meet and the resources available to conduct the program in your area.
- Gain approval from appointed and elected officials to use CERT as a means to prepare citizens to care for themselves during a disaster when services may not be adequate. This is an excellent opportunity for the government to be proactive in working with its constituency.

- Identify and recruit potential participants. Likely CERT candidates are community groups, business and industry workers, local government workers, and "Neighborhood Watch" groups.
    - Train a CERT instructor cadre.
    - Conduct CERT training sessions.
- Conduct refresher training and exercises with the CERTs. The CERT course is delivered in the community by a team of first responders who have the requisite knowledge and skills to instruct the sessions. It is suggested that the instructors complete a CERT Train-The-Trainer (TTT) course conducted by their state's training office for emergency management or the Emergency Management Institute in order to learn the training techniques that are successfully used by the LAFD.
- Obtain recognition for CERT members. Provide incentives for their continued participation.
- Train the emergency response community (EMS, fire, and law enforcement) to recognize, trust and communicate with the CERT members. Nothing will destroy CERT morale more than lack of involvement.

The CERT training for community groups is usually delivered in 2 ½-hour sessions, one evening a week over a seven-week period. The training consists of the following:

Session I: Disaster Preparedness. Addresses hazards to which people are vulnerable in their community. Materials cover actions that participants and their families take before, during, and after a disaster. As the session progresses, the instructor begins to explore an expanded response role for civilians in which they begin to consider themselves disaster workers. Since they will want to help their family members and neighbors, this training can help them operate in a safe and appropriate manner. The CERT concept and organization are discussed, as well as applicable laws governing volunteers in that jurisdiction.

Session II: Disaster Fire Suppression. Briefly covers fire chemistry, hazardous materials, fire hazards, and fire suppression strategies. However, the thrust of this session is the safe use of fire extinguishers, assessing the situation, controlling utilities, and extinguishing a small fire.

Session III: Disaster Medical Operations, Part 1. Participants practice diagnosing and treating airway obstruction, bleeding, and shock by using simple triage and rapid treatment techniques.

Session IV: Disaster Medical Operations, Part 2. Covers evaluating patients by doing a head-to-toe assessment, establishing a medical treatment area, performing basic first aid, and practicing in a safe and sanitary manner.

Session V: Light Search and Rescue Operations. Participants learn about search and rescue planning, assessment, search techniques, rescue techniques, and, most importantly, rescuer safety.

Session VI: Disaster Psychology and Team Organization. Covers signs and symptoms that might be experienced by the disaster victim and worker. It addresses CERT organization and management principles and the need for documentation.

Session VII: Course Review and Disaster Simulation. Participants review their answers from a take-home examination. Finally, they practice the skills that they have learned during the previous six sessions in a disaster activity.

During each session participants are required to bring safety equipment (gloves, goggles, mask) and disaster supplies (bandages, flashlight, dressings) which will be used during the session. By doing this for each session, participants are building a response kit of items that they will need during a disaster.

When participants have completed this training, it is important to keep them up to date and practiced in their skills. Trainers should offer periodic refresher sessions to reinforce what has been taught. Events such as drills, picnics, neighborhood clean-up, and disaster education fairs will help to keep them involved.

CERT members should receive recognition for completing their training. Communities may issue ID cards, vests, and helmets to graduates. These items become "badges of pride" for the program participants and serve to boost their morale significantly.

First responders need to be educated about the CERT and their value to the community. Using a CERT as a component of the response system when there are exercises for potential disasters can reinforce this idea. In addition, using CERTs to complement resources during other events such as mass gatherings and multi-venue operations, gives the members additional opportunities to practice their new skills and increase their morale.

FEMA supports CERTs by conducting or sponsoring TTTs for members of the fire, medical, and emergency management community. The objectives of the TTTs are to teach attendees to promote this training in their communities, conduct TTTs at their locations, conduct training sessions for neighborhoods, businesses and industries, and government groups; and organize teams with which first responders can interface following a major disaster.

CERT is about readiness, people helping people, rescuer safety, and doing the greatest good for the greatest number. CERT is a positive and realistic approach to emergency and disaster situations in which citizens will initially be on their own and able to make a difference. Through training, citizens can manage utilities and put out small fires; treat the three killers (opening airways, controlling bleeding, and treating for shock); provide basic medical aid; search for and rescue victims safely; and organize themselves and spontaneous volunteers to be effective.

If you have questions about the CERT program, contact:

> Emergency Management Institute
> 16825 South Seton Avenue
> Emmitsburg, Maryland 21727
> Phone: (301) 447-1071

In addition, a comprehensive CERT training program including Instructor Guide and Student Manuals can be downloaded from the FEMA electronic library at *www.fema.gov.*

DISASTER COMMUNITY HEALTH ASSISTANCE TEAMS (DCHATS)

Essentially, a DCHAT is a local or regional version of the DMAT provided by the National Disaster Medical System (NDMS). This is an effective approach to ensuring that routine medical health care can be maintained without overtaxing the EMS System. DCHATs are organized groups of local physicians, nurses, and public and mental health personnel that establish medical treatment locations throughout the community. These locations should be, if possible, pre-identified. The primary responsibility of a DCHAT is to provide routine medical assistance and nonemergent medical care. This serves to ensure that EMS remains available for life-threatening emergencies.

Resources provided by a DCHAT may include many of the same materials offered by a DMAT: medical supplies, communications equipment, surgical equipment, and X-ray capabilities. The resources available to a DCHAT should be identified in advance by local medical and public health officials.

One important lesson that has been learned by NDMS, DMATs, and DCHATs is the need for outreach capability. To simply place a medical resource into a disaster area will not ensure that necessary medical attention is delivered.

When families are affected by disasters, one primary concern is to return their lives to some level of order as quickly as possible. Clearing debris, repairing damages and securing livable conditions become victims' primary goals. Unfortunately, routine medical care for minor injuries is relegated to a low priority. Combine this change in priorities, with the fact that the environmental conditions after a major storm, flood, earthquake or other disaster creates a high potential for infections and diseases, and we can easily see the need for medical care to reach out to the victims.

Regardless of the mechanism used to provide routine medical care in times of disaster, that care may need to be taken *to* the victims. Only a coordinated effort between local health officials, EMS, and other resources like DMATs and DCHATs, as well as transportation capabilities will enable this outreach to occur. Although your primary medical care resources will be placed in a fixed location, you should be prepared to bring these resources to victims by any means necessary (for example, four-wheel drive vehicles and boats). The Toledo Area DMAT (TADMAT) has also gone as far as to develope a suggested "Disaster Outreach Severity Score" methodology which can be used by outreach teams to score the medical needs of victims.

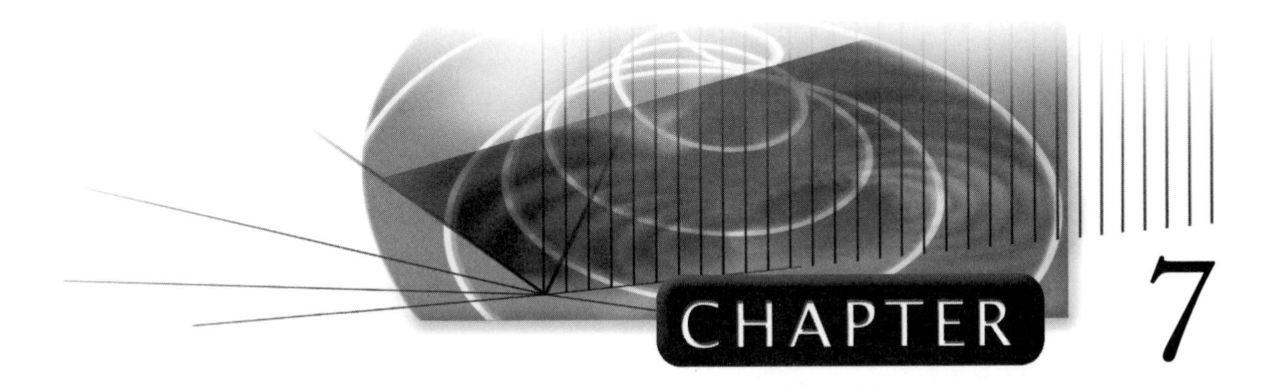

Federal Medical Resources

OVERVIEW

After reading this chapter you will be able to identify the federal consequence management responses that can be deployed to assist with the medical aspects of a terrorist event. You will learn to

- Describe the basic organizational concepts of the FRP
- Describe the integration of Emergency Support Function #8 (ESF#8)
- Identify the resources available under the National Disaster Medical System
- Describe the capabilities of federal urban search and rescue teams

DISASTER AND EMERGENCY CONDITIONS

A disaster or emergency may overwhelm the capabilities of a state and its local governments in providing a timely and effective response to meet the needs of the situation. For example, the occurrence of a large or catastrophic earthquake in a high-risk, high-population area will cause casualties, property loss, and disruption of normal life support systems and will impact the regional economic, physical, and social infrastructures.

Disasters and emergencies have the potential to cause substantial health and medical problems with hundreds or thousands of deaths and injuries, depending on factors such as time of occurrence, severity of impact, existing weather conditions, area demographics, and the nature of building construction. Deaths and injuries will occur principally from the collapse of man-made structures and from collateral events, such as fires and mudslides.

The economic and physical infrastructure may also be damaged by a disaster or emergency. An earthquake may trigger fires, floods, or other events that will multiply property losses and hinder the immediate emergency response effort. A hurricane may significantly damage or destroy highway, airport, railway, marine, communications, water, waste disposal, electrical power, natural gas, and petroleum transmission systems.

FEDERAL RESPONSE PLAN ASSUMPTIONS

The FRP, as seen in Figure 7–1, assumes that a disaster or emergency, such as an earthquake, may occur with little or no warning at any time of day that produces maximum casualties. The ERP also deals with other types of disasters, such as hurricanes, which could result in a large number of casualties and cause widespread damage, or with the consequences of any event in which federal response assistance under the authorities of the Stafford Act is required. In all cases, the ERP assumes that the response capability of an affected state will be quickly overwhelmed.

The large number of casualties and/or the heavy damage to buildings, structures, and the basic infrastructure will necessitate direct federal government assistance to aid local and state authorities in conducting life-saving and life-supporting efforts.

The immediate response of federal search and rescue personnel and equipment (and medical personnel and supplies for minimizing preventable deaths and disabilities among those injured or trapped in damaged buildings) is critical during the first 72 hours.

Federal departments and agencies may need to respond on short notice to provide effective and timely assistance to the state. Therefore, the FRP provides pre-assigned missions for the declaration process to be carried out under P.L. 93-288, as amended, and as prescribed in 44 C.F.R., Part 205. Based on the severity

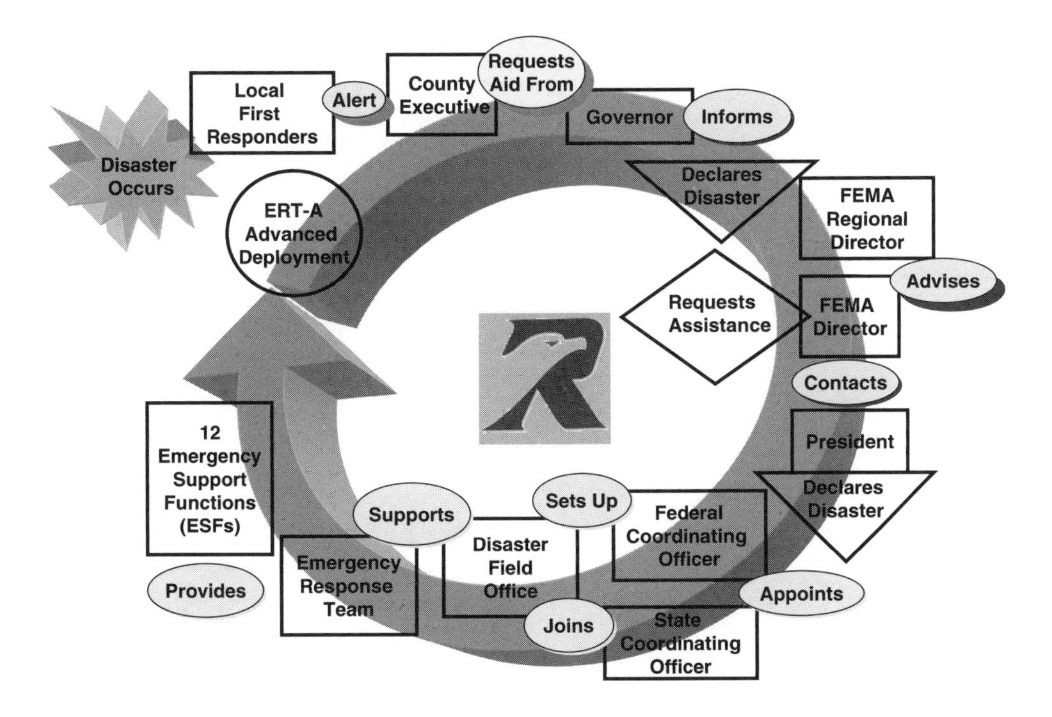

Figure 7–1 Federal Response Plan (FRP)

and magnitude of the situation, the governor will request the president to declare a major disaster or an emergency for the state, and the president will issue a declaration, as warranted. The president will also appoint a Federal Coordinating Officer (FCO) to oversee all the activities under the declaration.

For certain situations, the president may declare an emergency with or without a governor's request, as specified in Title V of P.L. 93-288, as amended. Under Title V, the president may direct the provision of emergency assistance, either at the request of a governor (Section 501.a), or upon determination by the president that an "emergency exists for which the primary responsibility for response rests with the United States..." (Section 501.b).

The American Red Cross (ARC) is deemed to be a federal agency for the purposes of the FRP. Though created by the United States Congress in 1905, the ARC is a private, charitable corporation whose primary functions include the alleviation of human suffering caused by disaster or other natural catastrophe.

THE PLAN'S CONCEPT OF OPERATIONS

GENERAL CONCEPT

During the period immediately following a major disaster or emergency requiring federal response, primary agencies, when directed by FEMA, will take actions to identify requirements, and mobilize and deploy resources to the affected area to assist the state in its life-saving and life-protecting response efforts.

Agencies have been grouped together under the functional ESFs to facilitate the provision of response assistance to the state. These functions are transportation, communications, public works and engineering, firefighting, information and planning, mass care, resource support, health and medical services, urban search and rescue, hazardous materials, food, and energy. If federal response assistance is required under the FRP, it will be provided using some or all of the ESFs, as necessary.

Each ESF has been assigned a number of missions to provide response assistance to the state. The designated primary agency, acting as the Federal Executive Agent, with the assistance of one or more support agencies, is responsible for managing the activities of the ESF and ensuring that the missions are accomplished. ESFs have the authority to execute response operations to directly support state needs. The primary and support agency assignments by each ESF are shown in Figure 7–2.

Specific ESF functional missions, organizational structures, response actions, and primary and support agency responsibilities are described in the Functional Annexes to the FRP. ESFs will coordinate directly with their functional counterpart state agencies to provide the assistance required by the state. Requests for assistance will be channeled from local jurisdictions through the designated state agencies for action. Based on state-identified response requirements, appropriate federal response assistance will be provided by an ESF to the state, or at the state's request, directly to an affected local jurisdiction.

A Federal Coordinating Office (FCO) will be appointed by the president to coordinate the federal activities in each declared state. The FCO will work with the State Coordinating Officer (SCO) to identify overall requirements, including unmet needs and evolving support requirements, and coordinate these requirements with the ESFs. The FCO will also coordinate public information, Congressional liaison, community liaison, outreach and donations activities and will facilitate the provision of information and reports to appropriate users.

The FCO will head a regional interagency Emergency Response Team (ERT), composed of ESF representatives and other support staff. The ERT provides initial response coordination with the affected state at the state's EOC or other designated state facility and supports the FCO and ESF operations in the field. The FCO will coordinate response activities with the ESF representatives on the ERT to ensure that federal resources are made available to meet the requirements identified by the state.

#	1	2	3	4	5	6	7	8	9	10	11	12
ESF / ORG	TRANSPORTATION	COMMUNICATIONS	PUBLIC WORKS AND ENGINEERING	FIREFIGHTING	INFORMATION AND PLANNING	MASS CARE	RESOURCE SUPPORT	HEALTH AND MEDICAL SERVICES	URBAN SEARCH AND RESCUE	HAZARDOUS MATERIALS	FOOD	ENERGY
USDA	S	S	S	P	S	S	S	S	S	S	P	S
DOC		S	S	S	S	S	S			S		
DOD	S	S	P	S	S	S	S	S	S	S	S	S
DOEd					S							
DOE	S		S		S		S			S		P
DHHS			S		S	S	S	P	S	S	S	
DHUD						S						
DOI		S	S	S	S					S		
DOJ					S			S		S		
DOL			S				S		S	S		
DOS	S									S		S
DOT	P	S	S		S	S	S	S	S	S	S	S
TREAS					S							
VA			S			S	S	S				
AID								S	S			
ARC					S	P		S			S	
EPA			S	S	S			S	S	P	S	
FCC		S										
FEMA		S		S	P	S	S	S	P	S	S	
GSA	S	S	S		S	S	P	S	S	S		S
ICC	S											
NASA					S							
NCS		P			S		S	S				S
NRC					S					S		S
OPM							S					
TVA	S		S									S
USPS	S					S		S				

P - Primary Agency: Responsible for Management of the ESF
S - Support Agency: Responsible for Supporting the Primary Agency

Figure 7–2 ESF assignment matrix

A national interagency Emergency Support Team (EST), composed of ESF representatives and other support staff, will operate at FEMA headquarters to provide support for the FCO and the ERT.

The Catastrophic Disaster Response Group (CDRG), composed of representatives from all departments and agencies under the FRP, will operate at the national level to provide guidance and policy direction on response coordination and operational issues arising from FCO and ESF response activities. The CDRG is also supported by the EST and will operate from FEMA Headquarters.

Activities under the FRP will be organized at various levels to provide partial response and recovery (utilizing selected ESFs) or to provide full response and recovery (utilizing all ESFs).

ORGANIZATION OF EFFORTS

The organization to implement the procedures under the FRP is composed of standard elements at the national and regional levels. The overall response structure is shown in Figure 7–3. It is designed to be flexible in order to accommodate the

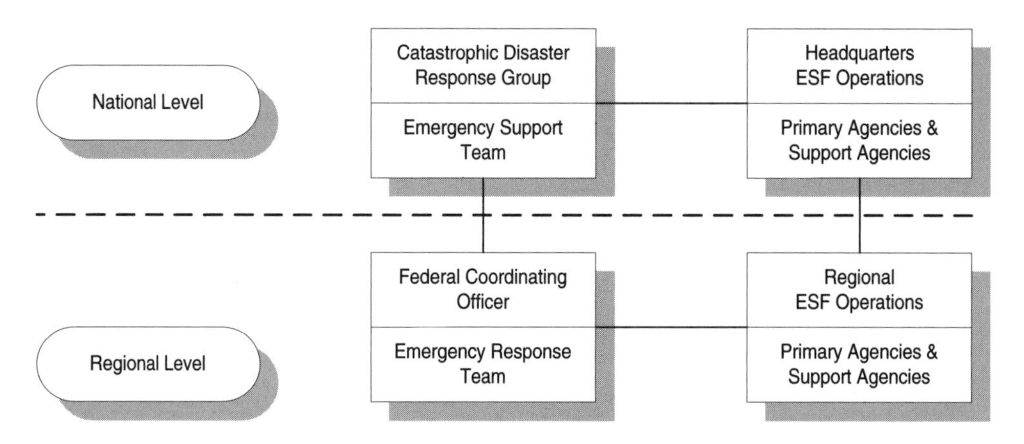

Figure 7–3 Federal response structure

response and recovery requirements specific to the disaster. The response structure shows the composition of the elements providing response coordination and response operations activities at the headquarters and regional levels, but does not necessarily represent lines of authority or reporting relationships. In general, national-level elements provide support to the regional level elements that implement the on-scene response operations in the field.

National-Level Response Structure

The national-level response structure is composed of national interagency coordination and operations support elements from the participating departments and agencies. Overall interagency coordination activities are supported by the CDRG and EST at FEMA Headquarters. These elements will be augmented by department and agency operations support elements at other locations. As shown in Figure 7–4, the national-level response structure is composed of the following specific elements:

Catastrophic Disaster Response Group. The CDRG is the headquarters-level coordinating group that addresses policy issues and support requirements from the FCO and ESF response elements in the field. It is chaired by the FEMA associate director, PT&E, and includes representatives from the federal departments and agencies that have responsibilities under the FRP. The CDRG addresses response issues and problems that require national-level decisions or policy direction. The CDRG may be augmented by officials from other organizations not listed in the FRP that have resources, capabilities, or expertise needed for the response effort.

The CDRG will meet on an as-needed basis at the request of the Disaster Recovery Center (DRC) chairperson. Meetings, unless otherwise indicated, will be held at the Emergency Information and Coordination Center (EICC), located in FEMA Headquarters in Washington, DC.

Emergency Support Team. The EST is an interagency group composed of representatives from each of the primary agencies, select support agencies, and FEMA Headquarters staff. It operates from FEMAs EICC. Detailed procedures regarding the EST organization and operations are found in the "EST Organization and Operational Procedures" document published by FEMA. The EST has the following functions:

- Supports the CDRG and assists in assuring interagency headquarters information and coordination support for response activities.
- Serves as the central source of information at the national level regarding the status of federal response activities and helps disseminate information through a Joint Informatin Center (JIC) to Congress, the media, and the general public.
- Provides interagency resource coordination support to the FCO and regional response operations. In this capacity, the EST provides coordination support for FCO, ERT and ESF activities, as necessary. ESF representatives from the primary agencies provide liaison between field operations, their respective EOCs (if applicable), and headquarters activities. The EST also coordinates donations, including unsolicited resources offered by various individuals and groups, with field elements for use in response operations.

To accomplish the resource coordination function, the EST

- Coordinates the acquisition of additional resources to support operations that an ESF is unable to obtain under its own authorities
- Advises the CDRG regarding resource conflict resolution between two or more ESFs that cannot be resolved in the affected region(s)

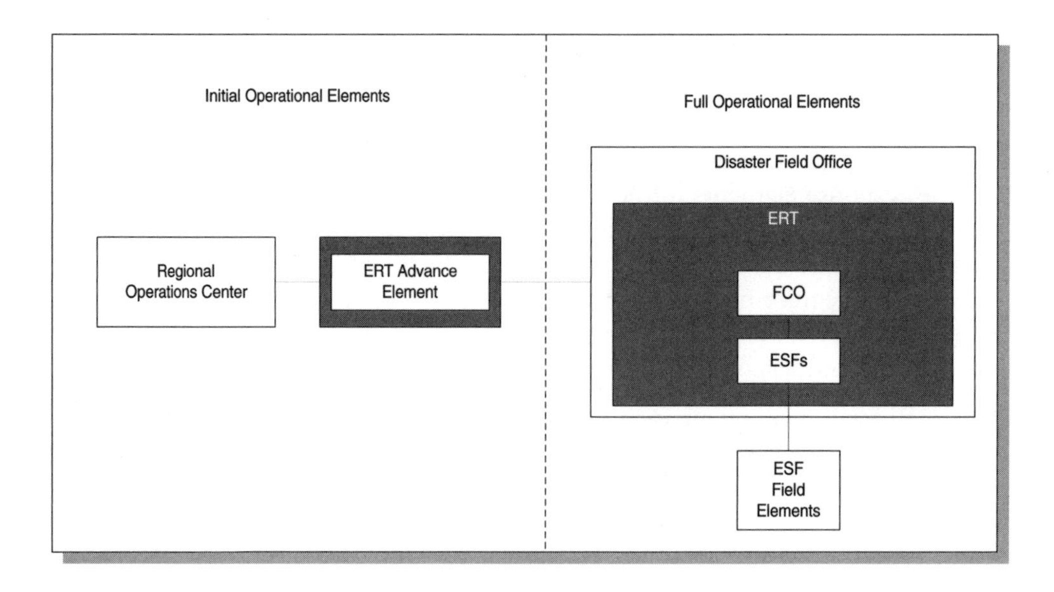

Figure 7–4 National-level response structure

- Supports coordination of resources for multi-state and multi-regional disaster response and recovery activities

AGENCY OPERATIONAL CENTERS

In addition to supporting EST activities at the FEMA EICC, headquarters departments and agencies will conduct national-level response activities at their own EOCs.

REGIONAL-LEVEL RESPONSE STRUCTURE

The regional-level response structure (see Figure 7–5) is composed of interagency elements operating from various locations. Initially, representatives from the ESFs and FEMA will assemble at the Regional Operations Center (ROC) located at the FEMA Regional Office (or Federal Regional Center). As needed, an Advance Element of the Emergency Response Team (ERT-A) will deploy to the field to assess or begin response operations as required. When fully operational, the regional-level response structure will include the FCO and ERT in a Disaster Field Office (DFO), with regional ESFs conducting response operations to provide assistance to each affected state.

REGIONAL OPERATIONS CENTER

The ROC is activated by the regional director at a FEMA regional office. It is staffed by FEMA and representatives from the primary agencies and other agencies, as needed, to initiate and support federal response activity. The ROC

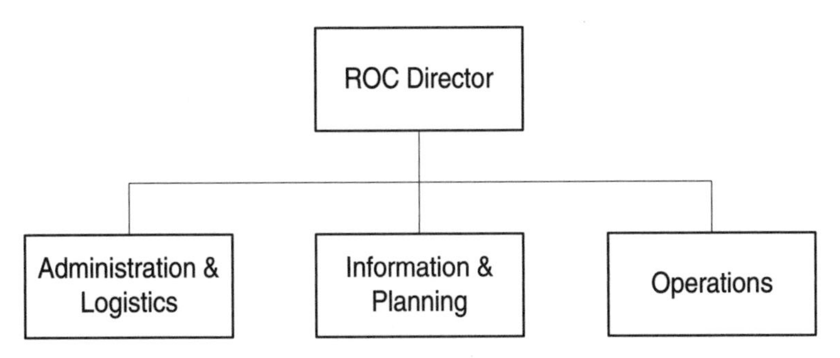

Figure 7–5 Regional-level response structure

- Gathers damage information regarding the affected area.
- Serves as a point-of-contact for the affected state(s), national EST, and federal agencies
- Establishes communications links with the affected state(s), national EST, and federal agencies
- Supports deployment of the ERT(s) to field locations
- Implements information and planning activities (under ESF#5)
- Serves as an initial coordination office for federal activity until the ERA is established in the DFO
- Supports coordination of resources for multi-state and multi-regional disaster response and recovery activities, as needed

The organization of the ROC is shown in Figure 7–6.

EMERGENCY RESPONSE TEAM

The ERT is the interagency group that provides administrative, logistical, and operational support to the regional response activities in the field. The ERT includes staff from FEMA and other agencies that support the FCO in carrying out interagency activities. The ERT also provides support for the dissemination of information to Congress, the media, and the general public. Each FEMA Regional Office is responsible for fostering an ERT and developing appropriate procedures for its notification and deployment.

Advance Element of the ERT

The ERT-A is the initial group to respond to an incident in the field. It is the nucleus of the full ERT that operates from the DFO. As shown in Figure 7–7 and Figure 7–8, the Advance Element is headed by a team leader from FEMA and is composed of FEMA program and support staff and representatives from selected ESF primary agencies. It is organized with Administration and Logistics, Informa-

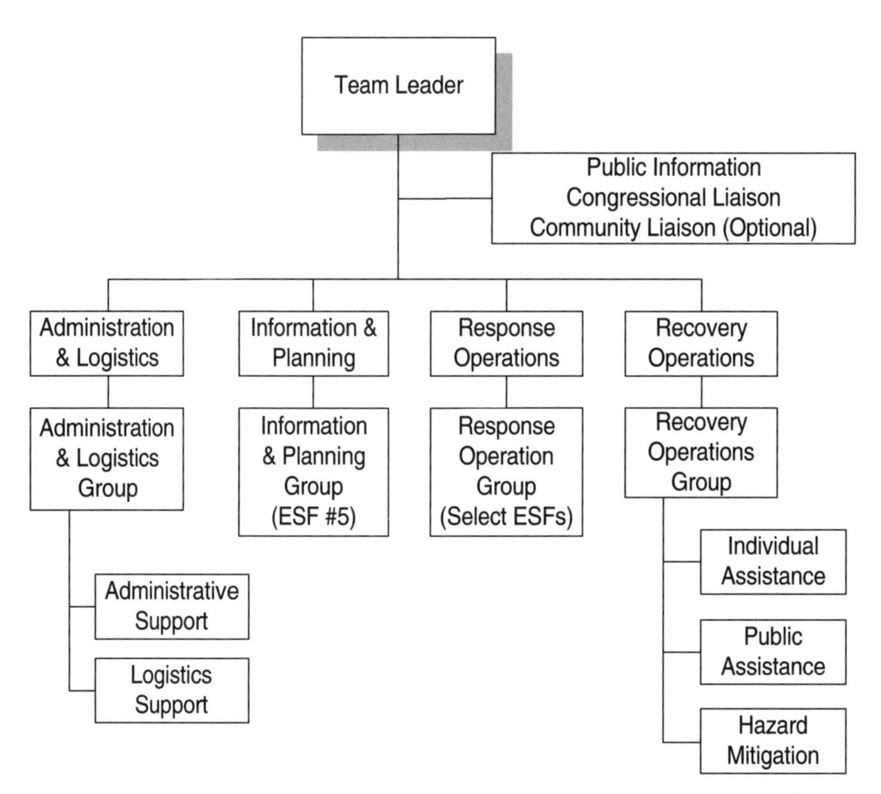

Figure 7–6 Regional operations center organization

tion and Planning, and Operations groups and includes staff for public information, congressional liaison, and community liaison activities, as required.

A part of the ERT-A will deploy to the state EOC or to other locations to work directly with the state to obtain information on the impact of the event and to begin identifying specific state requirements for federal response assistance.

Other members of the Advance Element, including leasing, communications, and procurement representatives, logistical and other support staff from FEMA and the General Services Administration (GSA), the Federal Emergency Communications Coordinator (ECU) or a representative, and the forest service, as required, will deploy directly to the disaster site to identify or verify the location for a DFO. They will also establish communications and set up operations, including the establishment of one or more mobilization centers, as required.

Structure of the ERT

As shown in Figure 7–9 and Figure 7–10, the ERT is headed by the FCO. The FCO is appointed on behalf of the president by the director of FEMA. The FCO

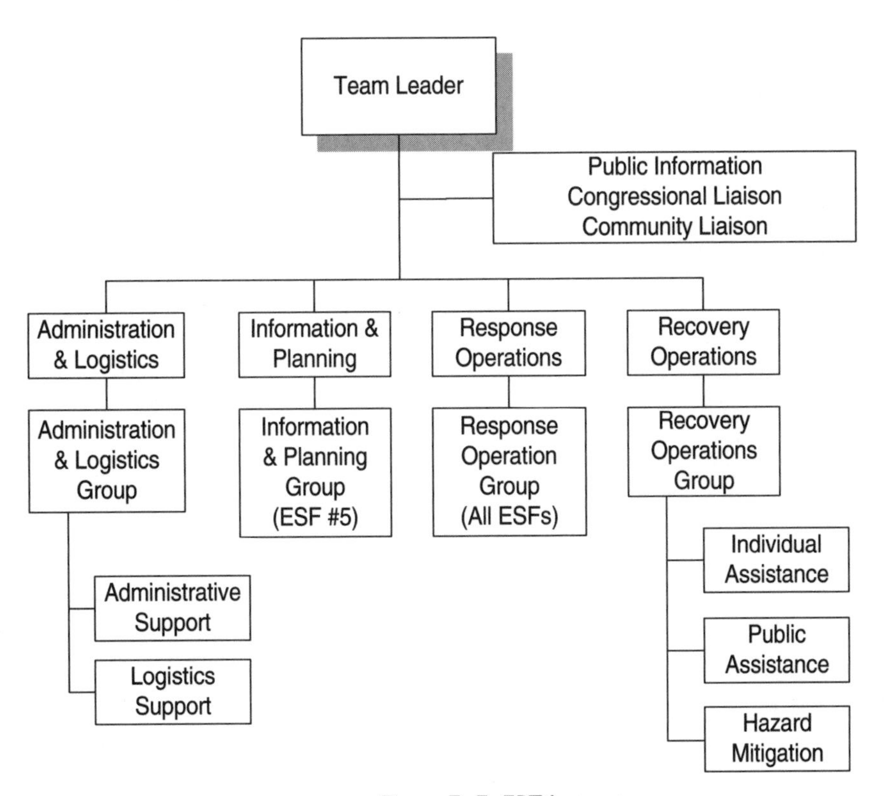

Figure 7–7 ERT-A structure

heads the ERT and is supported in the field by staff carrying out public information, congressional liaison, community relations, outreach (to disaster victims), and donation coordination activities. The FCO does the following:

- Coordinates overall response and recovery activities with the state
- Works with the SCO to determine state support requirements and to coordinate these requirements with the ESFs
- Tasks ESFs or any federal agency to perform missions in the FRP and to perform additional missions not specifically addressed in the FRP
- Coordinates response issues and problems with the CHUG that require national-level decisions or policy direction

Administration and Logistics

This element includes activities that provide facilities and services in support of response operations, as well as for recovery activities. It includes the DFO support functions of administrative services, fiscal services, computer support, and a message center.

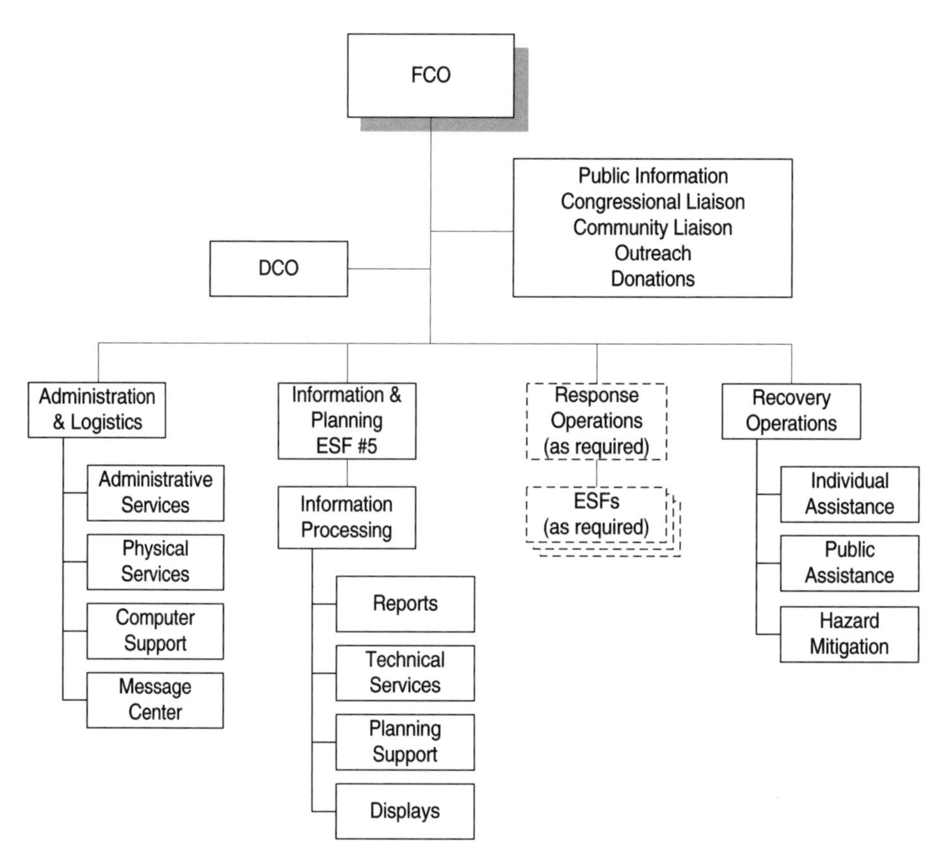

Figure 7–8 FEMA advanced element structure

Information and Planning

This element includes information and planning activities to support operations. It includes functions to collect and process information; develop information into briefings, reports, and other materials; display pertinent information on maps, charts and status boards; consolidate information for action planning; and provide technical services in the form of advice on specialized areas in support of operations.

Response Operations

This element includes the ESFs, which are activated to provide direct response assistance in support of state requirements. The functions include

- ESF#1 Transportation
- ESF#2 Communications
- ESF#3 Public Works and Engineering
- ESF#4 Firefighting

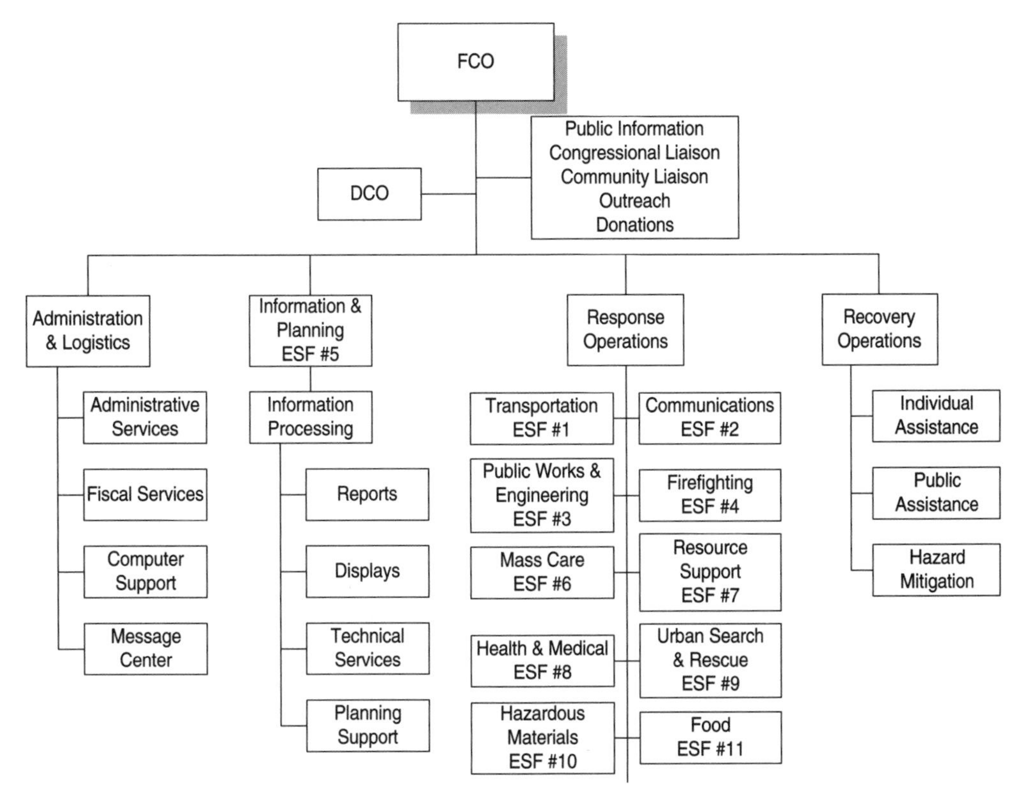

Figure 7–9 FCO structure

- ESF#6 Mass Care
- ESF#7 Resource Support
- ESF#8 Health and Medical Services

ESF#9 Urban Search and Rescue

ESF#10 Hazardous Materials

ESF#11 Food

ESF#12 Energy

Each ESF is responsible for assessing state-identified federal assistance requirements and resource requests, and for organizing and directing appropriate ESF response operations. The ESF's primary agency will identify the functional support requirements to be provided by itself, support agencies, and other ESFs.

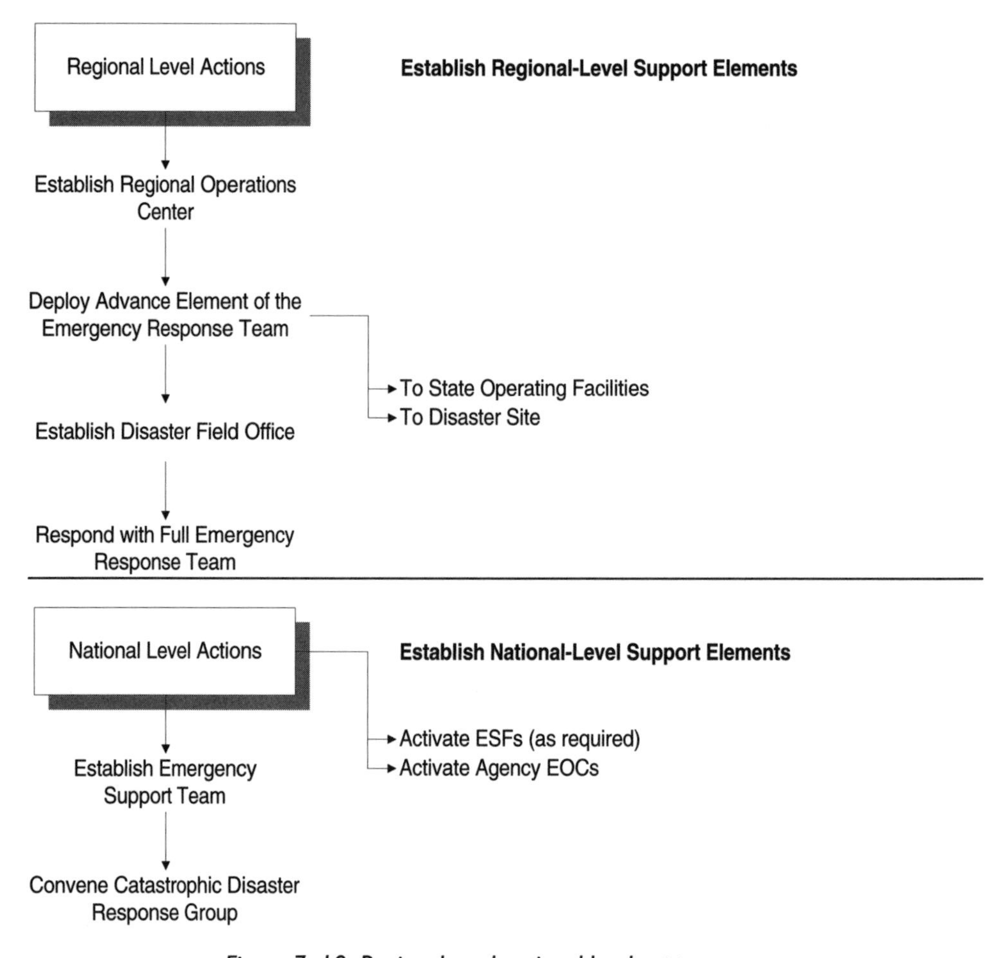

Figure 7–10 Regional- and national-level support structure

Recovery Operators

This element includes the program activities of FEMA and other federal agencies that provide disaster recovery assistance. This consists of these types of assistance. Individual Assistance includes temporary housing, grants, and loans to individuals, families, and businesses. Public Assistance includes debris clearance, the repair or replacement of roads, streets, and bridges, and the repair or replacement of water control facilities, public buildings, and related equipment, public utilities and the repair or restoration of recreational facilities and parks. Hazard Mitigation Assistance includes measures to lessen or avert the threat of future disasters.

Defense Coordinating Officer

The Defense Coordinating Officer (DCO) function is supported by the DoD. The DCO is provided by DoD to serve in the field as the point of contact to the FCO and the ESFs regarding requests for military assistance. The DCO and staff coordinate support and provide liaison to the ESFs.

NOTIFICATION

FEMA may receive initial notification or warning of a disaster from multiple sources, including the National Earthquake Information Service (NEIS) of the United States Geological Survey (USGS); the National Weather Service (NWS) (including the National Hurricane Center, the Severe Storms Forecast Center, and the River Forecast Center); the Office of Territorial Affairs of the Department of the Interior, the Nuclear Regulatory Commission Operations Center; the FEMA National Warning Center; a FEMA regional office; a state EOC; or the news media.

Upon the determination of the occurrence of a disaster or emergency, the FEMA National Emergency Coordination Center (NECC) will notify key FEMA headquarters and regional officials. If there is a need for activation of response structures of the FRP, the NECC will notify CDRG; and EST members at the national level, as required. The NECC will also notify the National Response Center, as appropriate. At the regional level, the appropriate regional director will notify members of the regional ERT.

Upon notification by FEMA, each agency is responsible for conducting its own internal national and regional notifications. CDRG members may be called to assemble at the FEMA EICC for an initial meeting. CDRG members or alternates must be available at the call of the CDRG chairperson to meet at any time during the initial response period, as necessary. Detailed federal headquarters and regional response notification actions are described in regional and headquarters procedures.

ACTIVATION

The FRP will be utilized to address particular requirements of a given disaster or emergency situation. Selected ESFs will be activated based on the nature and scope of the event and the level of federal resources required to support local and state response efforts.

Once a response requirement is identified, some or all of the structures of the FRP will be activated. This includes the establishment of the EST at headquarters level, the activation of some or all of the ESFs, and the deployment of an ERT from the regional office. The sequence of actions that will be taken at the national level and at the regional level upon activation of the FRP is shown in Figure 7–11.

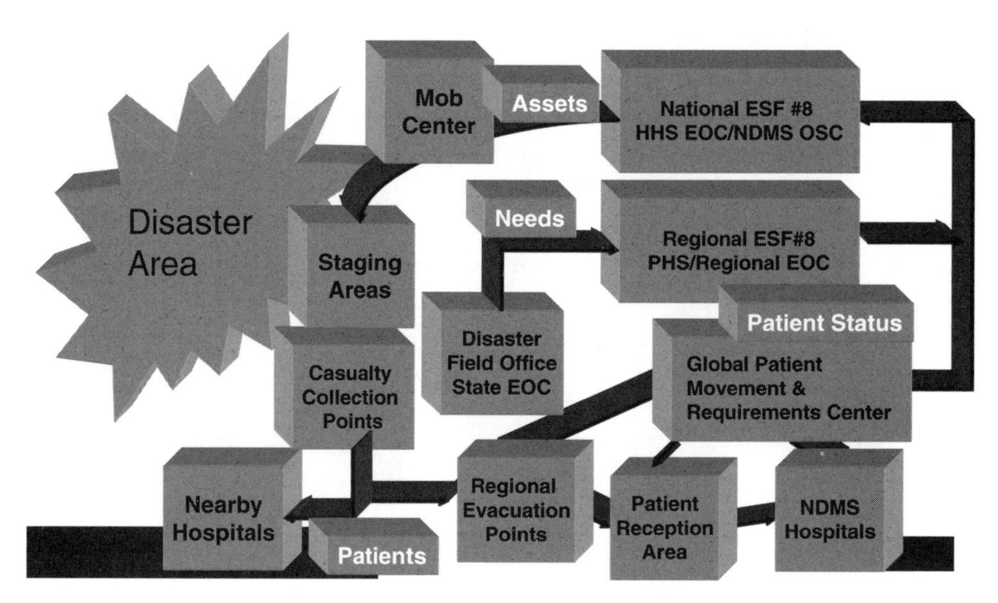

Figure 7–11 Sequence of national and regional actions upon FRP activation

At the national level, the FEMA associate director and PT&E, in consultation with the FEMA director, has the authority to activate part or all of the response structures at the headquarters level to address the specific situation.

At the regional level, a FEMA regional director, in consultation with the associate director, SLPS, and the FEMA director, also may activate part or all of the response structures of the FRP within the region for the purpose of providing response support to an affected state.

Based on requirements of the situation, FEMA headquarters and regional offices will notify federal departments and agencies regarding activation of some or all of the ESFs and other structures of the FRP. Priority for notification by FEMA will be given to contacting primary agencies.

DEPLOYMENT

When activated, ESFs and other operational elements will take actions to identify, mobilize, and deploy personnel and resources to support regional and national response operations, including the ROC and ERT activities in the regions and CDRG and EST activities in FEMA Headquarters.

RESPONSE ACTIONS

FEMA headquarters will take the following actions:

- The FEMA director will provide information on the requirements for federal response assistance to the White House and to senior-level federal government officials, as required. The FEMA associate director, SLPS7 will activate the EST and convene the CDRG, as appropriate. A JIC will be established, as required.

- The interagency EST will assemble in the FEMA EICC within two hours of notification to initiate headquarters interagency operations. The EST will provide support for regional response activities, as needed.
- At the call of the CDRG chairperson, the CDRG will convene in the FEMA EICC. Members will report on their agency deployment actions and initial activities in support of the ESFs.
- Federal departments and agencies may activate their headquarters' EICCs to provide coordination and direction to regional response elements in the field.
- FEMA will take the necessary actions to expedite the processing of a governor's request for a presidential major disaster or emergency declaration.

REGIONAL ACTIONS

- Upon the occurrence of an event that requires or may require a federal response, the FEMA regional director will initiate federal response activities from the regional office.
- FEMA and other federal agencies will activate a ROC and establish links with the affected state until the ERT is established in the field.
- The FEMA regional director, with the support of the ESFs, will initially deploy members of the ERT-A to the affected state for the purpose of assessing the impact of the situation, collecting damage information, and determining response requirements. The regional director will coordinate the federal support of state requirements until the FCO assumes those responsibilities. A JIC will be established, as required.
- ESFs will take actions to quickly determine the impact of the disaster on their own capabilities and will identify, mobilize, and deploy resources to support response activities in the affected state.

THE NATIONAL DISASTER MEDICAL SERVICE

Although the United States has never experienced a great disaster comparable in magnitude to the 1988 Armenian earthquake, the 1984 Bhopal, India toxic gas release, or the 1985 Mexico City earthquake, it is still susceptible to the kinds of catastrophic events that occur elsewhere. For example, the 1857 earthquake (Richter magnitude of 8+) that destroyed Fort Tejon, California, approximately 100 miles northwest of the center of Los Angeles, caused negligible casualties. Because the area has since become densely populated, authorities estimate that a modern recurrence could cause from 3,000 to 14,000 deaths and 12,000 to 55,000 injured persons requiring hospital treatment. Such an event in Los Angeles could cause 20,000 deaths and close to 100,000 serious injuries. There is also substantial risk of an earthquake in the central United States, which could devastate Memphis and St. Louis. No portion of the United States is free of risk from a major earthquake.

No single city or state can be fully prepared for such naturally occurring or man-made catastrophic events. Although many cities in the nation are well provided with health resources, those resources would be overwhelmed by a sudden surge of disaster injuries proportional to the population. The health resources of most states would be similarly overwhelmed. A system for dealing with disaster casualties must, therefore, provide for "mutual aid" among all parts of the nation and must be able to handle large numbers of patients that might result from a catastrophic incident.

In addition, in the event of a conventional overseas war involving American forces, the military medical system could be over whelmed by casualties returning to the United States for hospitalization. To meet the need, military casualties would be distributed among the federal and private hospitals for treatment. The National Disaster Medical System (NDMS) is a single system designed to care for large numbers of casualties from either a domestic disaster or conventional overseas war.

CONCEPT AND MISSION OF NDMS

The National Disaster Medical System (NDMS) is an organizational structure administered by the federal government to provide emergency medical assistance to states following a catastrophic disaster or other major emergencies. It is usually activated when the catastrophic disaster overwhelms both local and state resources. It is designed to supplement other resources and is oriented primarily to large-scale disasters in which local medical care capabilities are severely strained or overwhelmed. The NDMS has two primary missions:

1. To supplement local and state medical resources during major domestic natural and man-made catastrophic disasters and emergencies
2. To provide backup medical support to the DoD and Department of Veterans Affairs (VA) medical systems in providing care for United States Armed Forces personnel who become casualties during overseas conventional conflicts

Although the NDMS is administered as a partnership, the Department of Health and Human Services (DHHS) is charged with overall direction of the program. This responsibility is delegated to the DHHS Office of Emergency Preparedness (OEP). This office maintains operational control of the program during periods when it is not activated and during peacetime activation. Control of the program is transferred to the DoD in wartime support situations.

In peacetime activations that generally consist of domestic natural or man-made catastrophic disasters, the NDMS has three objectives:

1. To provide health, medical, and related social service response to a disaster area in the form of medical response units or teams, and medical supplies and equipment
2. To evacuate patients who cannot be cared for in the affected area to designated locations elsewhere in the nation
3. To provide hospitalization in federal hospitals and a voluntary network of non-federal acute-care hospitals that have agreed to accept patients in the event of a national emergency.

To carry out these three objectives, the NDMS has three sets of organizational resources:

1. *Disaster Medical Assistance Teams and medical professionals from the DHHS, VA and DoD.* DMATs are voluntary medical manpower units organized and equipped to provide austere medical care in a disaster area or medical services at transfer points or reception sites associated with patient evacuation. Hospitals, volunteer agencies, or health and medical organizations sponsor DMATs and recruit interested medical and paramedical personnel to participate. DMATs are classified into four readiness levels, as follows:

    Level One—DMATs that are fully deployable with standardized equipment and supply sets, are self sustaining for up to seventy-two hours, and are capable and willing to meet the following mission assignments as required: pre-hospital care, ambulatory care, in-patient care, medical transportation, patient disposition and evacuation, patient administration and processing, and collateral health and medical duties.

    Level Two—DMATs that are deployable with personnel and their personal equipment and supplies. Their primary mission is to augment on-ground Level One teams.

    Level Three—DMATs that have local response capability only.

    Level Four—DMATs with Memorandum of Understanding (MOU) executed in some stage of development but have no response capability.

    An activated DMAT generally consists of thirty-five members, and a team roster may include over one hundred individuals to ensure sufficient personnel on activation. Two or three DMATs may be combined to form emergency medical response units with larger treatment capabilities. See Figure 7–12.

    The OEP administers the DMAT program by maintaining memoranda of agreement with sponsors for approved teams, rating readiness levels of teams, maintaining personnel files with the credentials of team members, monitoring and approving training, and supplying teams with certain articles of equipment. Upon

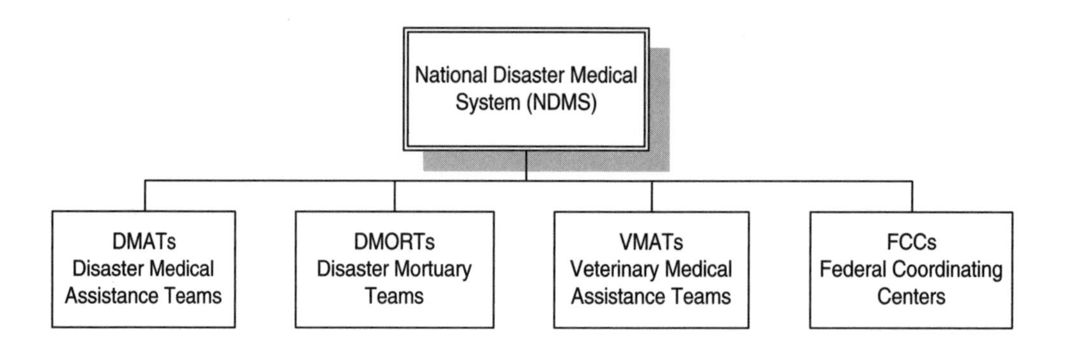

Figure 7–12 Location of current Level One DMATs

activation of the NDMS, the OEP establishes an NDMS Operations Support Center that calls up and deploys DMATs, federalizes DMAT personnel for perforate of medical tasks outside of their state of licensure, and maintains overall control of DMAT utilization. In the field, DMAT activities are supported and coordinated by Medical Support Units (MSUs). Besides general DMATs, specialized units have been formed for pediatric care, bum care, mental health, disposition of the deceased (Disaster Mortuary Services Teams; DMORTs), and assistance to Urban Search and Rescue units. In addition, field medical and related services are provided by federal health professionals from the DHHS, DoD, and VA.

2. *Casualty Evacuation System.* Movement of patients from disaster sites to locations where definitive medical care can be provided is administered through the DoD. Casualty tracking is conducted by the Armed Services Medical Regulating Office (ASMRO), and the United States Air Force provides airlift through the Air Mobility Command, which can be supplemented by civilian resources through the Civil Reserve Air Fleet (CRAF). Other types of transportation, such as specially outfitted AMTRAK trains, can be called into service through this system.
3. *Definitive Medical Care Network.* The NDMS has enrolled over 110,000 reserve beds in 1,818 participating civilian hospitals to receive casualties from disaster areas. The DoD and VA can provide additional beds, if required. Maintaining this network of hospitals in 107 metropolitan areas is the responsibility of the DoD and the Department of VA under the current concept of NDMS operations. This administration is performed through liaison offices in Federal Coordinating Centers (FCCs) in 72 DoD and VA facilities around the United States.

FCCs control patient distribution within their areas during catastrophic disaster casualty inception situations.

The entire NDMS, or selected components, can be activated in a number of ways. The governor of a state can request assistance from the president who, in turn, can either declare a disaster or order activation of federal assistance to that state. Cur-

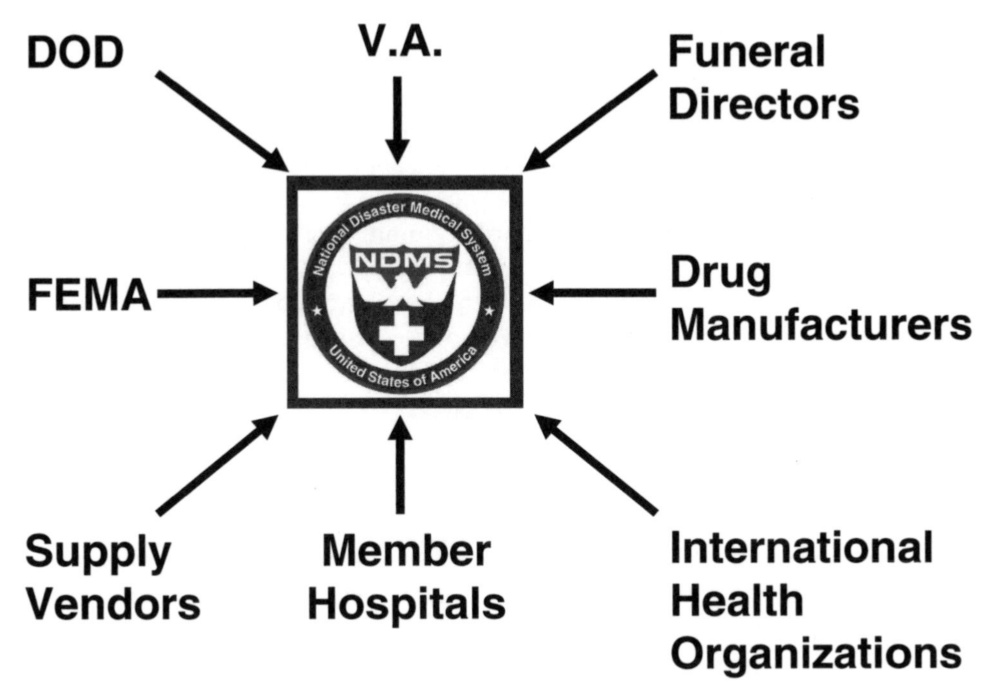

Figure 7–13 Participants to NDMS

rently, such activations are authorized under the Stafford Act of 1988 and administered through the FRP of 1992, which is coordinated by FEMA. The Public Health Service Act also authorizes the secretary of the DHHS to provide emergency medical assistance on request of local or state authorities, and the NDMS is an authorized vehicle for such assistance. The secretary of defense can also activate the NDMS in situations of national emergency. In practice, the director of the OEP would be the principal operating agent for the NDMS in all of these cases. Figure 7–13 shows participants in a National Disaster Medical System (NDMS).

HISTORICAL DEVELOPMENT OF THE NDMS

The NDMS is relatively young, having been established in 1980. During this period, the NDMS has grown considerably. Changes within the HHS and the concurrent development of the federal government's overall catastrophic disaster response structure has had a major impact on the program. Along with these recent developments, the NDMS still carries many of the traits it acquired in its inception that reflect federal policy for national emergencies dating back to the National Security Act of 1947. It has been only in the last five years that the system has been tested by activation in "real world" disasters. The most recent disasters have included team deployments during Hurricane Andrew (fifteen teams), Hurricane

Iniki (seven teams), Hurricane Emily (one team), and the Northridge Earthquake (ten teams). A Management Support Unit (MSU) was activated at each of these disasters.

Currently, the structure of the NDMS is as much a product of incremental change and evolution as it is of design. The program also appears to be coming into a period of increased utilization with modifications introduced from results of lessons learned from recent disasters. In this section, that background is reviewed in terms of three historical phases that the program has undergone in the last twenty-one years: the development of the basic idea for the NDMS, the institutionalization of the NDMS as part of the federal catastrophic disaster response structures, and the confrontation with real world catastrophic disasters.

SOURCE AND ESTABLISHMENT OF NDMS

Although authority for the secretary of the DHHS and its predecessor organizations to become involved with disaster medical services can be traced back to the National Security Act of 1947, the impetus for HEMS developed in the 1970s. The Disaster Relief Act of 1974 directed the secretary of the DHHS to provide medical response capability to catastrophic disasters. However, a specific program to carry out this responsibility did not begin until 1980. In that year, the DoD initiated the Civilian Military Contingency Hospital System (CMCHS), through which civilian nonfederal hospitals were enlisted to provide reserve beds to treat American military casualties of overseas conflicts if military bed capacity in the United States proved inadequate. At the time, the DoD was shifting some of its military medical capacity to contract sources. The CMCHS began to recruit hospitals in 1981 but ran into resistance from interest groups concerned that our government was preparing for nuclear war, even though the CMCHS was explicitly designed for non-nuclear, conventional conflicts.

At the same time, federal medical planners became increasingly concerned that the nation did not have an organized program at the federal level to deliver a medical response to catastrophic disasters. This concern led to a recommendation to establish a single national system that could provide backup support including medical services to the DoD for military contingencies and assist local and state authorities overwhelmed in domestic disasters. In December 1981, President Reagan established the Emergency Mobilization Preparedness Board with twelve working groups, one of which was the Principal Working Group on Health (PWGH). Over the next four years, the PWGH initiated actions that resulted in the announcement of the formation of the NDMS in 1984 and the operation of that system as a partnership among the DHHS, DoD, VA, and FEMA with the signing of an interagency memorandum in 1985.

The PWGH was dissolved in 1985, and in 1987, a new oversight structure was established that is essentially the one that exists today. The DHHS was designated as the lead agency in the partnership. Policy and oversight was placed in the hands of an NDMS Senior Policy Group (SPG) made up of the assistant secretary for health of the DHHS, the assistant secretary of defense for Health Affairs, the under secretary for health of the Department of VA, and the director of FEMA. Operational policy was delegated to an NDMS directorate composed of sub-secretaries from each of the partner agencies. Operations management was further delegated to an NDMS directorate staff consisting of a representative from each agency's office most directly involved with emergency medical response to catastrophic disasters. The director of the OEP within the DHHS was designated as the chair of the NDMS directorate staff.

By 1988, a series of executive orders and department directives authorized the NDMS to operate much as it does today. The program retains a commitment to the partnership concept, but responsibility for overall direction by the DHHS has been increased by actions such as the shift in the chair of the SPG from the DoD to the DHHS in 1992. This trend was reinforced by the initiation of an integrated federal response system for catastrophic disasters in 1988, which clearly designated the DHHS as the lead agency for health and medical support.

INTEGRATION OF THE NDMS INTO THE FRP

In 1988, Congress passed and the president signed the Robert T. Stafford Disaster Relief and Assistance Act (Public Law 100-707), which essentially established the integrated federal disaster response structure that we have today. It consolidated a series of existing authorities for specialized disaster assistance into one "all hazards" response system. It placed overall direction for federal disaster response in the hands of FEMA and firmly reinforced two principles of federal disaster assistance policy.

First, FEMA was to assist localities and states whose capacity proves inadequate to handle catastrophic disaster response. Along with this, there is a charge to assist and otherwise ensure that local and state preparedness is strengthened. Catastrophic disaster response, like public health, remains a state responsibility.

Second, federal catastrophic disaster response is aimed at using existing resources, not creating new ones, at the federal level. Federal agencies with existing responsibilities for disaster-related programs are networked under FEMA direction, rather than creating a new federal disaster agency. The DHHS was assigned responsibility to support FEMA for health, medical, and health-related human services. Some of these are within the NDMS. These services are components of Emergency Support Function #8 of the FRP.

Over the next four years, a federal structure for implementing the Stafford Act was worked out, resulting in the publication of the FRP in 1992. Under the FRP, twelve

ESFs were identified, with the DHHS assigned lead responsibility for ESF#8: health, medical, and related human services support. In the FRP, sixteen medical support activities are associated with ESF#8, three of which are explicitly identified as NDMS responsibilities and another three which are strongly associated with NDMS. In addition, NDMS partner agencies and other federal offices are designated support agencies to the DHHS in carrying out ESF#8 functions.

The FRP also prescribes an organizational structure for federal coordination and control of catastrophic disaster response situations. FEMA is the agency charged with overall direction of the system. Each agency that has lead responsibility for an ESF is represented on a series of coordination and control groups from the federal level to on-site at the catastrophic disaster location. This structure is primarily concerned with expeditious routing and response to requests for assistance, and the management and control of federal response units at all levels.

The NDMS, as it currently exists, finds its authority under the Stafford Act/Federal Response Plan, through interagency agreements between partner agencies, and through a series of orders and directives issued on behalf on DHHS. There are currently sixty-two DMATs at various stages of readiness, with approximately twenty-one rated Response Category One teams. Over 4,100 individual volunteers are represented by these teams. DoD and VA operate seventy-two FCCs, networking over 110,000 beds in 1,818 participating hospitals. The United States Air Force is responsible for long range transportation for the NDMS system on an "on demand" basis. Most of the growth in the system has taken place in the last few years. For instance, just four years ago, there were only 500 registered volunteers in ten active DMATs.

Part of this growth has been associated with experience in actual disasters. This experience, too, has primarily occurred in the last few years and has had a significant impact on the program and the formulation of this document.

RECENT EXPERIENCES OF THE NDMS

Since 1989, the NDMS has been activated in response to six "real world" disasters that were large enough to activate the FRP. It has also been alerted for possible activation in two other major disasters in this same period. There has also been a rise in the number of emergencies that have called for some federal response in recent years, from twenty-two in 1988 to sixty-three in 1993. These experiences have tested the concept and performance of the NDMS and have stimulated initiatives for improvement. In this section, the experience of the NDMS in its six major activations is reviewed.

HURRICANE HUGO

Hurricane Hugo first hit the east Caribbean Islands on September 17, 1989. On September 18th, the storm cut across the U.S. Virgin Islands (St. Thomas, St.

Croix, and St. John) with winds of up to 210 mph and caused extensive damage before heading toward Puerto Rico and the South Carolina coast. Although there was some damage on the United States mainland, the major damage was in the Caribbean, especially on St. Croix. Hurricane Hugo represented the first activation anal deployment of the NDMS

Beginning on September 29, 1989, two DMATs from New Mexico were deployed on St. Croix to staff a temporary emergency room, clinic, and inpatient care facility. These teams were airlifted to the island by the Air National Guard. An NDMS aeromedical evacuation system was established between the U.S. Virgin Islands and Tampa, St. Petersburg, Florida. An evacuation link was also established between St. Croix and San Juan, Puerto Rico. The DMAT used equipment for a 106-bed field hospital supplied by the Alabama National Guard. This facility took over the burden of patient care from the St. Crow Hospital, which was inoperable after the storm. During its week of deployment, the New Mexico DMATs treated 294 patients, admitted thirty-eight, and airlifted eight.

On October 7th, the New Mexico DMATs were replaced by two teams from the United States Public Health Service in Rockville and Bethesda, Maryland. The Maryland DMATs were supplemented and eventually replaced by other treatment teams activated under a number of different programs. In addition to DMATs, the DHHS OEP coordinated other medical assistance, including the delivery and control of medical supplies and assistance in reestablishing sanitary and water systems. In addition, the OEP established and operated its emergency command and control structure.

Besides being the first actual deployment of the NDMS, Hurricane Hugo provided lessons learned that would be reinforced by further experience with disasters in the years to come. These included learning to work in an environment where the normal health services system is severely damaged and assuming responsibility for many primary care functions that, although not a direct result of disaster injury, are left uncovered by the disruption in existing services. The NDMS also gained firsthand experience in coping with the complexities of coordinating service delivery with multiple agencies, maintaining communication in a catastrophic disaster environment, and providing transportation in a timely manner. In addition, the experience underlined the importance of locally focused prior planning for catastrophic disaster response and recovery. In spite of many problems, however, Hurricane Hugo proved that the NDMS was a viable tool for actual catastrophic disaster response.

HURRICANE ANDREW

The most extensive test of the NDMS to date came in August and September of 1992. Hurricane Andrew hit southern Florida with vigorous force in the early morning hours of Monday, August 24. Winds were sustained at 145 miles per hour with gusts up to 175 miles per hour, making Andrew a Category IV hurricane. The NDMS had already been alerted to the possibility of a disaster and on August 23rd had established a temporary EOC in a hotel in Oklahoma City where most key NDMS personnel were attending an NDMS national conference. Also, the OEP had dispatched an advanced element of its field command operations to Dade County and put several DMATs on alert. On the day the storm went through the area, Florida Emergency Medical Services asked for assistance, and two DMATs and an MSU were activated. They arrived in South Florida on August 25th and were followed over the next week by several other DMATs. ESF#8 was responsible for providing medical care for 1.9 million people below Kendal Avenue because Florida's resources were overwhelmed. By the end of September, a total of fifteen DMATs from eleven states were deployed to Florida. These teams included nearly 600 temporarily federalized volunteers who provided primary health care, emergency medical services, mental health services, and health and medical outreach to over 17,000 patients. Of those treated, only three deaths were reported.

Other elements of the NDMS were activated in the Florida hurricane. Some casualties were flown out of the area, and DMATs were flown in and out by military airlift. In addition, other ESF#8 activities were carried out in coordination with NDMS, such as vector control, sewage, public information, water contamination control, and health surveillance.

Hurricane Andrew represented a major test of the federal catastrophic disaster response system. The FRP had just been published, and virtually every element in it was put into operation in some way. All the player agencies in the NDMS—the DHHS, DoD, VA and FEMA—played a role both on-site in southern Florida and at the various levels of coordination including state, regional and national EOCs. Health and medical care was provided through a joint command under the leadership of the DHHS with medical assets from the Medical Brigade of the 82nd Airborne (DoD), VA mobile vans and hospitals, public health service personnel, and DMATs. A new innovation, the MSU, was introduced to manage ESF#8 functions including the NDMS.

Hurricane Andrew proved that the NDMS could work well in a complex response situation that included coordination with a large number of national, regional, state, and local organizations and programs. But it also underlined the need for condoning improvement in a number of areas. Coordination and communication between responding agencies and organizations at all levels from national to local were often strained. Federal assets could not be made available on as timely a basis as the situation often required. The Hurricane Andrew experience also demonstrated that the NDMS must be able to provide routine medical care as well as emergency medical services.

Hurricane Andrew introduced the need for a complex, integrated recovery effort that requires coordination at the local, state, and federal levels. Hurricane Andrew illustrated the complexity of the federal response structure and the need for training, standardization, and clear guidance to states and localities to make the response system effective. It also showed that the entire response structure works better if local areas are prepared and response resources are available locally to handle much of the burden of directing the response effort immediately after the catastrophic disaster event. The need for prioritizing risk areas for development of response capabilities was strongly indicated.

The OEP sponsored a major after-action conference to help identify some of the lessons learned from Hurricane Andrew and pinpoint areas where the NDMS could be improved. Recommendations from this after-action exercise contributed significantly to the development of this document.

HURRICANE INIKI

On September 11, 1992, Hurricane Iniki passed over the Hawaiian islands of Kauai and Oahu with winds of over 130 mph. The NDMS control system had already been activated, and an MSU had been dispatched to Hawaii. Between September 11th and October 2nd, when NDMS field operations were terminated, seven DMATs were deployed and treated 1,552 patients. In addition, DMATs participated in evacuation of casualties from Kauai to Oahu. The MSU coordinated the provision of a number of additional services such as back-up pharmacy support and assessments of health and hygiene conditions, including the potability of water supplies and environmental safety. DMATs also worked closely with deployed Armed Forces medical units, including relief of Air Force aeromedical personnel at Lihue Airport.

The NDMS capability deployed to Hawaii was coordinated with state disaster officials. Hawaii maintained firm control over recovery operations, so fewer NDMS-acquired resources needed to be used than otherwise might have been the case; and those that were used were deployed effectively. However, there was some delay in providing transportation for DMATs and in the management of other resources.

The experience with Iniki showed the importance of strong local preparedness and control of operations. It also showed that NDMS resources, including DMATs, should have a degree of flexibility. Local preparedness, propositioning of medical resources as close to potential disaster sites as possible, training and standardization, and flexibility were all underscored as important from the Hurricane Iniki experience.

MIDWEST FLOODS

During the summer of 1993, the Mississippi River carried a much higher volume of water than usual and produced significant flooding throughout the midwestern United States. The FRP was activated in June, and the director of the OEP at the

DHHS participated on a number of joint local, state, and federal committees to respond to the floods in the states that were impacted. Under the FRP, DFOs were established in nine states with ESF#8 representation.

All partners to the NDMS provided resources for the disaster response as part of the ESF#8 effort. These resources included military units and local VA assets. The DHHS provided assistance for water contamination control, environmental health and vector control, sanitary engineering, preventive medicine, and mental health services. Related DHHS social services included assistance in caring for the aged, food and drug supplies, and other support for primary care. A specialized DMORT was activated to assist with problems caused by flooding in cemetery areas.

The experience of the floods in the Midwest showed how catastrophic disasters can build and extend over time and emphasized the importance of integrating response with recovery plans and activities.

HURRICANE EMILY

Almost exactly one year after Hurricane Andrew devastated south Florida, Hurricane Emily threatened the east coast of the United States. The president activated the FRP on Sunday, August 29, 1993 and an EOC was established at Raleigh, North Carolina. All NDMS partners alerted their locally based assets and the DHHS dispatched an advanced party to North Carolina where the hurricane was expected to hit the Nags Head area. The VA mobilized emergency plans in its Richmond, North Carolina facility, the DHHS activated its North Carolina DMAT, and the DoD supplied air support from Fort Brag for the DHHS to assess the impact area. The NDMS also established an MSU at Richmond.

Although the storm did touch the Nags Head area and produced considerable flooding, there was little major damage and no loss of life. However, Hurricane Emily illustrated the importance of local, state, and federal cooperation, as well as speed of response. In addition, it showed the necessity for having locally positioned response assets and the significant role of the NDMS in maintaining such assets.

NORTHRIDGE EARTHQUAKE

On January 17, 1994 an earthquake occurred in Northridge, outside the city of Los Angeles. The president declared a disaster and FEMA activated the FRP. The OEP initiated activities under ESF#8, including the activation of elements of the NDMS. On the day of the earthquake, two DMATs in southern California were activated and deployed to assist emergency personnel handling the large number of people seeking emergency care at Northridge Hospital. An MSU was established and response elements of the VA were activated. Two additional DMATs were flown to March Air Force Base to be ready if needed.

All partners in the NDMS provided significant resources in the response to the Northridge earthquake. The DHHS activated a total of ten DMATs and a MSU.

These teams provided medical services to 5,676 disaster victims, including those identified in outreach activities. VA saw an additional 1,165 victims in its Mobile Health Clinics and 19,468 at Disaster Assistance Centers established at its fixed facilities.

The California earthquake illustrated how suddenly a disaster can arise. It also illustrated that many of the problems that have to be dealt with cannot be contended with immediately. In events like these, disaster response trails off into the recovery period.

OTHER RESOURCES FROM THE PUBLIC HEALTH SERVICE

There are numerous other resources available under the United States Public Health Service. In addition to being available under the activation of the FRP, these resources are also available to your local public health officials during non declared emergencies. Resources include the Centers for Disease Control and the Agency for Toxic Substances and Disease Registry. Additional information and resources can be found on the Internet at the addresses listed below:

www.emsa.ca.gov/dms2/download.htm

www.va.gov/emshg

www.cdc.gov/nceh/nps

www.bt.cdc.gov

www.hazmatforhealthcare.org

FEDERAL URBAN SEARCH AND RESCUE RESOURCES (US&R, USAR)

The Federal Emergency Management Agency has developed the Urban Search & Rescue (US&R) Response System as a component of ESF#9—Urban Search and Rescue—of the FRP. Within this framework, tactical response task forces will be mobilized for large-scale disaster mitigation on a nationwide basis.

The FEMA US&R Response System is based upon providing a coordinated response to disasters in an urban environment. Although US&R capabilities could be used for other operations, such as ground search and construction collapses, its primary mission is to locate and extricate victims trapped in collapsed buildings, especially of reinforced concrete construction. This is accomplished with specialized equipment, resources, and training that may be of limited additional value for other less substantial forms of construction.

The US&R Task Force comprises 56 persons and canines. The task force functional organization and associated terminology are predicated on, and will operate within, the National Interagency Incident Management System (NIIMS).

One of the unique precepts of the task force is that members are sufficiently crossed-trained in alternate functions to ensure depth of capability and integrated operations during search and rescue missions. Furthermore, there are two personnel assigned to each identified task-force position for rotation and relief of personnel. This practice allows for around-the-clock operations.

TASK FORCE LEADER

The task force leader is the central point of coordination of task force operations. The task force leader is responsible for receiving direction from both local and military sources, implementing strategic and tactical assignments, and providing the required information flow for proper disaster management. The task force leader is responsible for managing and supervising all search and rescue activities of the task force during a mission assignment.

The task force leader is prepared to report directly to the assigned DoD official during the mobilization and demobilization processes and to the appropriate local jurisdiction authority or IC when assigned to a mission location. Therefore, the task force becomes a resource to the IC within the operational procedures established for task force operations.

SEARCH TEAM

The first of four functional elements of the US&R Task Force is the search team. The primary focus of this team is to locate live victims trapped in collapsed buildings. The team must provide canine, electronic, and physical search strategies, tactics, and techniques, either separately or in an integrated fashion as required by the incident. The search team component of the task force is composed of

- Two search team managers
- Four canine search specialists
- Two technical search specialists

The search team managers are responsible for managing and supervising the search function of the task force during operations. This includes the employment of canine search operations, as well as technical searches utilizing specialized electronic equipment.

RESCUE TEAM

The rescue team is the second functional element of the task force. This team is further subdivided into four rescue squads, each comprising one officer and five specialists. The primary responsibilities of the rescue team are

- Evaluation of compromised areas
- Structural stabilization

- Breaching, site exploration
- Live victim extrication

The personnel comprising the rescue team and its squads are competent in

- Basic structural stability assessment
- Emergency shoring procedures
- Rigging and heavy lifting
- Breaking and breaching reinforced concrete
- Use of specialized tools

MEDICAL TEAM

The next functional component of the US&R Task Force is the medical team. This team is composed of two team managers and four medical specialists. It is very important to note that this medical team is *not* a free resource available for redeployment. The purpose of this medical team is to provide medical care for

- US&R personnel
- Victims directly encountered by the task force
- Task force search canines
- Others as practical, but not free-standing resource

A unique concern is the fact that these medical specialists are uniquely qualified to treat collapse victims. They are specifically trained in Crisis Incident Stress Debriefing (CISD), crush syndrome, respiratory injuries, and other care issues directly related to this type of disaster. These medical specialists will potentially be handing off victims to less specialized medical care once extrication and removal has occurred. This practice is considered standard under disaster circumstances. Additionally, the US&R medical team is considered a special category of the DMAT under the auspices of the NDMS. This DMAT designation will provide all requisite licensure and liability coverage for participating medical personnel.

The Medical Team Manager (MTM) is a licensed physician with substantial qualifications in emergency medicine as well as an extensive working knowledge of the medical implications of collapse victim care and rescue operations. The medical specialists are, at a minimum, paramedic certified with an accredited organization or municipality and meet the National Registry of EMT-Paramedic standards. If the specialist is a registered nurse, he or she meets all mobilization, airway maintenance, extrication skills, and vascular access abilities of the National Registry at the Paramedic Level.

TECHNICAL TEAM

The fourth functional element of the task force is the technical team. This team is composed of specialists that support the overall search and rescue mission of the task force. The primary responsibilities of the Technical Team are

- Evaluation of hazardous or compromised areas
- Structural assessment
- Stabilization advice
- Hazardous materials monitoring
- Liaison with local capabilities
- Communications and logistics

Headed by a team manager, the technical team comprises the following disciplines:

- **Structural specialist**—A certified structural engineer with a minimum of five years experience in structural design and analysis as sanctioned and approved by FEMA US&R Response System Steering Committee.
- **Hazardous materials specialist**—A currently certified hazardous materials responder technician level responder in accordance with NFPA 472 with at least three years of regular experience in hazardous materials response.
- **Heavy eequipment and rigging specialist**—A specialist with a minimum of three years of trade-related experience such as heavy equipment operation or crane operation. In addition, this specialist has extensive knowledge in lifting and rigging techniques.
- **Technical information specialist**—A specialist in the documentation and reporting requirements of the US&R Task Force operations.
- **Communications specialist**—A team member responsible for managing the US&R task-force communication system and equipment.
- **Logistics specialist**—A person responsible for the management of the equipment cache of the task force. The logistics specialist is knowledgeable in the capabilities and limitations of all equipment used by the task force, as well as aviation shipping and weights and balances requirements

The expertise and equipment cache that can be brought into action at a collapse rescue event through the use of the US&R capabilities of FEMA is enormous. Additionally, multiple US&R Task Forces can be deployed to any incident that demands their skills. As a general rule, the three closest US&R teams will be dispatched to an event as needed, and then other teams are utilized on a rotational basis.

Coordination with these task forces needs to be carried out in accordance with a well-organized Incident Command structure. Understanding the capabilities of a US&R

team—as well as its limitations—is key to the successful utilization of this resource. Figures 7-14 through 7-19 show Urban Search and Rescue within the ICS.

SUMMARY

This chapter is intended to assist in the understanding of the influx of resources that will arrive with or without your request at the local and state level during an event. It is important to remember, as a local member of the emergency service. That you must know how to obtain and manage the resources, not how each resource works. Leave that up to the specialists assigned to the task.

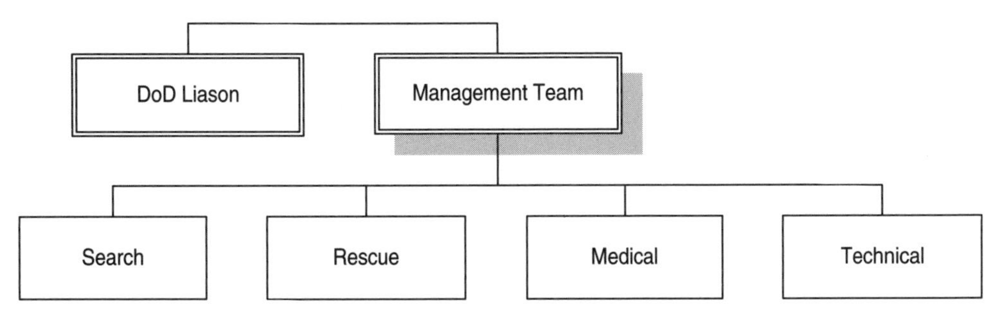

Figure 7–14 IC organizational structure

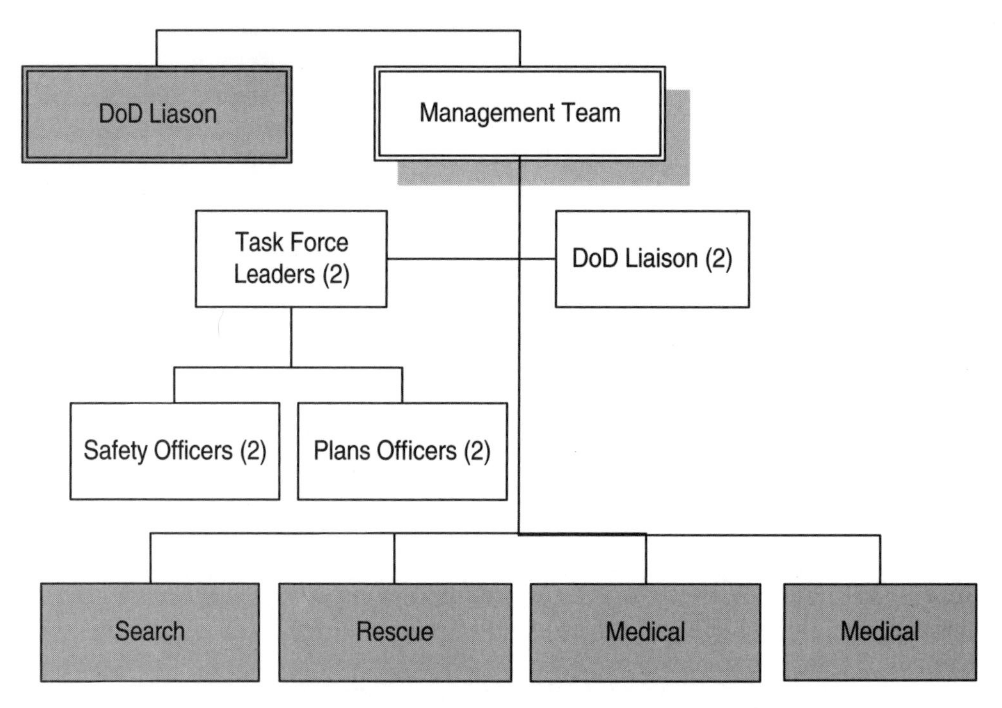

Figure 7–15 Management team—responsible for the overall management of task force operation including planning and safety

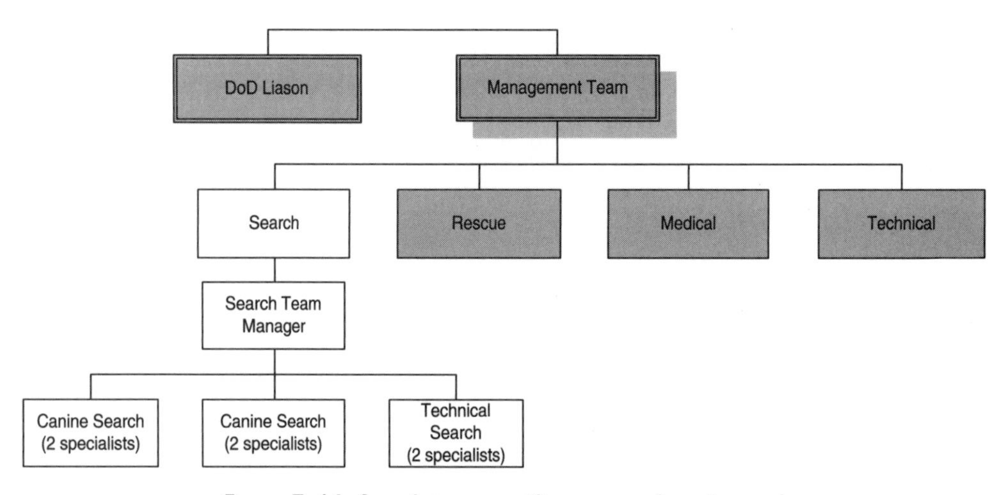

Figure 7–16 Search team—utilizes manual, canine and technical search technologies to locate trapped victims

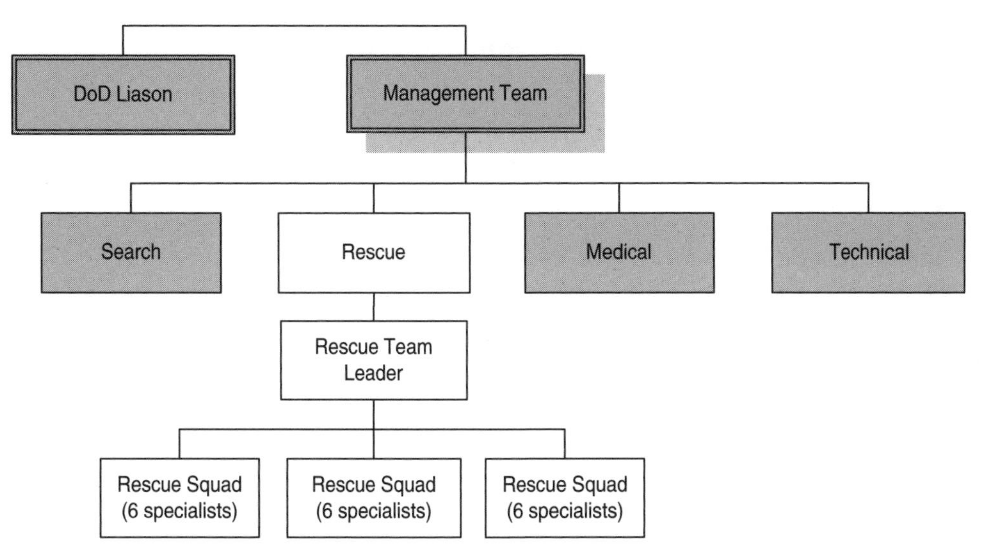

Figure 7–17 Rescue team—involves labor-intensive rescue

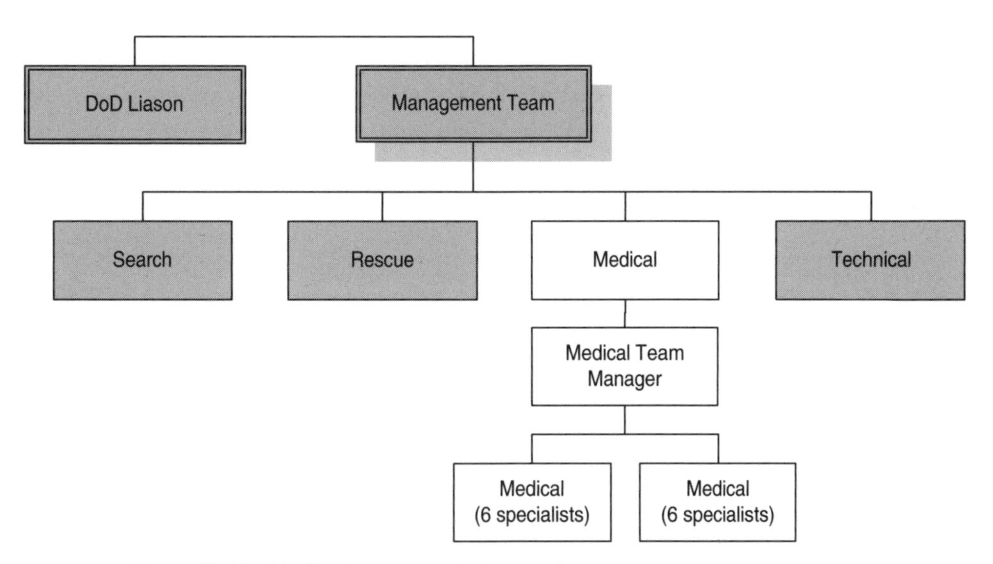

Figure 7–18 Medical team—includes pre-hospital team and victim care, specialized medicine, and limited HAZMAT medicine capabilities

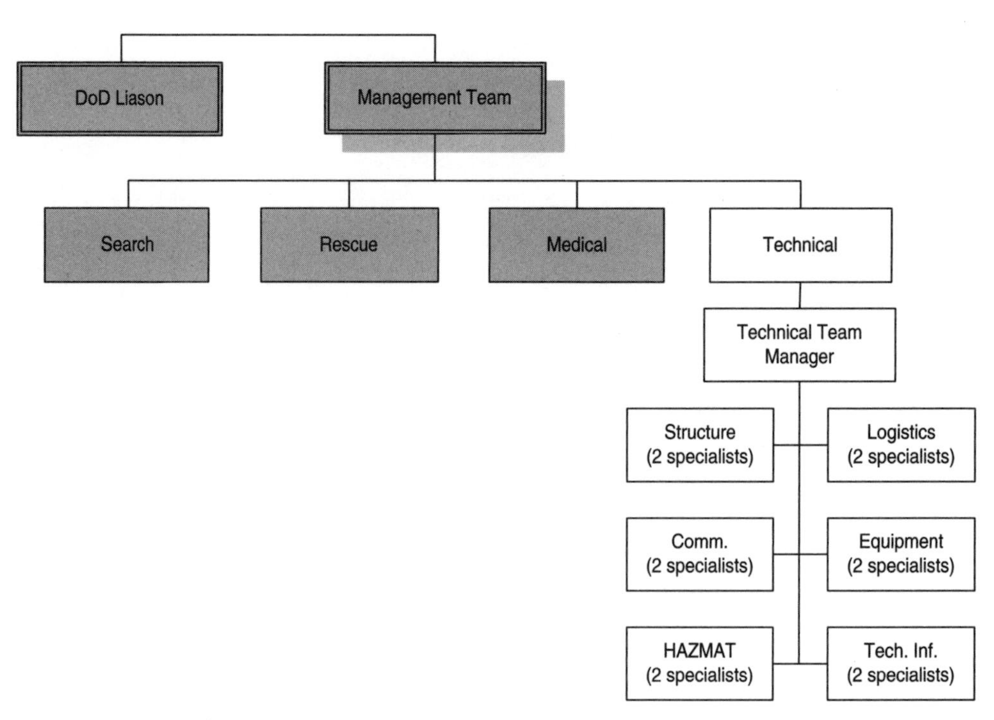

Figure 7–19 Technical team—this is an organized group of technical experts who provide specialized input covering operations

Medical Consequences

OVERVIEW

Contents presented in this chapter describe incident potentials and medical-system impacts of various WMD. After reading this chapter you will be able to

- Identify the medical-care implications of the various types of WMD
- Properly utilize existing medical management reference materials for specific types of WMD
- Identify potential roles of the local medical community with regards to major biological weapon events
- Identify potential roles of field medical treatment protocols upon the WMD event

INTRODUCTION

We have discussed in previous chapters the basic strategies for the overall management of the terrorist event and the strategies that must be considered for a safe response. Strategies that are essential to the safety of our personnel include

- Recognition of the terrorist event
- Isolation of the event
- Security measures
- Protection of responders and the public

Only when these strategies have been ensured can we consider beginning the medical treatment to the casualties of the terrorist event.

To ensure the proper care for the casualties, we will discuss in this chapter the appropriate sequence or mainstays of patient care, and the specific effects and treatments appropriate to the weapon used (physiological effects and care).

B-NICE

One acronym you may have heard is B-NICE, which stands for Biological, Nuclear, Incendiary, Chemical, and Explosive. All of the weapons we will be discussing can be placed into one of these categories. This is not to discount the times when the terrorist may use an improvised explosive device to disseminate a chemical weapon, but rather, it simply means that the terrorist has used two weapon classes. In this chapter, we will discuss the different physiological effects and patient-care issues concerning each of the classes of weapons. In the event the terrorist has used a combination weapon, you will know what the primary hazards and patient-care issues are and how to look at the whole picture of the patient. This situation would be much like when you encounter patients with different ongoing disease processes; you prioritize the presenting problems and then begin patient care.

The chapter is designed to examine the B-NICE concept for organization, and commonality between programs. A more basic introduction to these weapons can be found in the "Emergency Response to Terrorism: Self Study Guide and Basic Concepts" course. Go to www.usamriid.army.mil/education/bluebook.html for access to an electronic, printable version of both the U.S. Army Research Institue of Infectious Diseases (USAMRIID) and the U.S. Army Medial Research Institute of Chemical Disease (USAMRIC) books. These are guides to the care of biological and chemical properties. Although we will be making frequent references to them for suggested courses of treatment, use of these books does not in any way substitute for your local protocols and decisions of your medical director.

PATIENT CARE MAINSTAYS

The "mainstays of patient care" form a logical and sequential approach to caring for the victims of a terrorist event that take into consideration the protection of the rescuers, the patient, and the downstream medical community while ensuring that the patient receives life-saving medical intervention as soon as safely possible. Using the mainstays concept, broad strategic goals regarding patient care are defined and then further refined based upon the weapon involved and the health and safety threats present.

Prior to discussing the actual mainstays of patient care, we must first identify the "patient." We must remember that the primary goal of our operations at a terrorist event is to minimize suffering and the loss of life while ensuring the safety of our responders. With that goal in mind, it becomes apparent that no patient-care actions can be taken until we have properly protected our personnel against unreasonable risks. The following definitions will aid you in identifying casualties and patients:

> **Casualty (victim)**—A person who has been wounded (traumatically, psychologically, or by illness) or killed as a result of the event.
>
> **Patient**—Those casualties that are (1) viable to medical intervention without (2) causing unreasonable risk of permanent injury or death of the rescuer.

In earlier chapters, we discussed the potential harms that are likely to be present at terrorist events through the use of the TEAM CPR acronym. These potential harms are the *hazards* that face our personnel. Based upon these hazards, we must take appropriate protective measures to provide reasonable assurances that the risks to our personnel are minimized. Risk is the likelihood that actual harm will occur. Appropriate protective measures include

- Administratively preventing exposure to the hazards by isolating the hazardous area and denying entry.
- Eliminating the hazard by removal. Examples of elimination might include a bomb disposal team rendering safe an explosive device, or ventilating work areas of toxic gases or vapors.
- Securing the hazard. The hazard is still present but its potential effects upon personnel are minimized.
- Protecting personnel from exposure to the hazard by having them wear PPE designed to prevent or minimize direct exposure to the hazard.

These four protective measures can be used singularly or in any combination to ensure a work environment that is reasonably safe for our responders. Ideally, we would like to totally prevent exposure by isolating the area and denying entry. Initially, such actions will be taken until we can develop other protective measures to minimize risks. But we all know that isolation and denial of entry will not be the final answer to the safety concern. Actions within the hazard area will be required, and therefore, we must implement appropriate measures to minimize the risks.

MAINSTAY #1: APPROPRIATE PROTECTIVE MEASURES

First and foremost is the initiation of any one or combination of the four protective measures listed above to reduce the risks to our personnel to an acceptable level, based upon the viability of the casualties and the urgency of the situation. The decisions with regard to risk/benefit operations must be made based upon an awareness of both the potential hazards (TEAM CPR) and the capabilities and limitations of the four protective measures that might be utilized.

MAINSTAY #2: PREVENT FURTHER EXPOSURE TO THE HAZARD

Depending upon the weapon used, measures must be undertaken to prevent the viable casualties' further exposure to life-jeopardizing hazards. In the case of a collapsed structure resulting from a bombing event, for example, this might involve rapid extrication or even respiratory protection from dust and airborne hazards during the extrication process. In the case of an NBC event, it will most likely involve rapid removal and decontamination.

MAINSTAY #3: PROVIDE SUPPORTIVE MEDICAL CARE AS SOON AS SAFELY POSSIBLE

The basic ABCs of medical care should be initiated as soon as it is safe for both the patient and rescuer to do so. The point at which such care begins will depend upon the circumstances of the event. Those viable casualties who can be rapidly removed from the highly hazardous areas should not have interventions initiated (except very basic immobilization techniques) until they have been removed. Those viable casualties whose extraction is impeded might have basic treatment measures started in the hazard area as long as such action does not unreasonably jeopardize the health and safety of the rescuers or other viable casualties. Other interventions that may be unique to the NBC environment at this time might include the use of self-injectors designed for antidotal treatment of specific chemical weapons.

MAINSTAY #4: DECONTAMINATE

Again, depending upon the weapon used and the circumstances surrounding the event, decontamination may be required. Therefore, all patients should be assessed for decontamination needs, and the appropriate decontamination measures must be instituted to protect the casualty, responders, and downstream medical community.

NBC events will require decontamination of some form. Remember, however, that persons exposed only to vapors and gases present little risk of secondary contamination once clothing is removed. Those contaminated by or exposed to radiological materials will more than likely require extensive decontamination and monitoring for decontamination effectiveness. Those contaminated by or exposed to biological or chemical agents will require decontamination with soap and water. Check your Standard Operating

Procedure (SOP) and protocols for inhanced decontamination procedures using 0.5% hypochlorite solution.

Decontamination steps include gross decontamination involving the removal or inactivation of gross or bulk contaminants, and secondary decontamination which is the thorough washing, rinsing, and irrigation of the body. Tertiary decontamination measures may also be necessary at a medical facility.

MAINSTAY #5: MAINTAIN PERSONAL PROTECTIVE MEASURES

Assess potential for contagious disease processes and take appropriate measures. If the terrorist event involves the use of biological weapons, the potential for transmission of the disease to others must be assessed and appropriate isolation and personal protective measures must be instituted.

Generally speaking, the majority of weaponized contagious diseases are transmitted through the respiratory tract, and this mode of transmission is easily protected against with the proper use of High Efficiency Particulate Arresting (HEPA) filter masks or PPE/SCBA and standard universal precautions. A small number of other biological weapons are contracted through cutaneous or gastrointestinal contact, but the majority of these are not transmittable from human to human. When in doubt, the use of positive-pressure SCBA or HEPA filters, in conjunction with standard universal precautions, provides a very high level of protection against the *majority* of biological agents.

MAINSTAY #6: BLS/ALS INTERVENTION

Initiate Basic Life Support (BLS) and Advanced Life Support (ALS) medical intervention. Ideally, ALS intervention measures should not be started until after at least gross-decontamination procedures have been completed.

Remember that during a mass casualty event, the pulseless or apneic patient would not be considered a viable patient. Committing excessive resources to an arrested casualty can impede our system's ability to render life-saving care to more viable patients. Many times, ALS measures will be supportive in nature. In these cases, ALS care is directed at supporting respiratory and circulatory systems, maintaining proper perfusion, and enhancing the elimination of any chemical products. Many terrorist events utilize weapons that require specific medical treatment considerations in addition to the supportive ALS treatments. These treatment considerations may include

- Nerve agent treatment protocols directed toward preventing stimulation of acetylcholine receptors
- Pulmonary effects created by respiratory irritants and vesicants
- Potential crush syndrome
- Cyanide poisoning by reversing the effects upon the oxidative phosphorylation process

- Traumatic injuries resulting from explosions
- Burn injuries resulting from the use of incendiary devices

MAINSTAY #7: MEDICAL RECEIVING FACILITY (TRADITIONAL AND NONTRADITIONAL)

Obviously, patients will need to be transported to a receiving medical facility that is capable and prepared to deal with such events. These medical facilities can either be the well-prepared traditional medical facilities or nontraditional medical facilities that have been designated or established specifically to handle patients of the unique medical nature involved in the incident.

Efforts must be made to keep the local medical community intact during and after a terrorist event. Therefore, the destination facilities should be prepared or established to handle the number and types of patients that will be received without impeding the community's ability to provide continued health care. Therefore, to help keep the local medical infrastructure intact, nontraditional receiving facilities may need to be established. Such nontraditional receiving facilities could include the utilization of non-emergency medical "walk-in" and other health-care facilities for minor injuries and walking wounded, and the establishment of Casualty Collection Points (CCPs) and Patient Staging Areas (PSAs) to aid in the control of patient flow to traditional hospital emergency rooms.

CCPs and PSAs have been developed so that patients can be collected and treated in nontraditional locations. This method of patient care should continue until time permits patients to be placed into the local community's traditional health-care system while minimizing the potential for overwhelming or contaminating any single receiving medical facility.

A CCP is a predefined location at which patients can be collected, triaged, and provided with initial medical care. Generally, the CCP is staffed during the initial stages of the event by EMS personnel and other medical-care providers that are sent into the field. Once initial treatment has been received, those requiring continued medical care are then transferred to either a PSA or a local or regional hospital.

If the local or regional medical capabilities have been overwhelmed or directly impacted by the event, then PSAs can be further developed to hold the patients for long periods of time. This greater involvement of the PSAs will generally require the additional support of resources, such as DMATs, or specialty resources such as National Medical Response Teams (NMRTs), which are a specialty NBC DMAT available through the NDMS.

MAINSTAY #8: TRANSFER AT THE RECEIVING FACILITY

Regardless of whether the destination facility is traditional or nontraditional, the patient must be transferred in such a manner to ensure that any potential cross-contamination is minimized. This process takes place outside the doors of the receiving facility through a "clean-team transfer."

During the clean-team transfer process, field units avoid entering the medical facility with the patient. Instead, a clean medical team meets the transporting unit outside and transfers the patient from the litter onto a clean bed. Any equipment, clothing, or personal articles are left with the transporting team or securely packaged prior to allowing the items to enter the hospital. Direct contact between the field unit personnel and hospital personnel must be avoided.

Medical facilities need to be prepared for such events. The EMS system must urge local medical facilities to prepare for contaminated patients regardless of terrorist attack or hazardous materials accidents in the community. The EMS system should also be prepared to assist in that effort.

As a result of the Tokyo sarin attack on March 20, 1995, a local hospital, St. Luke's, was inundated with 640 patients within a two-hour period. Of the 640 patients seen by St. Luke's Hospital Emergency Department, sixty-four arrived by ambulance and thirty-five by EMS minivans. The remaining 541 victims, or eighty-four percent of those seen by St. Luke's, arrived by private vehicle and were not seen by EMS field units. Had it not been for the implementation of the Hospital Disaster Plan that required the stopping of any routine or nonemergency surgeries and the suspension of all outpatient care procedures, the hospital would have been unable to cope with the massive onslaught of victims. However, because the hospital was prepared with a plan, it was able to quickly reallocate over 100 physicians and 300 nurses and volunteers to care for the patients.

Sarin, when used in a manner as it was in the Tokyo subway, produced a vapor that caused the illness of several thousand victims. Had direct contact with the sarin occurred in the Tokyo incident, the likelihood of viable victims would have been greatly reduced. Additionally, if agents other than sarin had been used that might present a significant risk of secondary contamination, St. Luke's Emergency Department as well as any other hospital could have been shut down easily. Hospital emergency departments must therefore be prepared to control the inward flow of patients as well as to decontaminate those who arrive.

PHYSIOLOGICAL EFFECTS OF TERRORIST WEAPONS

In this section we will discuss a variety of different toxicological processes regarding WMD. This information is provided for those wishing to understand, in more detail, how these materials actually affect the body.

The overriding emphasis of this section is to instill the importance of specialized resources needed for these events, as well as to better understand how to recognize particular weapons based upon signs and symptoms.

Through the next several sections we will discuss

- The nervous system and the effects of nerve agents and neurological toxins
- The metabolic pathways and the effects of "blood agents" or cyanides
- The blood and hemoglobin and the effects of methemoglobinemia
- The respiratory system and how it affects weapon use and dissemination

NERVOUS SYSTEM

INTRODUCTION

To fully understand the effects of nerve agents upon the body, we must review and build upon some of the basic nervous-system anatomy and physiology that we learned in our basic medical training. Whether we are first responders, EMTs, or paramedics, an understanding of these basic concepts is essential to

- Identify potential agents that have been used
- Implement decisions regarding patient care

Not only nerve agents but also certain biological weapons (neurotoxins) can have pronounced effects upon the nervous system.

A simplistic example that integrates the concepts concerning all of the functions of the nervous system can be addressed by discussing what occurs when we accidentally touch a hot stove. Our nervous system (1) detects changes and gathers sensations, (2) causes the appropriate response to begin, and (3) organizes information for future recall or immediate use and interpretation. When you place your hand on the hot stove, the nervous system immediately gathers the information ("touching this stove burns my hand, which is not a good thing"), causes a reflex response ("move the hand"), and organizes thought and future recall ("this is going to hurt"). In the matter of a split second our nervous system has sensed, acted upon and "cataloged" the event.

The two major subdivisions of the nervous system that allow this event to be assessed and acted upon are the central and peripheral nervous systems. The Central Nervous System (CNS) lies within the brain and spinal cord, and the Peripheral Nervous System (PNS) comprises the cranial and spinal nerves. Our focus will be on the PNS.

NEURONS AND NERVES

The cells of the body that make up nerves are called *neurons*. All neurons have common components regardless of their location in the nervous system. Each neuron has a cell body that contains the nucleus, mitochondria, and other organelles essential to the life and function of the neuron cell. In addition to the body, each neuron cell has *dendrites* processes that extend from the cell body and are used to gather stimuli and transmit impulses towards the cell body. The length and arrangement of the dendrites varies with the specific neuron and its function. Furthermore, each neuron has another process called an *axon* that is used to transmit signals away from the body of the neuron cell to the next neuron in the series or to the muscle or gland that the nerve affects.

This entire process of gathering and transmitting signals from the dendrites to the body and subsequently down the axon is carried out by a nerve impulse called *wave depolarization*. Simply put, this electrical impulse occurs when the neuron has a greater concentration of sodium ions (Na^+) outside of the cell and a greater concentration of potassium ions (K^+) inside the cell. This results in a positive charge outside of the cell membrane relative to the negative charge stored in it. When a stimulus is received by the dendrites (such as in the case of a neurotransmitter substance), the sodium permeability of the membrane is greatly increased through chemical change. This allows the sodium (Na^+) to rush into the cell, causing depolarization. Immediately after this depolarization occurs, membrane permeability changes to allow the potassium (K^+) to move to the outside, thus restoring a positive charge on the outside resulting in repolarization. Then, the sodium/potassium pumping mechanism moves the ions to their original location and concentration gradients so that the cell is ready to transmit another impulse. This entire process—depolarization, repolarization, and pumping—occurs very rapidly and is measured in milliseconds. A neuron cell is capable of carrying out this process hundreds of times each second.

In neurons with axons that are several inches long, this process is enhanced by myelin sheaths that increase the impulse velocity and serve as insulatation from neighboring neurons to prevent "short circuiting." Neurons are grouped together in bundles to form nerves.

SYNAPSES

Neurons are not in direct contact with the next neuron in the series because of a physical space or *synapse* between the axon of one neuron and the dendrites or body of the next. Therefore, electrical conductivity or wave depolarization does not directly transmit from one nerve to the next. Instead, a chemical *neurotransmitter* substance is released by the axon of the transmitting neuron. This neurotransmitter substance crosses the synapse to receptor sites on the receiving neuron. Once the transmitter is received, the sodium permeability of the membrane for the receiving neuron is changed, and the wave depolarization occurs.

Depending upon the area of the nervous system, the neurotransmitter substance will vary, as we will discuss later.

AUTONOMIC NERVOUS SYSTEM

We established that the nervous system has two major subdivisions: the CNS, which includes the brain and spinal cord, and the PNS, which lies outside of the central. The *autonomic, or involuntary* nervous system is physically a part of the peripheral division.

The autonomic nervous system innervates smooth muscles, such as those of the GI tract, to increase or decrease activity; cardiac muscle, to increase or decrease its rate and/or depth of contractions; and the glands of the endocrine system, to increase or decrease the release of secretions or hormones.

The autonomic nervous system has two subdivisions: *sympathetic* and *parasympathetics*. Many organs are innervated by both of these systems, and the final effects can be described as a "net sum effect" of the stimulation that the organ (gland, or smooth or cardiac muscle) receives from both. This net sum effect is extremely important to the concepts of neurotoxins and nerve agents. If either the sympathetic or parasympathetic systems become overly stimulated or inhibited, the net sum effects upon that organ will shift from either sympathetic or parasympathetic predominance.

A *ganglion* is a point in which the end of one set of neurons meets the start or bodies of the next neurons in the chain. This "bulge" of neurons is actually the grouping of the bodies of the receiving nerve. Within the transmission pathways of both the sympathetic and parasympathetic systems there are two stages: *preganglionic* and *postganglionic*. The impulses of both the sympathetic and parasympathetic nervous systems begin at their respective starting point in either cranial or spinal nerves, then move through the preganglionic fibers to the ganglion. At the ganglion, a neurotransmitter substance carries the signal across the synapse to the postganglionic fibers, and the impulse is carried down the axons to their termination. At the termination, the impulse must again cross a synapse through the same or a new neurotransmitter substance to the actual tissue that it will effect—the *effector organ.*

SYMPATHETIC NERVOUS SYSTEM

The sympathetic nervous system begins in the thoracic or lumbar areas of the spinal column, where preganglionic axons end at the sympathetic ganglia just outside and lateral to the spine. One preganglionic sympathetic neuron often synapses with numerous postganglionic neurons that effect many organs. This arrangement results in an important physiological finding—that the sympathetic division brings about a widespread systemic response.

At the ganglia, the neurotransmitter substance is *acetylcholine* (ACH). This substance is released from the preganglionic axons and then stimulates the neuron receptors that further carry the impulse to the effector organs. Once this stimulation has occurred, the ACH is deactivated to prevent restimulation of the postreceiving neuron by a chemical substance called *acetylcholinesterase* (ACHE). Sympathetic transmission at the ganglia uses ACH and is *cholinergic.* After the impulse crosses the synapse, it continues down the postganglionic fibers to the next synapse at the effector site. At this synapse, the neurotransmitter substance changes to *norephi-*

nephrine (NE), which causes the intended physiological response., postganglionic sympathetic transmission uses NE and is called *adrenergic.* Unlike ACH, about fifty to fifty-eight percent of NE is taken back up by the transmitting neuron. The remaining NE rapidly diffuses into the blood stream and does not require a chemical deactivator. Also, the effects of NE are very short lived. Therefore, to continue a short-lived response, the sympathetic nervous system stimulates the adrenal gland to release more NE and epinephrine (adrenaline) to help sustain its effects.

The effects of sympathetic nervous system stimulation are commonly referred to as "fight or flight." This term originates from ancient humans' need to protect themselves and stimulate activities needed for self-preservation. Therefore, fight or flight would encompass those activities needed to respond to stress. The organs that are innervated by the sympathetic nervous system are listed in Figure 8–1, along with the resulting effects of the stimulation. As you review this list, consider the implications for an individual's ability to cope with stress that might be experienced when engaging in a fight or fleeing from danger.

PARASYMPATHETIC NERVOUS SYSTEM

The other division of the autonomic nervous system is the *parasympathetic nervous system,* which starts in the cranial and sacral areas of the CNS. Once the neurons leave these areas, their long axons travel great distances throughout the body compared with the short distances traveled by the preganglionic nerves of the sympathetic division. This results in the ganglia being placed very close to the effector organs and also results in less branching of the postganglionic fibers. Therefore, parasympathetic effects are very specific rather than systemic, as seen in the sympathetic division.

The impulse transmission of the parasympathetic system starts by the wave depolarization of the preganglionic neurons down their long axons to the ganglia located in proximity to their effector organs. At the ganglion, as with the sympathetic division, the neurotransmitter substance is ACH. But unlike the sympathetic division in which the neurotransmitter changes to NE in the postganglionic fiber, the neurotransmitter remains ACH. To summarize, ACH is the neurotransmitter that is preganglionic on both the sympathetic and parasympathetic divisions but is only found postganglionic on the parasympathetic side. If you recall from our discussion of the sympathetic system, the postganglionic neurotransmitter substance was NE. In the parasympathetic nervous system, the neurotransmitter remains ACH. Therefore, any chemical substance that increases the effect of the ACH release within the body will initially cause stimulation of both the sympathetic and parasympathetic divisions.

The parasympathetic nervous system is responsible for negative functions, and its effects are not as system-wide as the sympathetic stimulation because of the limited-branching

Figure 8–1 Sympathetic effects on organs

Organ	Effect of Sympathetic Stimulation	Considerations
Eyes		
Pupils	Dilation	Allows for greater light and vision.
Ciliary muscle	Slight relaxation	Allows far vision.
Glands		
Nasal	Reduction of secretion	Need for digestion is reduced during the fight or flight. The body will be using energy stores within the body and will not desire food intake.
Lacrimal	Reduction of secretion	
Gastric	Reduction of secretion	
Salivary	Reduction of secretion	
Sweat	Increased secretion	Sweating will increase in order to promote cooling.
Blood vessels	Peripheral: constrict	Needs to shunt additional blood to vital organs needed to deal with the stressor.
Heart	Increased rate Increased depth	Increases blood and oxygen supply to tissue to support energy production (metabolism).
Lungs	Bronchial dilation	Increases oxygenation necessary for metabolism.
Stomach	Decreased activity	Not a time to eat. The body is working off of stored energy.
Liver	Conversion of glycogen to glucose for additional energy	Needs to meet high energy demand.
Kidney	Decreased output	
Bladder	Contents released	
Basal metabolism	Increased as much as 100%	Production of additional energy.
Adrenal	Epinephrine release stimulated	Sustains sympathetic response.
Mental activity	Increased activity	

postganglia. Many have called the parasympathetic effects the "feed or breed" effects, meaning they relate mainly to the restoration of energy stores within the body.

Lets go back for a moment to our discussion of what occurs when a person is experiencing sympathetic stimulation in the fight-or-flight mode. Heart rate, respiratory rate, and metabolic rates are all increased to respond to a stress. Since metabolic rates have been increased so dramatically, much of the available glucose for metabolism may have been consumed. Therefore, after the stressor has ended, the body will lessen sympathetic stimulation and the parasympathetic division will increase stimulation, which . brings the body to rest. In an effort to restore nutrient levels necessary for metabolism, the "feed" functions are increased. The effects of parasympathetic stimulation are summarized in Figure 8–2.

Figure 8–2 Parasympathetic effects upon organs

Organ	Effect of Parasympathetic Stimulation	Considerations
Eyes		
Pupils	Constrict	Greater available light not needed. Near vision.
Ciliary muscle	Constrict	
Glands		
Nasal	Stimulation of secretion	Stimulation of digestive tract secretions is increased to facilitate the "feed" effect.
Gastric	Stimulation of secretion	
Salivary	Stimulation of secretion	
Sweat	Sweating on palms of hands	Since metabolic rate is slowing, less cooling is required.
Blood vessels	Little to no effect	
Heart	Slowing of rate and force	Less perfusion required.
Lungs	Constriction	Decreased oxygen demand.
Stomach	Increased peristalsis and tone Sphincter relaxes (mostly)	Increases digestive activity.
Liver	Slight glycogen synthesis to glucose	
Kidney	No effect	
Basal metabolism	No effect	
Adrenal	No effect	No need for adrenaline dump.
Mental activity	No effect	

MUSCARINIC AND NICOTINIC EFFECTS

ACH activates two types of receptors: *muscarinic* and *nicotinic.* Muscarine, a poison from many mushrooms, activates only the muscarinic receptors and will not activate nicotinic receptors. Conversely, nicotine will activate only the nicotinic receptors.

Muscarinic receptors are found in all effector cells stimulated by postganglionic neurons of the parasympathetic nervous system.

Not only is ACH used in the autonomic nervous system, but it is also found in many nerve endings of the involuntary nervous system. For example, it is the neurotransmitter found in the membranes of skeletal muscle fibers at the neuromuscular junction. However, the receptors are nicotinic rather than muscarinic.

ACETYLCHOLINESTERASE CRISIS

Remember that ACH requires the use of ACHE, a chemical compound, for deactivation. Failure to deactivate the ACH with ACHE will allow ACH to accumulate in the synapse, causing constant restimulation of the postsynaptic neuron. The nerve agents, common organophosphate, and carbamate insecticides act by creating an *acetylcholinesterase crisis*. This medical condition results when the organophosphate compound binds to the ACHE, preventing it from deactivating the ACH. Therefore, continuous stimulation of the postsynaptic cells occurs. Remember that ACH is the preganglionic neurotransmitter on both the sympathetic and parasympathetic sides of the autonomic nervous system. Also, ACH is the neurotransmitter to the skeletal muscle fibers. Therefore, initially, there will be stimulation of the sympathetic and parasympathetic nervous system. The patient will present early with fight-or-flight characteristics such as tachycardia, tachypnea, and sweating. In fact, epinephrine will dump into the system from the adrenal gland to support this. However, since ACH is also the neurotransmitter substance in the postganglionic parasympathetic side, the parasympathetic muscarinic effects will begin to show. The characteristic Salivation, Lacrimation, Urination, Defecation, Gastric Distress, and Emesis (SLUDGE) will later begin to show, accompanied by the most common early sign of nerve agent exposure, *miosis* (pinpoint pupils) and visual disturbances. In fact, visual disturbances such as blurred or darkened vision are one of the most common signs of slight to moderate nerve agent exposure.

A review of the medical implications of nerve agent exposure will underscore the relevance of this discussion of the nervous system.

THE METABOLIC PATHWAYS

To understand the effects of cyanides (blood agents) upon the body, it is first necessary to understand the metabolic pathways that are found in virtually every cell in the body. The military term "blood agent" is, in fact, inappropriate regarding the cyanides because its toxic effect has nothing to do with the blood or the hemoglobin in the red blood cells. Rather, like other commonly encountered industrial chemicals or products of natural decomposition (e.g., hydrogen sulfide), the cyanides affect the cell's ability to properly utilize oxygen during the energy-producing metabolic process. Therefore, if the cell cannot produce usable energy, it cannot function, and death occurs rapidly.

ADENOSINE TRIPHOSPHATE

Energy is stored in the form of a chemical compound called adenosine triphosphate, or ATP, in all the cells that require energy to function, with the exception of red blood cells. Without ATP, the cells would not be able to carry out their basic functions.

The ATP molecule is made up of the chemical compound adenosine, which has attached to it three phosphate radicals. This compound contains the energy needed within the two high-energy bonds between each of the three phosphate radicals. When energy is needed for a particular function in the cells, the bond between the second and third phosphate is broken, or hydrolized, and the large amount of energy stored within the bond is released to supply the needed cellular process. After this occurs the result is adenosine diphosphate (ADP), which is adenosine with two phosphate radicals attached (ATP – P = ADP).

Once ATP has been hydrolized to perform a function within the cell, the resulting ADP must have another phosphate radical reattached so that it again can be used in an energy-requiring process (ADP + P = ATP). Therefore, the use of ATP is cyclical. ADP is combined with another phosphate radical to yield the high-energy-containing ATP. The ATP proceeds to an area of the cell requiring energy and is hydrolized back to ADP, and its stored energy is released and used in a cellular process such as the construction of proteins or the sodium-potassium pumping process. The ADP is then reassembled with another phosphate radical to again form ATP, and the process is repeated.

If we prevent the manufacturing of ATP, the cell can no longer function because it has no useable energy store. Cyanides attack the body by shutting down a process within the cell that is responsible for assembly of greater than eighty percent of the cell's ATPs. This fact provides a basic understanding of the effects of cyanides.

By shutting down ATP production, none of the cells within the body except the red blood cells can continue to perform their functions. The neurons within the brain have a very high energy demand and are among the first to be affected. Weakness sets in, the level of consciousness drops rapidly, seizures occur because of the CNS effects of highly irritated neurons, and the medulla oblongata of the brain, which controls cardiac activity and respiratory drive, shuts down. Therefore, respiratory arrest occurs very rapidly. But since the red blood cells do not require the production of ATP, they continue to function properly and cyanosis does not occur. The patient will look like a hypoxic or anoxic patient, depending upon the dose received, without being cyanotic. In fact, the term "blood agent" was coined because it was originally thought that the cyanides affected the blood's ability to carry oxygen. Nothing could be further from the truth.

To truly appreciate what is occurring in the cell, we must examine how ATP is manufactured and what ATP building process is being affected. In order to do this, we will discuss the three metabolic pathways that make ATP: glycolysis, the TCA (Krebs) cycle, and the oxidative phosphorylation process (also called the ion transport chain).

GLYCOLYSIS

The first metabolic pathway is that of glycolysis. "Glyco-" means glucose, and "-lysis" means to split apart, therefore, glycolysis is the splitting of the glucose molecule. During this process, the glucose molecule goes through a series of chemical reactions to produce two molecules of pyruvic acid. The energy from two molecules of ATP is required to carry out these reactions, but in themselves, the reactions yield an output of four molecules of ATP for an overall net gain of two ATP molecules. These molecules of ATP can then be used elsewhere in the cell for energy requiring processes.

The key factors about glycolysis are (1) there is a net gain of two ATPs and, as we will see, this is a very small amount; (2) the glycolysis process does not require oxygen and therefore is called anaerobic (without oxygen) metabolism; and (3) pyruvic acid is converted to lactic acid, resulting in lactic acidosis, unless the next two metabolic pathways are properly functioning.

TCA (KREBS) CYCLE

Once the pyruvic acid leaves the glycolysis process, it is prepared for entry into what is called the Krebs cycle. This is the start of the aerobic metabolism process.

The two pyruvic acid molecules created in glycolysis give up carbon dioxide, and what remains, the acetyl group, is combined with an enzyme called coenzyme-A. The resulting molecule, called acetyl coenzyme-A, enters the Krebs cycle. For our purposes, the Krebs cycle is another series of chemical reactions that takes place in an area of the cell called the mitochondria.

During these reactions, the acetyl coenzyme-A is acted upon to produce more carbon dioxide and to bind the enzymes NAD and FAD to hydrogen atoms. These enzymes, with the attached hydrogen atoms, are essential to the third metabolic pathway. Therefore, the Krebs cycle's major role is to get chemicals ready for the ion transport chain. However, as an added benefit, the Krebs cycle also manufacturers another two ATPs.

There are several families of chemical compounds that can adversely affect the Krebs cycle. If the cycle is damaged, the important NADH + H and FADH + H are never delivered to the ion transport chain.

Fluoroacetate is one such chemical that can interrupt the Krebs cycle. In the field, patient care for exposures to such a chemical would be supportive in nature, that is, to prevent further exposure by decontaminating and gastric binding with activated charcoal, and to provide Advanced Life Support (ALS) care.

What is important to remember about the Krebs cycle is that it prepares essential enzymes and hydrogen ions for use in the ion transport chain. Failure to deliver these ions to the ion transport chain will result in that metabolic pathway's failure. In addition, if the Krebs cycle is shut down, the pyruvic acid released from the glycolysis process is not consumed and therefore will convert to lactic acid. Lactic acidosis will result. As a by-product of the Krebs cycle, carbon dioxide and two molecules of ATP are generated.

OXIDATIVE PHOSPHORYLATION (THE ION TRANSPORT CHAIN)

The Oxidative Phosphorylation (OP) process also occurs in the mitochondria of the cell. During this process, a tremendous amount of usable energy in the form of ATP molecules is generated. Compared to glycolysis, which produces a net gain of two ATPs; and the Krebs cycle, which produces an additional two ATPs; the OP process generates twenty-six to thirty-two ATPs or more than eighty percent of the cell's usable energy.

During the OP process, hydrogen ions are brought to the start of the ion transport chain. The ion transport chain is a series of chemical compounds called cytochromes that can be oxidized to give up electrons and reduced to take on electrons. For each of these cytochromes, the chemical compound composing it is reduced by taking on hydrogen ions (H+ and e-) and then oxidized by giving up these hydrogen ions. These hydrogen ions are then used to reduce the next cytochrome in the chain. This series of oxidation/reduction reactions occurs until the hydrogen ion is moved to the bottom of the transport chain or the "terminal cytochrome." Once the hydrogen ion reaches the terminal (A3) cytochrome , it is then released to awaiting oxygen to form water. If oxygen is not present, the hydrogen ion will remain on the terminal (A3) cytochrome. If the ion remains there, then the ion from the previous cytochrome cannot move down to the open acceptor site and the transportation of hydrogen ions stops in the entire chain.

The series of chemical reactions in each of the cytochromes in the transport chain releases energy every time one of the cytochromes is oxidized and gives up the hydrogen ion. This energy is then utilized by a "coupled" chemical reaction called phosphorylation that assembles a third phosphate to the ADP molecule (ADP + P = ATP). Therefore, the assembly of ATP from ADP and P is directly dependent upon the release of energy from the transport chain, and that release of energy is directly dependent upon the movement of hydrogen ions down the chain. If oxygen is removed from the bottom of the reaction or something else causes that ion transport process to stop, such as filling the acceptor sites on the terminal cytochrome with another chemical other than hydrogen, then energy is not released. Without that energy, the usable form of cellular energy (ATP) cannot be generated.

The conditions and toxins that affect the ion transport chain are

- **Anoxia**—Lack of oxygen would stop the process because the hydrogen has no place to go once it reaches the bottom of the chain.
- **Cyanides and hydrogen sulfide**—They bind to the terminal (A3) cytochrome, preventing the hydrogen ion from moving down the chain. Therefore, the oxidation/reduction reactions stop, and the coupling of ADP to another P stops.
- **Phosphorylation uncouplers (a form of commercial pesticide)**—These prevent the addition of the third "P" to the ADP molecule. The ion transport chain works, but the associated reaction to make ATPs stops.

BLOOD, THE MEDIATOR OF INFLAMMATION

Without going into excessive detail, the blood is another physiological system that we must discuss during the study of the effects of WMD. We have already put to sleep the myth that cyanides attack the blood. However, other commonly found industrial materials that could be weaponized through improvised means, as well as some agents that are already weaponized, have specific effects on the blood or on functions in which the blood plays a vital role.

ROLES OF THE BLOOD

The blood, in fact, is an organ unto itself. It is the only organ in the body that exists as a liquid, and it is responsible for numerous functions:

- Transport of nutrients to the cells of the body (e.g., glucose, oxygen)
- Removal of wastes from the cells (e.g., carbon dioxide, urea)
- Regulation of function (e.g., movement of hormones, maintaining pH balance, maintaining cellular hydration)
- Defense through clotting, controlling mutated cells and fighting infection

We will focus on only two functions: the transport of nutrients, specifically oxygen, and the defense mechanism.

TRANSPORT OF OXYGEN

The Red Blood Cells (RBCs) are responsible for carrying oxygen through the tissues of the body. The RBCs are bioconcave hockey puck shaped-disks that are able to slightly deform themselves in order to fit into the tight passages of the capillaries. There are between five and six billion RBCs in every cubic centimeter of blood.

The RBCs are unique in that they do not have a nucleus or a mitochondria. And, if you recall from our discussion about the metabolic pathways, the absence of mitochondria would indicate that the RBCs do not produce their own energy. Furthermore, the absence of a nucleus that contains the DNA blueprint necessary for reproduction would indicate that they cannot reproduce themselves. Therefore, their production is carried out by the marrow of the bone. And any chemical or material that destroys the bone marrow would stop blood cell production. This is one of the major effects of ionizing radiation. The syndrome that results from exposure to such radiation, called Acute Radiation Syndrome (ARS), suppresses the blood cell production capabilities of the marrow. As RBCs die off after 90 to 120 days on average, they are not replaced, and the patient becomes anemic. But even before this anemia develops, the white blood cells and platelets, having a life span of thirty days or less, die off even faster, and the patient in ARS becomes very prone to infections and bleeding.

The bulk of the weight of a RBC is derived from a molecule called hemoglobin (Hb). Each hemoglobin molecule in the RBC is composed of four peptide chains, or "globins," that are intertwined. Within each RBC, there are approximately 250 million hemoglobin molecules. At the end of each of the four peptide globins in the molecule is attached a "heme" group that contains iron, hence the name "hemoglobin." It is to this iron or heme group to which oxygen attaches to be carried throughout the body. When the heme group is oxygenated, we refer to the hemoglobin as oxyhemoglobin.

Oxygen, as we know, is brought into the lungs through respiration. Once inside the alveoli, the oxygen diffuses through the very thin alveoli wall, the capillary wall, and the membrane of the red blood cell and attaches itself to the heme groups of the hemoglobin molecule. Once oxygen is attached to the hemoglobin and forms oxyhemoglobin, it is carried to the tissues of the body and then released by the heme. The oxygen then diffuses through the membrane of the red blood cell again, through the wall of the capillary, into the interstitial spaces between the cells and then through the membrane wall of the cell requiring oxygen. Once inside the cell, the oxygen is carried to the end of the ion transport chain we discussed earlier and awaits a hydrogen ion to be released from the terminal cytochrome. Two hydrogen ions then bind to oxygen to form water as one of the byproducts of cellular respiration.

On the RBC's return trip, the heme group picks up a bicarbonate ion (HCO_3-) that is holding the waste carbon dioxide that was generated by the glycolysis process and the Krebs cycle. The bicarbonate radical carrying carbon dioxide, when added to the hemoglobin, changes the hemoglobin to carbaminohemoglobin for the return trip to the lungs. At the interface between the capillary and the alveoli, the heme group releases the carbon dioxide and it is diffused into the alveoli.

As you can see, the heme group plays a vital role in cellular respiration by making sure the oxygen moves to where it is needed and that carbon dioxide (in the form of bicarbonate) is removed. Therefore, properly functioning hemoglobin is essential to the ion transport chain process and the resulting production of ATP.

Several industrial compounds could conceivably cause drastic changes in the RBC and its ability to carry oxygen:

- **Arsine**—Arsine, which can be found in various chemical compounds and as arsine gas, passes through the lungs and into the RBC. Once inside the RBC, arsine causes the production of hydrogen peroxide within the cell, which in turn "splits" the red blood cells in massive numbers (hemolysis) and allows the hemoglobin to be released into the free plasma of the blood stream. Not only is the oxygen-carrying capacity of the RBC destroyed, but the hemoglobin in the plasma causes severe damage to the kidneys.
- **Methemoglobin formers**—Methemoglobinemia is a condition in which the iron of the heme group is changed so that it can no longer carry oxygen. In order for oxygen to be carried by the heme group, the iron within the heme group must be in what is referred to as the ferrous state (Fe+2). If a chemical enters the RBC that causes the iron to change from the ferrous to the ferric (Fe+3) state, the hemoglobin can no longer carry oxygen, and the effects will be similar to those of suffocating in an oxygen-deficient atmosphere. When methemoglobin is formed, the iron in the heme group is actually being oxidized (think of it as "rusted") by the causative agent. Therefore, since oxyhemoglobin is responsible for the characteristic red color of the blood, methemoglobin gives the blood a characteristically rust or chocolate color. And, since methemoglobin will not change back to the ferrous state by itself, the blood will remain that color when exposed to air (as would the blood from someone who was just simply in an oxygen-deficient atmosphere). The most striking sign of methemoglobinemia is this brownish cyanosis or cyanosis that is refractory to good airway control and oxygenation attempts.

  The compounds that can cause this medical problem are nitrate and nitrite compounds, aniline, and nitrobenzene, to name a few. The treatment for methemoglobinemia is the administration of methylene blue, which allows the hemoglobin to return to its natural state.
- **Carbon monoxide**—In basic firefighter or medical training we learn that carbon monoxide binds much more easily to the heme group than does oxygen. This results in a reversible condition called carboxyhemoglobin. The treatment for carbon monoxide poisoning is, of course, high flow oxygen and definitively, treatment in a hyperbaric chamber.

One last point we need to address before examining the mediators of inflammation: the use of pulse-oximetery for methemoglobinemia and carbon-monoxide poisoning. Pulse-oximetery works of the principle of measuring the color characteristics of the hemoglobin molecule. Carboxyhemoglobin looks very much the same color as regular oxygenated blood, and therefore, pulse-oximetery readings in the presence of carbon monoxide poisoning will be normal or high. In the case of methemoglobinemia, the hemoglobin changes to a color that is darker. Therefore, the corresponding pulse-oximetery reading will be low. The bottom line is that in either of these cases, pulse-oximetery is of little value.

BLOOD AS A MEDIATOR OF INFLAMMATION

The inflammatory response is a combination of natural processes that attempts to minimize tissue injury and and protect the body from chemical irritants, bacteria, mechanical injuries, cuts and even burns. The inflammatory response may also accompany immune system reactions.

When an irritation occurs, the site that is affected releases inflammation mediators such as histamines that initiate the inflammatory response. During the response, three specific things begin to happen:

1. The vascular system in the affected area dilates to allow a greater flow of blood. This results in characteristic redness in the area and allows for white blood cells to move into the area.
2. The inflammation mediators also cause the vascular system in the area to increase its permeability. This allows fluids and white blood cells to leave the capillaries and come into direct contact with the tissues that have been irritated or injured. As fluid permeates the tissues, swelling occurs and the white blood cells hold the affected area in check so that the offending material cannot spread. This results in inflammatory exudate or pus formation.
3. Chemotaxis is brought about to consume the damaged cells and bacteria through phagocytosis (or "amoebalike eating"). This results in some collateral damage to area tissues, scarring, and the formation of fibrotic tissues, but the irritant is ultimately destroyed and removed.

Why is this important? Some of the very common warfare agents are choking agents. In fact, many of the most commonly manufactured industrial materials in the world act upon the body in very much the same way as choking agents. The inflammatory response to the agents is what makes them deadly.

RESPIRATORY SYSTEM CONSIDERATIONS

The respiratory tract is uniquely susceptible to materials that can make their way to the lower respiratory passages. Materials that are between 0.5 and 3 microns in size can make their way all the way down to the alveoli, where gas exchange takes place with the RBCs. The body's initial response to these irritants is to cough in order to get the offending materials out of the respiratory passageways. In addition, the bronchioles will begin to close, causing shortness of breath and wheezing in an attempt to further minimize exposure of the delicate membranes. However, if the gas or vapor that was inhaled is water soluble, it will quickly mix with the moisture in the lungs to possibly form corrosive and irritating liquids. Once the irritation occurs, particularly in the terminal bronchioles and alveoli, the inflammatory response begins.

Solubility and insolubility of the agent has a direct bearing on the level of effect and damage to the lung tissue. Typically, the lungs are lined with a mucus barrier. At the upper level of the respiratory tree the mucus is a water-based material. Thus, a water-soluble chemical will affect this area. An example is chlorine, which is water-soluble. When inhaled, the water-based mucus and chlorine mix, causing irritation at this site. Lower-level lung damage is limited except under high concentrations and exposure for a long duration.

Non-water-soluble chemicals have a tendency to bypass the upper portion of the respiratory tree and attack the lower portion which is lipophilic (surfactant), causing damage in the fine bronchioles and alveoli. The water-based mucus in the upper airways tends to repel the non-water-soluble products, sending them deeper into the respiratory tree. Phosgene, a non-water-soluble chemical, is a good example of this type of product.

In both cases of water-soluble and non-water-soluble chemicals, the length of time and degree of concentration has a direct effect on the lung tissue. If a water-soluble chemical is present in high concentrations over a period of time, deep injuries can occur. If a non-water-soluble chemical in high concentrations reaches the lower regions of the lungs, the damage is devastating; an example is sulfur mustard.

Vapor pressure also has a hand in this equation. The higher the vapor pressure, the higher the velocity, and thus, the deeper the penetration of the chemical. All these principles work together, in reference to respiratory injuries.

As the alveoli become irritated, the inflammatory mediators are released, and the three phases of the inflammatory response begin. The capillaries that are in direct contact with the outside of the alveoli dilate, allowing greater blood flow to the area to occur. The permeability of the capillaries increases to allow the fluid and the white blood cells to come into direct contact with the outer wall of the alveoli. When this occurs, fluid begins to build up between the capillary wall and the alveolar wall, and pressure begins to develop. Since the pressure will follow the path of least resistance, it will begin to leak into the alveoli and pulmonary edema (fluid in the lungs) begins to develop. Lastly, phagocytosis begins, and the damaged cells are destroyed along with some neighboring undamaged cells. This results in long-term damage and scarring of the lung tissues. Depending upon the extent of damage, it may lead to long-term adverse health effects if the patient survives the pulmonary edema.

Noncardiogenic pulmonary edema is the form of edema that was discussed above. It is dramatically different from the cardiogenic pulmonary edema that we normally see as a result of left heart failure.

Remember that cardiogenic pulmonary edema is a result of damage to the left side of the heart resulting in increased pulmonary vascular pressures. In that case, the pressure is coming from within the blood vessels, and we control this by trying to increase vascular volumes through the use of nitroglycerine and morphine sulfate (in addition to 100 percent oxygen therapy). However, the pressure being developed in noncardiogenic pulmonary edema is not coming from within the vascular space but rather from within the potential space that exists between the alveolar wall and the capillary wall. As a result, diffusion of oxygen across the space is decreased, resulting in a more profound state of hypoxia; and the administration of nitrates and morphine will not reduce the fluid in the lungs.

Patients with a noncardiogenic pulmonary edema require aggressive airway control, the use of Positive End Expiratory Pressure (PEEP or CPAP) to minimize inward leakage of the fluids, and possibly the use of steroid therapy to minimize the inflammatory effects and to stabilize the cell membranes.

The most common chemicals in the world, not to mention choking agents, cause respiratory irritation and noncardiogenic pulmonary edema. Chlorine, sulfur dioxide, anhydrous ammonia, the mixture of household bleach and ammonia, and even vapors from acids in sufficient doses can result in this syndrome. Everyone has tons of these chemicals in his or her community. Special consideration must be given to agents that do not have to be manufactured and hidden but are so common that they can be found and handled easily and openly.

MASS PATIENT DECONTAMINATION

One of the unique considerations of the terrorist events is the potential need for mass patient decontamination. Primarily, the decontamination of large numbers of patients will be related to the biological, nuclear radiation, and chemical events. However, even in explosive and incendiary events, mass patient decontamination may be required because of the agents involved and the spread of bodily fluids.

Historically, fire and EMS operations have focused decontamination on one to several individuals. However, now we must develop systems to decontaminate potentially several hundred people in a short period of time.

Regardless of the system your community establishes, the basic concept of decontamination applies to the mass patient decontamination arena; gross decontamination must occur to remove the bulk of the contaminants. This stage involves stripping and flushing the patient. If the situation permits, efforts should be made to protect the victims' modesty. However, when chemicals (particularly the vesicants) are involved, rapid decontamination is essential for the minimization of toxic effects. Under no circumstances should contaminated persons be allowed to leave the scene without at least undergoing gross decontamination.

After gross decontamination, various levels of secondary decontamination will be required. This process involves the systematic washing and rinsing of all body surface areas in order to remove, to the extent possible, any remaining contaminates.

Further decontamination may or may not be required. Tertiary decontamination is generally performed at a medical facility and may involve actions such as gastric decontamination and further wound cleansing.

When conducting decontamination there are two objectives: (1) prevent further extension of the incident (gross or secondary contamination), and (2) minimize adverse health effects of continued contact. In order to determine the extent of decontamination that is required, several factors should be considered.

- **Physical state of the agent involved**—Is the agent a solid, liquid, gas, or vapor? Solids and liquids on the surface of the body require more aggressive decontamination efforts, whereas exposure to only gases or vapor generally results in minimal residual contamination. Therefore, those only exposed to gases and vapor present little risk of secondary contamination once clothing is removed. The safety level for both the patient and responder is greatly enhanced by simply flushing with water. If solids or liquids are involved, stripping and flushing may or may not be fully effective. In these cases, the next two items should be considered.
- **Water solubility of the agent**—If the solid or liquid is water soluble, then flushing with water is very effective. However, if the material is not hydrophilic (water loving), then some form of emulsification will be required. Thorough washing of all surface areas with soap and water would be indicated. Plain soap is very effective, however many references recommend the use of "tincture of green soap." This type of soap is available from medical supply houses or can be mixed by many pharmacies.
- **Vapor pressure of the agent**—The vapor pressure of the liquid agent must also be considered. The higher the vapor pressure, the less persistent the material is and the faster it will evaporate. For example, cyanide gas has a vapor pressure greater than 1 atomsphere and therefore will immediately evaporate. However VX, with a vapor pressure of 0.0007, will remain for days. (In the case of VX, though, the patient that you are trying to decontaminate from liquid contact will generally not be a viable one.)

There are numerous layouts and systems for the field decontamination of mass casualties. We have determined three major types of layouts:

1. Crowd emergency decontamination
2. Two-corridor decontamination
3. Three-corridor decontamination

Crowd emergency decontamination generally involves the use of existing fire department resources to achieve the goal of stripping and flushing patients. In this type of operation, handlines and master-stream devices operating at approximately 50 to 60 psi nozzle pressure in a narrow fog pattern are used to surround and flush the patients. In other cases, fire apparatus are positioned side by side to create an area that provides the patients with a logical direction of movement as well as some very basic modesty protection. The use of ladder trucks with elevated streams can further assist this process.

Two- and three-corridor decontamination systems allow for a greater level of protection for the patients (from both the elements and observation). These systems generally require greater resources specifically designed or adapted for this process. Two corridor decontamination systems permit an area for ambulatory and non-ambulatory patients, and three corridors allow male and female ambulatory patients to have separate areas. This may seem extensive for an emergency situation, however, the more modesty we provide, the better the chance that decontamination will occur. Patients will be more likely to shower more thoroughly if they are not placed into embarrassing situations.

In addition to the design of the system, other equipment can be helpful in decontaminating patients. Items such as mesh stretchers (such as the Raven® stretcher or wire stokes baskets), tents, stretcher basins, and specially designed decontamination trailers can be helpful.

NOTE TO READER
This chapter has presented a basic understanding of how terrorist weapons might affect the body, and combined with the information in Appendixes D and H, forms the entire content of patient care.

Figure 8–3 Agent chart

Agent Civilian (Military) Name	Route of Exposure	Onset of Symptoms	Signs and Symptoms	Decontamination	Prehospital Care
Nerve Tabun (GA) Sarin (GB) Soman (GD) VX	R, S	Seconds to 18 hours	Miosis, difficulty breathing, headache, muscular twitching, salivation, lacrimation, urination, defecation, gastrointestinal distress, emesis (SLUDGE), seizures, coma, death	Universal body-fluid precautions for responders, decontamination with 0.5% hypochlorite solution, then soap and water wash.	Respiratory support; Atropine 2 mg IM repeated until atropinization occurs; Pralidoxime chloride (2-PAM Cl) 600 mg IM, maximum prehospital dose 1800 mg; Diazepam 10 mg IM repeated according to local protocol
Vesicants (Blister) Nitrogen Mustard (HN1)(HN2)(HN3) Lewisite (L) Phosgene Oxime (CX)	R, S	2 to 24 hours	Tearing or burning eyes, runny nose, sneezing, cough, nosebleed, redness on skin followed by blisters. *Symptoms are delayed but tissue damage occurs within minutes of contamination.*	Universal body-fluid precautions for responders, decontamination with 0.5% hypochlorite solution, then soap and water wash.	Aggressive burn management; airway support; and for Lewisite, British Anti-Lewisite. The blister fluid will not contain vesicant agent.
Blood Hydrogen Cyanide (AC) Cyanogen Chloride (CK)	R, I	15 seconds to 2 minutes	Increased respirations, loss of conscious, seizures, death. *All rapid onset.*	Universal body-fluid precautions for responders, decontamination with soap and water wash.	Supportive care and respiratory and circulatory support, use of cyanide antidote kit or IV of sodium nitrate and sodium thiosulfate.
Pulmonary (Choking) Phosgene (CG) Chlorine	R	20 minutes to 24 hours	Eye and airway irritation, dyspnea, chest tightness, bronchospasm, delayed noncardiogenic pulmonary edema.	Universal body-fluid precautions for responders, decontamination with soap and water wash.	Supportive care and respiratory and circulatory support. Aggressive airway management, use of intubation with PEEP, use of ACLS pulmonary edema medications may be ineffective.
Incapacitating (Riot Control) Mace (CN) (CS) Pepper Spray Adamsite (DM)	R, S, I	Seconds	Burning pain on mucous membranes, skin and eyes, tearing, burning in nostrils, elevated blood pressure, irregular respiration. Has been fatal in confined spaces. Symptoms will usually resolve in 15 to 20 minutes after removal to fresh air.	Eyes, flush with water or saline. Skin, flush with copious amounts of water and alkaline soap, or mild alkaline solution, sodium bicarbonate or sodium carbonate. *Do not use hypochlorite, which will worsen skin symptoms.*	Supportive care, as the effects are usually self-limiting. May trigger asthma attacks in sensitive patients in a confined space.

General Biological Terrorism Concepts

OVERVIEW

This section defines the framework for developing and sustaining a comprehensive and integrated approach to addressing terrorism. It is a blueprint and serves as a guideline for the development of operational-area efforts for responding to and combating biological terrorism.

The objective of a biological terrorism plan is to

- Augment the existing comprehensive emergency management plan for your area of responsibility
- Assist in the formidable task of delivering medical care
- Assist with implementing public-health strategies for combating the devastating physical, psychological, and sociological impact of mass illness and/or death

BIOLOGICAL EMERGENCIES

Biological emergencies are an actual or imminent set of conditions in which biological agents are intentionally introduced. The Area of Responsibility (AOR)[1] is defined as the comported and unincorporated area of your agency. Specialized response and management capabilities are required to effectively mitigate the impact of a biological terrorism situation.

Biological incidents can involve the release of warfare agents (known as Weapons of Mass Destruction or WMD) or the intentional release of locally produced left- or right-wing groups (see Chapter 1 for definitions) or laboratory and industrial agents.

BIOLOGICAL RESPONSE

Overall responsibility for responding to and managing the consequences of a terrorist incident is the responsibility of the local jurisdiction. Command responsibility for incident response falls with the designated Incident Commander or Unified Command Structure. Because of the complex multidisciplinary and multijurisdictional issues involved in managing a significant terrorist event, the formation of a Unified Command among responding agencies is the preferred management structure. The Unified Command begins at the field level and is concerned with all aspects contributing to the resolution of the incident. In addition to the typical response assets available to an Incident Commander, specialized resources are also available to the Unified Command to support field response efforts. These resources will perform mission-specific tasks as assigned by the Unified Command and should operate within the incident's Operations Section.

For agencies that have it, the Metropolitan Medical Response System (MMRS) will directly perform field-level response efforts for WMD consequences. This group forms the technical nucleus of a comprehensive response capability to bioterrorism. It includes specialized personnel to direct and coordinate immediate response, mitigation, and recovery operations at the incident scene. Your system is already in place and is most likely to be in operation.

For agencies that are not equipped with MMRS or who are developing a system, this book will assist you in formulating plans and acquiring resources for your comprehensive biological terrorism response system. Medical, EMS, and public health personnel; local, state, and federal law enforcement; and hazardous materials (HAZMAT),environmental, and epidemiological specialists each participate in some form within this system.

As it is developed, the system should be based on the federal MMRS model provided by the DHHS.[2] If your agencies do not have an MMRS and an incident

1. You agency or department may call this Area of Operation, Operational Area, etc.
2. http://www.mmrs.hhs.gov/

occurs, assistance will come in the form of the MMRS. It is better to plan using that system now than to learn on the disaster site.[1]

The MMRS is designed to work alone or in conjunction with the local agencies and with the National Medical Response Team for Weapons of Mass Destruction (NMRT-WMD). Response activities include rescue, decontamination, and immediate treatment of the injured. Mitigation efforts include the deployment of detection and containment equipment to limit or contain the incident. Recovery activities include efforts to mobilize recovery resources, to clean and remove the agents, and to declare the area safe for restoration of normal activities.

WMD incidents are too large and complex for agencies to manage. If and when a WMD incident occurs, state and federal assistance will arrive automatically to support the local agencies. It is important for state and local law enforcement and fire service to develop groups such as Area Command Teams (ACTs)[2] and Unified Local Command Teams (ULC). I used Florida as a benchmark, and the State of Florida's Fire-Rescue Disaster Response Plan as a reference.[3] Your state may reference this as well as other titles. The teams can be activated to support the management of law enforcement and fire service activities at terrorist incidents. Typical duties for these teams, which are also known as incident management teams, would include supporting law enforcement or fire service activities outside of the directly impacted Hot/Warm Zone of bio incidents. These teams could contribute to effective overall incident operations by providing a skilled staff to support the command-and-control activities of law enforcement operations, including security, traffic control, evacuations, and fire/rescue operations at a terrorist incident scene. Each ACT should be self-contained with intelligence, operations, and logistics specialists to support complex law enforcement and fire service operations. The individual ACT commanders can also provide staff support for the "unified command" among law enforcement, fire service, and other responding entities.

INTEGRATING FEDERAL, STATE, COUNTY, AND LOCAL RESPONSES

Integrating resources from many levels requires the commitment of agencies to mitigate against, prepare for, respond to, and recover from an incident of biological terrorism. A comprehensive approach to terrorism requires coordination with a variety of federal agencies. For example, the FBI is the lead federal agency for crisis management and terrorism investigations. Similarly, the president has designated FEMA as the lead agency for consequence-management efforts. The Terrorism Incident Annex of the FRP implements Presidential Decision Directive 39 (PDD-39) and defines federal roles in terrorist incidents. Despite these designated

1. http://www.ndms.dhhs.gov/CT_Program/MMRS/mmrs.html
2. For the purpose of this book I will use ACT for outside command teams.
3. This plan is managed by the Florida Fire Chief's Association

roles, local agencies will be in the forefront of emergency response to terrorist incidents. Local jurisdictions have the responsibility to manage the consequences of terrorist incidents occurring within their borders. With the increased awareness of cyber-terrorism and continuity of government, refer to PDD 62 and 63 for additional roles in this area.

Agencies at the local level should establish a committee for the planning and management of incident response. At the local level, this committee can address themselves as they see fit. For example, groups around the country call themselves Terrorism Stakeholders Group (TSG) or Terrorism Task Force (TTF). For our purposes, I will refer to this committee as the TSG, which should comprise but is not limited to:

- Local and regional federal agencies and departments such as the FBI, FEMA, HHS, and DoD
- State agencies and departments of emergency management, and health and fire service
- Local agencies and departments of health, fire service, and law enforcement; churches; airports; schools; port services; the coroner, and GIS personnel

The TSG's role should be in planning and training development.[1] The TSG should coordinate inside and outside agencies for technical assistance in developing measures to address biological terrorism. During an incident, federal agencies and departments will augment these efforts, particularly in regard to crisis management. In the areas of intelligence sharing and situation monitoring, the TSG should coordinate the lead local law enforcement agency. Local law enforcement should monitor open source information for trends and potential criminal intelligence operations to assess the capabilities and intentions of known terrorist groups. Local and regional law enforcement should coordinate their activities with the FBI.

In the area of response and consequence management, local law enforcement should be the lead point of contact for marshalling during consequence management efforts at the AOR. The fire service will have direct consequence-management roles. The EOC, which is responsible for strategic resource management, will coordinate support to local consequence management activities with FEMA through the state's Division of Emergency Management (CDEM).

In the realm of investigations, the FBI is the lead investigative agency. Local law enforcement investigations will be the primary supporting investigative entity, coordinating investigative efforts with the FBI. The representative of local law enforce-

1. See Appendix L, Sample Domestic Preparedness Training Courses

ment assigned to the EOC should coordinate law enforcement support to the incident site or AOR with state law enforcement.

The strategic activities of all of the response efforts will be coordinated through the local EOC. The state's DEM[1] will support incident site or AOR efforts by helping to obtain the resources necessary to manage an incident that exceeds local capabilities. DEM will generally act as the point of contact for obtaining routine mutual aid or resource support from within the state and federal government. Due to the FBI's direct federal crisis management responsibilities, national security considerations, and the need for rapid access to federal or military resources to manage a WMD incident, the local division of the FBI may directly request federal resources for biological incident(s) independently, or at the request of the EOC. The involved process by which the FBI requests technical (or other) assistance from supporting federal agencies or departments is coordinated by FBI headquarters at the request of the local FBI division and with the concurrence of the FBI director and the attorney general. At such a time, the FBI will establish a Joint Operations Center (JOC). For more information, refer to the FBI's WMDICP. In these cases, the state's DEM will be immediately notified to minimize duplication of effort. Figure 9–1 shows the sequence for requesting aid for a WMD event.

THE FIRE SERVICE AND/OR METROPOLITAN MEDICAL RESPONSE SYSTEM (MMRS)

If you are planning to develop an MMRS system, you can contact agencies within an MMRS system for assistance in preparing a Standard Operating Procedure (SOP).

The MMRS or the system you design will require a unique response asset specifically designed to address terrorist incidents involving WMD. It will augment field medical, HAZMAT, and mass decontamination operations and provide medical information and consultation to the incident response organization and hospitals within the AOR. An MMRS or comparable system should address its efforts on

- Agent identification
- Safe extraction and decontamination of victims
- Antidote and administration
- Triage and primary care
- Forward movement of victims for further care
- Enhancing the capacity to provide definitive care

---

1. States address this agency as Department, Division, or Office of Emergency Management. Check with your state for the correct title.

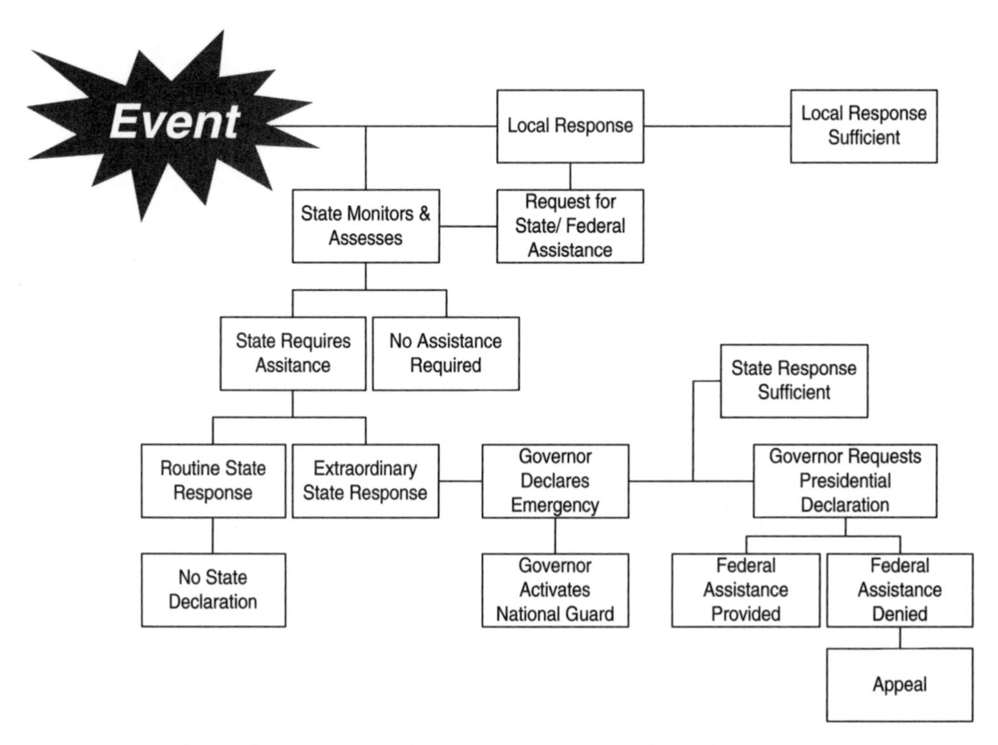

Figure 9–1 Sequence for requesting local and federal assistance

- Appropriate disposition of the deceased
- Providing advice on site recovery and decontamination issues

The MMRS is a specially trained, interdisciplinary NBC consequence management response team. When deployed, the MMRS is composed of forty-nine persons, allocated into five elements: (1) Command & Control; (2) Plans, Intelligence, and Research; (3) Law Enforcement/Security Operations; (4) Logistics; and (5) Intervention.[1] In order to ensure availability of personnel for multiple operational periods, the MMRS will be staffed three times its normal capacity for a total of 147 persons. Within smaller agencies, you may want to develop the system in a regional faction to augment local personnel and resources. All team members will benefit from a high degree of cross-training and frequent emergency mobilization and crisis rehearsal drills.

COMMAND AND CONTROL

Overall management of the consequences of all actual or threatened terrorist incidents is the responsibility of the affected local jurisdiction. At the federal level, the

1. See Figure 9–5 Biological cross function action chart on page 184

FBI is the lead agency and the on-scene manager for the United States Government. Command and control of all incident activities remains with the jurisdictional Incident Commander (or Unified Command). Strategic resource coordination and the acquisition and deployment of all ARO or mutual-aid resources will be coordinated by the EOC.

Field elements should be under the command of a Unified Incident Command. The MMRS will be under the direct command and control of the incident's Operations Section Chief (OSC). If the NMRT-WMD is deployed to augment the MMRS, the NMRT-WMD commander and medical system will establish a liaison and a closely coordinated response. The NMRT-WMD will also coordinate its field activities through the MMRS's deputy leader, who is typically a physician. Coordinated and collaborative medical control and treatment priorities will be established. Transfer of decontaminated patients to definitive care will be coordinated through the hospital coordination leader in the MMRS's Intervention Element utilizing the Medical Alert Center (MAC) at the Department of Health Services.

INTELLIGENCE AND EARLY WARNING

The local level TSG is a cooperative mechanism for sustaining a comprehensive and integrated early-warning capability to address threats of terrorism. The TSG supports the efforts of all anti- and counter-terrorist crisis and consequence management teams. It also serves as a focal point for analyzing strategic and operational information and intelligence needed to respond to and combat terrorism. Special emphasis will be placed on early detection of emerging threats, including terrorist acts employing WMD (such as NBC devices or agents).

PREVENTION

Prevention efforts include intelligence gathering, physical security, and Random Antiterrorist Measures (RAM) at key facilities and infrastructure. Prevention measures will be defined in confidential target-specific outlines whose development will be coordinated through the TSG.

ONGOING PREPAREDNESS EFFORTS

This planning guide provides a starting point for biological-terrorism preparedness and response efforts. In order to achieve an optimal reasonable level of preparedness for potential acts of terrorism involving WMD, all emergency-response agencies need to become familiar with the contents of the plan, train their personnel to meet the threats, and develop an agency-specific plan to reinforce and amplify the operational area plan. Like all emergency plans, this is a living document that will be revised and refined as we gain more knowledge about these threats and as new capabilities are developed to manage and mitigate the consequences of terrorist acts. Ongoing efforts to build and enhance preparedness must include regular drills and exercises involving all agencies that will be involved in an actual response. These drills and exercises, together with after-action reports from actual terrorist incidents, will be evaluated to provide a basis for future updates of your plan.

CONCEPT OF OPERATIONAL MANAGEMENT ISSUES

OPERATIONAL CONTEXT FOR RESPONDING TO AND COMBATING TERRORISM

An integrated operational context that will embrace a cohesive command, control, and intelligence architecture should be essential to manage a terrorist event. To achieve optimal effectiveness, all emergency response elements within an AOR must be integrated into a comprehensive terrorism response and management structure. Key elements in a plan must include the EOC, a permanent TSG or similar committee, coordinated law enforcement, the county health department, a medical response system such a MMRS, and fire area command (or incident management command) teams.

Management of terrorist incidents is divided into two interrelated phases known as crisis and consequence management. These phases were expressed in PDD-39 to describe the division of responsibilities among federal agencies. (See Federal Response Plan, Terrorism Annex.) This terminology is used to ensure smooth coordination with federal terrorism-management efforts. Crisis management describes measures to resolve the hostile situation, including law enforcement efforts aimed at prevention, interdiction, and threat management; as well as efforts to support investigation and prosecution. Consequence management describes efforts to respond to and mitigate the impact of a terrorist incident on people and the infrastructure. It involves measures to treat the injured, protect public health and safety, restore essential services, and provide emergency relief.

EMERGENCY OPERATIONS CENTER

The EOC is a critical component of effective management of emergencies in the operational area. It will become the centerpiece of efforts to manage terrorist attacks as well. Since incident management of terrorist incidents is complex, involving multi-agency, multidisciplinary efforts; effective management of mutual-aid resources is essential. The EOC will act as a focal point for coordinating strategic issues and resource allocation.

CRISIS MANAGEMENT

Crisis management or activities to assess, manage, or interdict a terrorist threat,[1] as well as activities related to criminal intelligence, investigations, and preparation for prosecution are law enforcement activities. Each affected jurisdiction will have a role in crisis-management activities. Generally, the biological TSG (health and law enforcement) will be the centerpiece of operational area crisis-management activi-

1. Refer to the FBI's Weapons of Mass Destruction Incident Contingency Plan (WMDICP), unclassified edition, which is available to official agencies.

ties at the local level. The TSG will coordinate its crisis-management activities with operational area agencies responsible for consequence-management efforts to ensure optimal incident management. In all incidents involving threatened use, confirmed presence, or actual use of WMD, the FBI will be immediately notified.

SITUATION AND THREAT BACKGROUND

The events of September 11, 2001 make NBC emergencies a plausible scenario necessitating detailed contingency planning and preparation of emergency responders in order to protect the civilian populace in the United States. Among the events calling attention to the threat level are the bombing of New York's World Trade Center on February 26, 1993 in New York City's World Trade Center and repeated in September of 2001, the sarin gas attack on the Tokyo subway on March 20, 1995, and the attempted hydrogen-cyanide assault on the Tokyo subway on May 5, 1995. The presence of sodium cyanide residue in the debris of the World Trade Center bombing in New York City heightened domestic concern. In the biological arena, incidents of note include the blending of ricin by a Minnesota antigovernment, tax-protest group whose members were convicted for violating the Biological Weapons Anti-terrorism Act in 1996. Nuclear terrorism surfaced in Moscow when Chechen insurgents placed radiological waste in Moscow parks to further their cause.

Biological agents can be delivered by a variety of means including dispersal via explosive devices such as bio-bombs or dirty bombs, mechanical devices such as crop-dusting aircraft, mosquito-control trucks, garden spray devices, dispersal through a building's water or ventilation system, or in the food system. Additionally, conventional and stand-off attack on bio-engineering facilities, or goods in transit (i.e., an intentional HAZMAT incident) must be considered. Such routine shipments offer a prime target, either for hijacking to appropriate the materials or a more conventional attack aimed at extortion or the release of the agents.

A bio-attack would most likely involve aerosol dispersal and would afford no easily discernible signature (i.e., the on-set of symptoms will typically occur within days to weeks after the initial exposure making detection difficult). The medical and emergency management concerns are complex, since little experience exists in coping with the impact of biotoxins on a large scale.

Sports arenas, concert halls, department stores and malls, transportation terminals, and office buildings—in short, any location capable of housing a large or small number of people in an enclosed or nearly enclosed space—are amenable to aerosol dispersal and thus are at greatest risk in a bioterrorism incident.

INCIDENT TYPES: PACKAGE, COVERT RELEASE, THREAT

One challenge in responding to a biological terrorism incident is the nature of the attack. An attack using biological agents can present in four different ways:

1. **Package suspected of containing a biological agent**—Terrorists may deliver an envelope or package that claims to contain a biological agent. Even if these packages contain only inert material, the response must assume that a hazard exists. Furthermore, it is not prudent to rely upon the veracity of the terrorists and assume that the package is accurately labeled. Therefore, these packages must be assessed for other hazards, e.g., hazardous chemicals, radioactive materials, and other biological agents.
2. **Covert release**—A second form of attack is a covert release. While technically the most difficult to detect, this is one of the most effective uses of a biological weapon. Disseminating a biological agent without warning reduces the time available to successfully identify an agent and treat the victims, and thereby increases the overall impact. The scope of a covert incident may be as small as a location on a street or in a building, or as large as the whole city. Determining that a biological agent has been released, identifying the agent or agents, and treating the victims will require the coordinated implementation of a comprehensive plan.
3. **Threat**—The threatened release of biological agents will require careful analysis of the credibility of that threat. Because of the catastrophic consequences of a full-scale attack with biological agents, officials may be asked to assess the threat and make difficult decisions regarding the deployment of assets. Your plan should detail operational options for such a scenario.
4. **Actual release**—The actual release of a biological agent will require a major response and the use of large amount of resources from local, state, and federal agents along with the private sector.

THREATENED USE

A threat evaluation is essential in the case of a threatened release. The first step of the evaluation is a review of the threat information. Threats should be evaluated at the local level in consultation with the local fire service, Department of Health (DOH), and members of the local office of the FBI who will conduct and perform an in-depth comprehensive federal threat assessment. Factors that should be considered by the decision makers while determining a course of action include a (health) risk evaluation, and a determination of potential impact. If the threat is deemed credible and/or feasible, the decision makers should immediately initiate Incident Action/Operational Planning.

Essential Elements of Information (EEIs) for threatened acts include

- Does the threatened act threaten a populated area?
- Does the threatened act threaten critical resources?
- Is there any reason to suggest a hoax?
- Is the act feasible/technically possible?

- Could the threat be a diversion?
- Is the threat similar to previous threats?

CONFIRMED PRESENCE

In the event that an actual biological agent is discovered prior to actual release, the EOC will be immediately activated. Upon consultation between the incident commander, the FBI on-scene commander (or Unified Command), and the HAZMAT or MMRS leader, and any necessary specialized, federal response capabilities will be requested. Examples of federal resources include the Nuclear Emergency Search Team or military C/B response teams such as the Chemical Biological Quick Response Force, the Chemical Biological Incident Response Force, or the Tech Escort Unit. The MMRS leader(s) will immediately provide advice on suggested protective actions (i.e., evacuation or shelter-in-place protection). The confirmed presence of an explosive device or WMD capable of causing a significant destructive event prior to actual injury or property loss is considered a significant threat.

ACTUAL RELEASE

Notification and response needs are similar to those described under confirmed presence. The actual release of an NBC agent constitutes a complex emergency. In the event of an actual release, the following incident objectives are imperative:

- Secure a perimeter and designate zones of operational control and ID agent released
- Rescue, decontaminate, triage, treat and transport affected persons
- Move uninvolved crowds/persons to safe zones
- Stabilize the incident
- Mitigate new and second cases

THE AGENTS

Biological terrorism agents are germs. They are living organisms, such as bacteria or viruses, which can cause disease and are more formally known as pathogens. However, not all pathogens can be successfully weaponized. The CDC has compiled a list of agents that currently pose the greatest threat (see Chapter 10) and has prepared detailed treatment options for each of these highest-risk agents.

Planning for biological terrorism requires that each aspect of the threat receives intensive analysis in order to develop a highly focused, yet well-integrated, response. Appendix D presents a number of treatment approaches in order to maintain flexibility and the ability to respond to a wide range of possible agents.

PLANNING STRATEGIES

The following strategies—Threat Response Management, Recognition and Evaluation, Site Management, Casualty Management, Fatality Management, Logistics and Direction, and Control as found in the incident command system—form the substance of an agency plan and outline the steps that governmental and nongovernmental agencies will have to take to manage the consequences of biological terrorism. These strategies, and the actions described therein, are not necessarily time-sequenced with one following another during an event. Rather, many functions are integrated into daily operations while others will be carried out simultaneously during an event in order to ensure the most efficient resolution of the incident.

THREAT RESPONSE MANAGEMENT

The law enforcement and intelligence agencies in the United States have proven very adept at identifying and interdicting terrorists. When a biological terrorism threat against a jurisdiction is identified, defining whether the threat is credible will guide the response. The FBI will be the lead federal agency for crisis management. Local law enforcement will be the lead contact agency and should work closely with FBI on the federal level to interdict the terrorists and prevent the release of any biological materials.

The FBI, local law enforcement, emergency management, and DOH (local and state) should form an Interagency Task Force to define the nature of the threat. This task force, using the following strategies can determine the level of resource mobilization needed to counter the threat to the population. Depending on the nature and seriousness of the threat, officials may

- Convene an emergency interagency meeting of appropriate city, state, and federal agencies
- Call upon the local TSG
- Access the National Threat Assessment Group through the FBI
- Enhance public health surveillance and recognition activities throughout local agencies
- Alert hospitals and health-care providers of the threat
- Prepare to open points of distribution for medications in conjunction with state and federal agencies
- Prepare to open alternate care facilities
- Prepare to open alternate morgue sites
- Request pre-positioning of federal assets
- Activate the local and state EOC
- Declare a local state of emergency

Review you state's activation requirements.

RECOGNITION AND EVALUATION

One of the primary challenges associated with biological weapons is that their release may be covert: there may be no "K-boom" involved. The first sign that a biological agent has been released may be the illness it causes in the exposed population (see passive and active surveillance in Chapter 10). As a result, the most important component of an effective response to the release of a biological agent is recognition. Early recognition of an event creates a larger "window of opportunity" during which effective treatments and public-health measures, such as quarantine[1] or emergency public education, can be implemented.

An additional challenge posed by biological terrorism is that the first signs of an attack are similar to those associated with nonspecific Influenza-Like Illness (ILI): fever, cough, body aches, and sore throat. By the time victims exposed to some biological agents, such as anthrax, develop symptoms, it will be too late to effectively treat them.

ADMINISTRATION

TRAINING AND EXERCISING[2]

Special skills and requirements identified by local agencies should be taken advantage of and incorporated into your existing training and exercising program offered by state and federal agencies. The local health departments, in conjunction with the state and federal governments, should assume the lead in getting the medical community trained. Local fire services, with the DOH, the CDC, and the United States Public Health Services, should assist in the assessment of total training needs. Gaps that may be identified by the TSG group will serve as the basis for determining how to develop and deliver supplementary training.

Normally, exercises are used to evaluate levels of preparedness and readiness and to test the effectiveness of training programs. Operating under this premise, the DOH should work with the fire service to ensure the disaster exercise includes appropriate goals and objectives, including but not limited to ESF#8 (see Chapter 10).

PLAN MAINTENANCE

Your plan is a living document that should be reviewed annually to determine if it needs to be updated. At a minimum, it should be revised every three to four years, or as required by the emergency management planning guidelines set forth in your local or state laws.

---

1. Check with you state's department of health for additional information on requirements for quarantine.
2. See Appendix L, Sample Domestic Preparedness Training Courses

OUTLINE OF BIOLOGICAL RESPONSE OPERATIONS

DETECT EVENT

Detection depends on notification from local and state health surveillance or from the law enforcement community for plan activation (see Chapter 10, Passive Surveillance).

UNANNOUNCED

The detection of an unannounced event will begin with an irregularity in one or more key local indicators during passive medical and public health surveillance and will be reported to the DOH (see Chapter 10, Detection). This irregularity may range from a single case of an unknown illness or a suspicious increase in the number of people reporting common illnesses such as the flu. An event may be detected at local health-care provider level, at the state or district public health departments, or at the CDC. It is recommended that the local and state health department be designated as a central clearinghouse for medical reports during the incident to design, execute, and maintain appropriate records (see Chapter 10, Notification). When the local surveillance system detects an event, it should notify the state health departments.

ANNOUNCED

For an announced event, it is assumed that notice of a recent or potential release of an agent will have been received and investigated by local, state, and/or federal agencies, and found to be credible.

CONDUCT INTERNAL NOTIFICATION

Upon any notification of an event by local, state and federal agencies, the local health department should immediately commence a preliminary epidemiological review (see Chapter 10, Active Surveillance).

This should include analyzing the data to

- Verify that the report is valid
- Determine if the event is, in fact, suspicious
- Determine what course of action will be necessary to follow up on the alert and/or bring it to closure (see Chapter 10, Diagnosis)

DETERMINE THE COURSE OF ACTION

Once officials have formed an opinion on the emerging crisis, the emergency management plan should be implemented either in its entirety or in the appropriate sections. For further information on activation of a plan, see the training operations of the United States FRP.

ACTION PLANNING

The following points should be considered for action planning:

- The initial response will be from the local fire and law enforcement.
- DEM, through the EOC, will execute a bioterrorism emergency response system designed in conjunction with all related agencies from local, state, and federal governments.
- DEM should inform local and state elected officials of the decisions that have been taken and the background of the bioterrorism emergency. The governor of your state and the locally elected officials should jointly inform federal officials and congressional representatives.
- As necessary, the jurisdictions will initiate mass prophylaxis/immunization (see Chapter 10, Medical Prophylaxis).
- Local hospitals, neighborhood health clinics, and urgent care centers, whether freestanding or hospital-based, should execute their strategies once the decision to initiate mass prophylaxis is made, based on the advice and direction of the TSG.
- Pharmaceuticals will initially be distributed through existing medical institutions. In underserved or heavily populated areas, alternative methods to distribute medications may have to be used.
- As necessary, mass patient care operations will be initiated.
- If the number of victims rises and begins to overwhelm the ability of the local medical institutions to deliver care, the EMA, in concert with the local health department, will begin implementing various phases of the Medical Response Expansion Program (MRF-P)(this is not the official name of this system). (see Chapter 10, Medical Response Expansion Program).
- The MRFP will expand to include Alternative Medical Centers (AMCs), in addition to Outpatient Treatment Centers (OPTCs), as the crisis evolves and medical facilities become overwhelmed, based on the recommendations of the EMS and/or TSG.
- The EOC should designate EMS resources, in coordination and cooperation with fire and EMS agencies, to provide stand-by and transport services in and around clusters of hospitals, clinics, and AMCs (see Chapter 10, EMS Resources).
- Hospitals should make every attempt to discharge stable patients and transfer patients not affected by the biological incident to comparable outlying facilities in order to make bed space available for critically ill victims (see Chapter 10, Hospital Plan).

- ESF#8 (Health and Medical) in the EOC[1] will serve as the medical command-and-control element (see Chapter 10, ESF#8).
- Amateur Radio Emergency Services (ARES or RACES), personnel can be positioned at each hospital. OPTC and AMC can conduct communications among and between hospitals, AMCS, and the EOC by telephone, fax, and email whenever possible. As a backup, alternate radio links should be established using ARES or RACES.
- As necessary, mass fatality management operations should be initiated (see Chapter 10, Mass Fatality Management Implementation).
- The TSG will consider fatality issues in its analysis of the evolving situation.
- Once the TSG has formed an opinion on the emerging mass fatality crisis, they should take all matters into consideration and recommend implementation of the medical examiner Expansion Program (MEEP).
- The Office of the Chief Medical Examiner (OCME), in close cooperation and coordination with the EMA and the lead law enforcement agency, should execute an emergency response (Response Action Plan) designed to meet the recommendations of the TSG, the EOC once the plan is implemented.
- As it is deemed necessary, DOH should coordinate MEEP operations (see Chapter 10, MEEP Operations).
- As necessary, environmental surety measures should be implemented.
- According to the agent used in the incident, Fire and Rescue will determine the specific steps required for reentry (see Chapter 10, Agent Surety).
- The local DOH and/or local Department of Environmental Services, with the assistance of entry teams, should perform environmental sampling to determine the extent and level of contamination.
- The local DOH should examine any animals found in the area of the incident for disease and identify, log, and dispose of all dead animal carcasses.
- The local DOH should establish measures to control vectors.
- The local DOH should establish measures to limit access and decontaminate heavily contaminated areas.
- The local DOH should identify and assess buildings used to house or treat infected persons.
- The local DOH in conjunction with the local EOC should initiate a public information program to inform residents of residual dangers and protective measures.

---

1. If your EOC is setup using Federal Response Plan.

To determine when the bioterrorism emergency has ended, the local and state DOH should continue to monitor conditions in the affected areas. They will maintain communications with each other and with the EOC. Upon request, the local DOH will make the regional teleconference system available for emergency response to affected parties.

CONDUCT PUBLIC NOTIFICATIONS

The FBI's Joint Information Center (JIC) will receive and handle all media inquiries. The JIC leader may, however, authorize the JIC at a facility to release information once it is cleared and verified to be consistent with JIC information. In any interaction with the media, the goal is to ensure that information is accurate, timely, and consistent.

JIC will report their decisions to local and state Public Information Offices (PIOs). The content of the information distributed to the public should be coordinated regionally, although jurisdictions may opt to communicate on an individual basis.

EMS should establish and maintain an Internet-based Web page to release timely information and information on locations for assistance. A telephone hotline should be established for those members of the public who do not have Internet access.

END EVENT

Based on the continuous monitoring of the situation, the local DOH will determine when the emergency has ended. These officials will inform the JIC and EOC, who will inform local elected officials that the emergency is over and will direct the PIOs to report the ending of the emergency to the public.

STAGES OF SEVERITY

A sample system is designed to handle three stages of severity of a bioterrorism event. These levels are defined as follows:

Stage I: Developing Public Health Crisis—0 to 100 patients

Stage II: Public Health Disaster—101 to 1,000 patients

Stage III: Catastrophic Public Health Event—1,001 to 10,000 or more patients

The patient numbers associated with each stage are to be considered guidelines since the release of a highly contagious agent may require Stage III procedures even if the number of patients is far less than 1,001.

STRATEGIES AND ACTIONS

Each stage of severity requires a different strategic response that is implemented through a suitable general action plan. The following is a list of strategies and general actions for each stage of severity:

Stage I: Developing Public Health Crisis—0 to 100 patients
Strategy: Assess needs
General Action: Utilize existing response structure

Stage II: Public Health Disaster—101 to 1,000 patients
Strategy: Establish alternate response structure
General Action: Augment resources within the region

Stage III: Catastrophic Public Health Event—1,001 to 10,000 or more patients
Strategy: Utilize all available federal and state resources
General Action: Integrate resources into response structure

DECISION FACTORS

Decision makers must anticipate when or if a bioterrorism event will progress from one stage of severity to the next. By anticipating this progression, resources in place can be redirected, or additional resources can be requested in a timely, progressive manner. Your agency should develop decision factors to help decision makers anticipate increasing response requirements. When any decision factor is reached, leaders must look at initiating actions suitable for the next higher stage of severity. In addition to these specific decision factors, leaders must be aware that implementing any functional action normally associated with a stage of severity requires analyzing all other functional areas to determine if the next level of action is necessary in those areas.

The following is a list of sample decision factors associated with each severity stage. Decision makers must be alert to the possibility that information provided through critical information channels may present additional decision factors at any time.

Stage I: Developing Public Health Crisis—0 to 100 patients

Strategy: Assess needs
General Action: Utilize existing response structure
Decision Factors:

- More than 100 patients probable
- Hospitals approaching maximum capacity
- Hospital triage approaching maximum capacity

Stage II: Public Health Disaster—101 to 1,000 patients

Strategy: Establish alternate response structure
General Action: Augment resources within the region
Decision Factors:

- More than 1,000 patients probable
- Cohort hospital system approaching maximum capacity
- Casualty Collection Points ((CCPs) approaching capacity
- Logistics utilizing all regional staff, security, or transportation resources

Stage III: Catastrophic Public Health Event—1,001 to 10,000 or more patients

Strategy: Utilize all available federal and state resources
General Action: Integrate resources into response structure
Decision Factors:

- Alternate Care Facilities (ACFs) at maximum capacity
- Federal resources delayed for more than twenty-four hours

SPECIFIC TREATMENT MODELS

Figure 9–2 through 9–4 represent the specific treatment plans that will be implemented in the event of a bioterrorism incident.

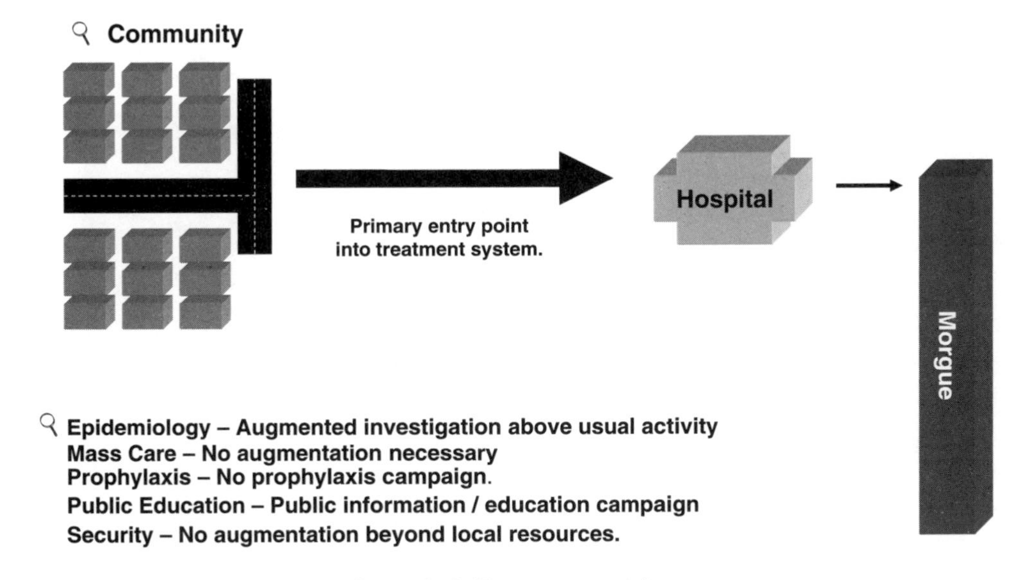

Figure 9–2 Treatment model one

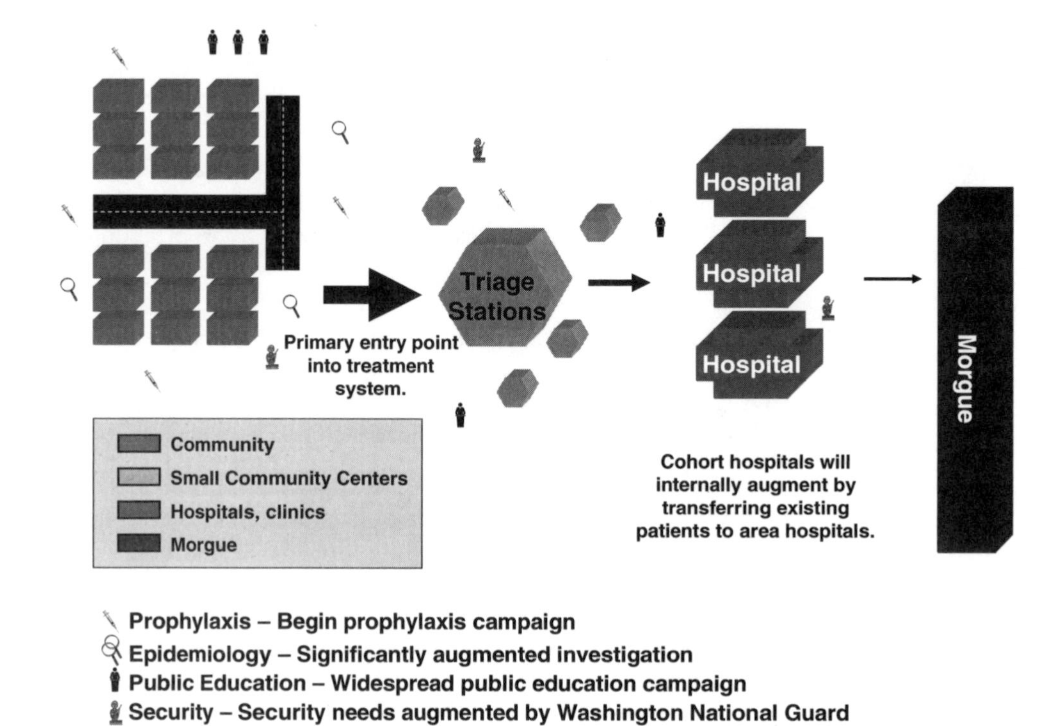

Figure 9–3 Treatment model two

BIOLOGICAL TERRORISM OPERATIONS

BIOLOGICAL WARFARE (BW) OR INTENTIONAL RELEASE OF BIOLOGIC AGENTS

Bioterrorism incidents fall into three broad classes:

- Threatened use
- Confirmed presence
- Actual release

SUSPICIOUS OUTBREAK OF DISEASE NOTIFICATIONS

In all cases of suspicious outbreak of disease, special notifications are necessary. Local law enforcement, because of its responsibility for marshalling consequence management resources at the AOR, is responsible for activating and notifying all affected response agencies and operational elements, such as the MMRS. The JIC is also responsible for disseminating Alerts and Advisories on behalf of the TSG. In all incidents involving threatened use, confirmed presence, or actual use of a WMD, the FBI will be immediately notified.

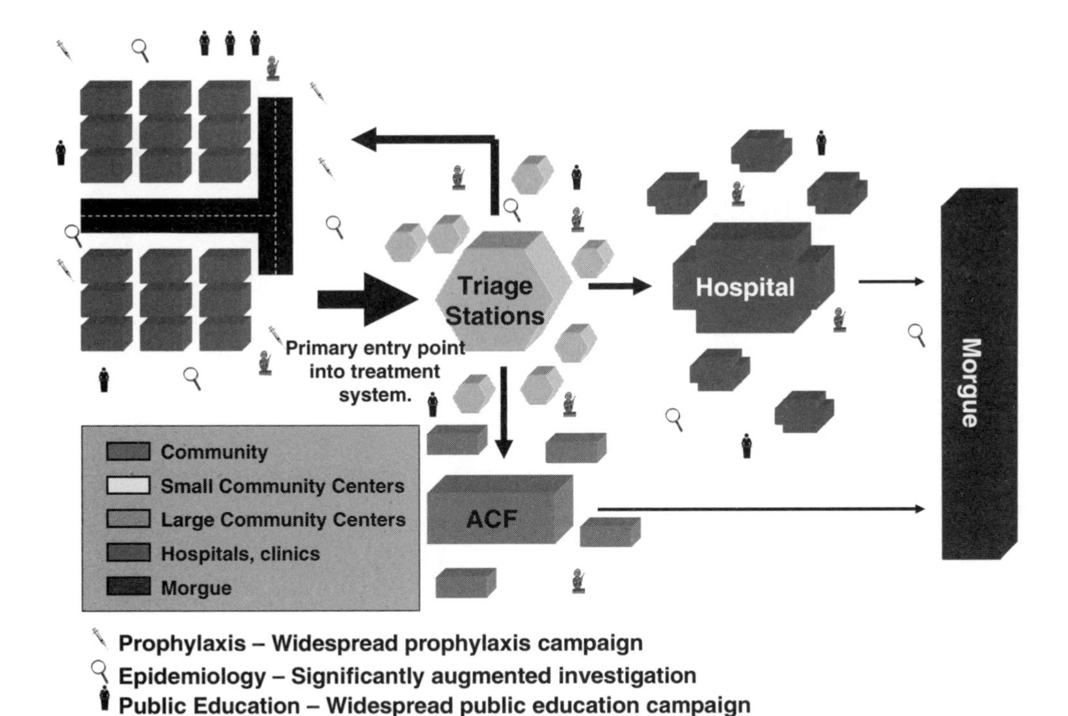

Figure 9–4 Treatment model three

IN THE EVENT OF AN ALERT

Local officials should monitor known threat situations and issues and advise the fire service, the emergency manager, MMRS leader(s), the NMRT-WMD commander, the CDC, and the FBI of the situation. Technical specialists should be activated as needed to initiate advanced planning.

IN THE EVENT OF A WARNING

A known impending threat has the potential to directly impact the AOR. If information exists to indicate a specific threat or a heightened threat of terrorist activity within the AOR, advise the fire service, the emergency manager, MMRS leader(s), the NMRT-WMD commander, the CDC, the FBI, and all affected agencies of the situation. The fire service may activate the MMRS(s) in advance and initiate incident action planning, threat-specific training, and response rehearsal.

	Command and Control	Early Recognition/ Threat Assessment	Initial Actions	Medical Care (Mass Care Treatment)	Prevention Prophylaxis/ Vaccination	Logistics	Fatality Management	Environmental Clean Up
Stage I: Developing Public Health Crisis **0–100 Patients** **Strategy: Assess needs** **Action: Utilize existing resources**	City EOC activated. Health department's biological response committee activated. Biological terrorism stakeholders group activated	Notification of expanded (active) epidemiological surveillance. Identify agent.	Identify hot zone. Rescue, isolate, and decon. Transport sample to lab. Transport patients.	Hospital control assesses need for activation of disaster plan and monitors status of hospital capacity.	Assess need for prophylaxis. Issue health recommendations for public. Consider need for forward movement of CDC stockpile.	Assess need for system personnel internal staffing increase. Assess security needs. Assess need for inter-hospital transfers.	Assess need for temporary morgue facilities.	Identify environmental hazards and assess ability for clean up with local resources.
Stage II: Public Health Disaster **101–1,000 Patients** **Strategy: Establish alternate response structure** **Action; Augment resources within the region and state**	Establish unified command. Single-source public information stream. State EOC activated. Initiate requests for federal resources.	Establish epidemiological investigation. Request CDC and law enforcement assistance.	Crime scene investigation. Laboratory identification.	ESF 8, hospital disaster actions activated Hospice care established. Casualty Collection Points established (CCP[a] centralized triage) outside treatment facilities.	Targeted population prophylaxis. CDC stockpile requested. Manage of worried well. Control Measures: Target population airborne precautions.	Increased staffing through regional call-up. Consider widespread security needs. Alternate transportation resources utilized.	Establish temporary morgue facilities.	Clean-up requiring state and federal assistance.
Stage III: Catastrophic Public Health Event **1,001–10,000+ Patients** **Strategy: Utilize all federal and state resources** **Action: Integrate resources into response structure**	Full integration of federal resources.			Establish ACF[b]. Establish home care. Casualty collection points decentralized.	Mass prophylaxis to include neighborhood distribution. Quarantine expanded.	Federal/ National staffing assistance. Integrate DoD and federal security personnel. Transport non-incident related patients out of region.	Mass fatality disposition: -burial -cremation	

a. CCP (Causality Collection Points): Centralized triage points that distribute suspect patients to hospitals, hospices, or back t home care
b. ACF (Alternative Care Facilities)—improvised medical stations: Sports center, schools, public buildings

Figure 9–5 Biological cross function action chart

IN THE EVENT OF AN ACTUAL INCIDENT (RESPONSE)

The TSG, together with the local DOH, will provide the Incident Commander (or Unified Command) with a needs assessment of the situation geared toward enhancing effective incident management.

The fire service will provide medical or MMRS responses to the threatened act, a found device, or actual completed NBC attack. The fire service and MMRS personnel should assist in incident action planning and provide threat-specific briefings including statements of response capability, logistical availability, and command relationships. See 9–5 for a biological cross-function action chart.

THREATENED USE

Local law enforcement should conduct threat evaluation with state and federal agencies to:

- Review threat information
- Designate the "Decision Authority" for determining a course of action

The local AOR in cases of bio-threats is the local law enforcement official in consultation with area jurisdictional law enforcement official outside its AOR, if any. In addition, the FBI conducts and performs an in-depth, comprehensive federal threat assessment. The federal threat-assessment process is integrated with the local jurisdiction process through the local office or division of the FBI.

Factors to be considered by the Decision Authority while determining a course of action should include

- Performing risk evaluation
- Determining potential impact

If the treat is deemed credible and/or feasible, immediately initiate Incident Action Planning and/or Operational Planning.

Essential Elements of Information (EEIs) for threatened acts include the following questions:

- What is the validity/reliability of the threat?
- What is the threat?
- Does the threatened act affect a densely populated area?
- Does the threatened act endanger critical resources?
- Is there any reason to suggest a hoax?
- Is the act feasible/technically possible?
- Could the threat be a diversion?

- Is the threat similar to previous threats?
- What is the agent type, quantity, and potency?
- Is the weather a factor?
- What response resources are available (<15 minute, 30 minute, 1 hour, 2 hours, >2 hours)?
- What is the course of action?
- What is the worst case initial response plan?

CONFIRMED PRESENCE

The confirmed presence of weaponized biologic agents, explosive devices, or WMDs capable of causing a significant destructive event prior to actual injury or property loss is considered a significant threat. In the event of a confirmed presence, local officials must:

- Make essential notifications
- Ensure FBI notification
- Advise the local health department director and related officials
- Request bomb and HAZMAT squads
- Request MMRS or similar medical support

MMRS BIOLOGICAL AGENTS

The MMRS uses the five agents in Figure 9–6 as its base line. However, training and preparedness is not limited to these, which represent a wide range of agents. See Figure 5–3 for a map of MMST locations.

THE HISTORICAL BIOLOGICAL AGENT SCENARIOS AS A BASELINE

A small number of the biological agents of concern are transmitted from person-to-person through airborne droplets. These agents are highly infectious and present the greatest possibility of creating a self-perpetuating epidemic (e.g., smallpox, pneumonic plaque). The key to limiting the spread of disease and breaking the chain of infection includes the rapid implementation of public health measures, use of proper Personal Protective Equipment (PPE), and adherence to appropriate infection control practices.

However, many organisms that might be used by terrorist are not transmitted from person-to-person (e.g., anthrax) but still may require a robust response and clean up. When properly managed, these biological agents will not be spread beyond the initial exposed area or population.

HHS SENDS EMERGENCY MEDICAL SUPPLIES TO NEW YORK CITY IN FIRST-EVER USE OF NATIONAL PHARMACEUTICAL STOCKPILE

HHS Secretary Tommy G. Thompson authorized the first emergency use of the National Pharmaceutical Stockpile, drawing on the two-year-old rapid-response program to provide special aid to New York City after September 11th. The decision will result in delivery of substantial supplies to support medical personnel caring for victims of the airplane attack on the World Trade Center. "Our emergency resources stand ready to be provided quickly to those who need them," Secretary Thompson said. "We are taking this step to support the medical personnel and facilities in New York, so that they can deliver the best possible care to those who have been injured." The CDC is releasing one of the eight "12-Hour Push Packages" that are maintained in prepackaged, prepositioned caches in secure storage facilities around the country. The packages are designed to be deliverable to any area of the continental United States within twelve hours of deployment, with substantial supplies to address a wide variety of potential needs. Push Packages contain pharmaceuticals, intravenous supplies, airway supplies, emergency medication, bandages and dressings, and other materials to cover a spectrum of medical needs. Each package involves several truckloads of materials and is intended to be sufficient to respond to an emergency involving mass casualties. In addition to the Push Packages, CDC will provide 84,000 bags of intravenous fluid and other intravenous supplies, as well as 350 portable ventilators and 250 stationary ventilators. The additional IVs and other materials are being sent in addition to the Push Packages to respond to the particular expected needs in New York.

Figure 9–6 MMRS baseline agents

Agents Disease	Characteristics
Smallpox (*Variola virus*)	Person-to-Person Transmission Transmitted by aerosols from infected victims
Plague (*Yersinia pestis*)	Pneumonic form—Person-to-Person Transmission (Transmitted by aerosols within six feet of victim) Bubonic form—No Person-to-Person Transmission
Anthrax (*Bacillus anthracis*)	No Person-to-Person Transmission High Morbidity/High Mortality
Tularemia (*Francisella tularensis*)	No Person-to-Person Transmission High Morbidity/Low Mortality
Brucellosis (*Brucella sp.*)	No Person-to-Person Transmission High Morbidity/Low Mortality

How serious can a bioterrorism incident be? There is no simple answer. For many, the 1918 Spanish Influenza pandemic is the catastrophe against which all modern pandemics are measured. This pandemic caused widespread illness, death, and social disruption with over 500,000 deaths in the United States and twenty to forty percent of the worldwide population becoming severely ill. Since our world is vastly more populated now than during previous pandemics and people travel the globe with ease, the spread of a pandemic could occur much more rapidly today.

Prepandemic planning, therefore, is essential if the bioterrorism-related morbidity, mortality, and social disruption are to be minimized. Indecisiveness or delays in recognizing an epidemic could cause, by the hour, enormous damage. In the 1918 influenza pandemic, many people who felt well in the morning became sick by noon and were dead by nightfall.

Whether a bioterrorism incident is announced or unannounced, or the organism used is contagious or noncontagious, minimizing the impact of the incident will depend on rapid coordination and effective communication between first responders, public health officials, health-care providers, emergency management, law enforcement, and political leaders.

TREATMENT VS. PROPHYLAXIS

Among the responder community, the terms "treatment" and "prophylaxis" are often confused. For the purpose of this book, "treatment" refers to treatment of victims who are already ill or symptomatic, while "prophylaxis" refers to the provision of pharmaceuticals or vaccines to persons who are not ill or symptomatic in the hopes of preventing disease.

DISCOVERY OF AGENT PRIOR TO RELEASE

In the event an actual NBC agent is discovered prior to actual release, the following actions need to occur.

ACTIVATE THE METROPOLITAN MEDICAL RESPONSE SYSTEM (MMRS)

The MMRS will augment field medical, HAZMAT, and mass decontamination[1] operations. It can also provide medical information and consultation to the incident response organization and hospitals within your AOR. The MMRS will focus its effort on

- Agent identification
- Safe extraction and decontamination of victims
- Antidote administration
- Triage and primary care

1. See Appendix K for a sample fixed mass decontamination system.

- Forward movement of victims for further care
- Capacity to provide definitive care
- Appropriate disposition of the deceased
- Advice on site recovery and decontamination issues

REQUEST SPECIALIZED RESOURCES

Upon consultation between the Incident Commander and the FBI on-scene commander (or Unified Command), and the Emergency Medical officer (Fire Service) and MMRS leader, the NMRT-WMD and any necessary specialized federal response capabilities will be requested. Examples of federal resources include the Nuclear Emergency Search Team (NEST), or military C/B response teams such as the Chemical Biological Rapid Response Force (CBRRF), the Chemical Biological Incident Response Force (CBIRF), and the Tech Escort Unit (TEU).

EVALUATE THE DEVICE/RELEASE POTENTIAL

- Type of agent/vector(s) involved
- Location of device(s)
- Type of device(s) (dispersal, explosive, etc.)
- Number of devices
- Device size and configuration (estimated capacity in liters)

CONSIDER INCIDENT CONSEQUENCES

- Evaluate the current situation
- Consider the range of consequences/alternative scenarios
- Determine whether single attack or campaign
- Establish whether multiple sites or venues at risk
- Identify potential for in-depth decontamination (multiple sites)
- Consider downwind/downstream potentials
- Consider secondary and systematic cascading effects
- Review target (response information) folder or develop spontaneous target folder

DEVELOP A COURSE OF ACTION

- Contain the situation
- Consider options
- Evaluate resource availability and capability
- Define response resources available (<15 minutes, 30 minutes, 1 hour, 2 hours, >2 hours)

- Consider response rehearsal
- Consult technical specialists/utilize virtual reach back capabilities
- Consider multiple operational periods
- Identify and consider performance decrements
- Balance consequence management and investigative needs

EVALUATE AND IMPLEMENT PROTECTIVE ACTIONS

- Consider PPE needs
- Determine potential protective actions
- Consult with the MMRS leader and technical specialists for advice on suggested protective actions (i.e., evacuation or in-place protection)

ADDRESS POTENTIAL FOR SECONDARY ATTACK

- Conduct search for secondary devices
- Clear site support areas (CP), (S), (LZ), (D), (T) for secondary, tertiary devices

DEVELOP A SITE SECURITY/FORCE PROTECTION PLAN

- Develop an incident traffic plan

ACTUAL RELEASE

The actual release of an NBC agent constitutes a complex emergency. Factors to be considered include

- Key indicators
- Observed indicator of biological agent use
- Wind direction and weather conditions at scene
- Plume direction
- Topographic influences

The threat of bioterrorism includes the intentional release of biological agents, as well as the deliberate use of weapons of biowarfare. Biological Warfare (BW) agents can be introduced via aerosol devices (munitions, sprayers, or aerosol generators), breaking containers, the release of a vector, or covert dissemination.

Biological agents have the potential to be more lethal than chemical agents and are primarily deployed through aerosol spray or introduction into a water system.

General indicators of possible BW usage include

- Unusual dead or dying animals
- Sick or dying animals, people, or fish
- Unusual casualties

- Unusual illness for the region
- Definite pattern inconsistent with natural disease
- Unusual liquid, spray or vapor
- Suspicious devices or packages
- Unusual swarms of insects

In the event of an actual release of an NBC agent, the following actions must occur:

MAKE ESSENTIAL NOTIFICATIONS

- FBI
- Local DOH
- Bomb squad and HAZMAT squads (if appropriate)
- Medical alert center

DEFINE INCIDENT OBJECTIVES

- Secure a perimeter and designate zones of operation
- Control and ID agent release
- Rescue, decontaminate, triage, treat, and transport affected persons
- Move uninvolved crowds/persons to safe zones
- Stabilize the incident
- Avoid secondary contamination
- Secure evidence and crime scene
- Protect against secondary attack

ACTIVATE THE METROPOLITAN MEDICAL RESPONSE SYSTEM (MMRS)

- Agent identification
- Safe extraction and decontamination of victims
- Antidote and prophylaxis administration
- Triage and primary care
- Forward movement of victims for further care
- Appropriate disposition of the deceased
- Advice on site recovery and decontamination issues

EVALUATE DEVICE/RELEASE POTENTIAL

- Location of device(s)
- Type of device/vector(s) (dispersal, explosive, etc.)
- Number of devices

- Device size and configuration (estimated capacity in liters)
- Time and duration of release
- Release characteristics (interior, exterior, limited, continuous, etc.)

DEFINE THE DISEASE/AGENT

- Identify agents employed and mechanism of delivery
- Consider predictability of effect on target population
- Assess vulnerability of target population to exposure
- Assess rapidity of effect or define expected incubation period
- Assess human-to-human transmissibility
- Identify appropriate prophylaxis (antimicrobial agents or vaccines)
- Review procedures outlined in "Suspicious Outbreak of Disease" on page 195 for potential biological problem-solving and differential diagnosis criteria

CONSIDER INCIDENT CONSEQUENCES

- Evaluate current situation
- Consider the range of consequences/alternative scenarios (hybrid attack, multiple vectors, etc.)
- Determine whether single attack or campaign
- Establish whether multiple sites or venues are impacted
- Identify the potential for in-depth decontamination (multiple sites)
- Consider downwind/downstream potentials
- Consider secondary and systematic cascading effects
- Review target (response information) folder or develop spontaneous target folder

DEVELOP A COURSE OF ACTION

- Contain the situation
- Consider options/evaluate intervention
- Evaluate prospects (practicality and availability) for prophylaxsis (antimicrobial agents or vaccines)
- Evaluate resource availability and capability
- Consider response rehearsal
- Consult technical specialists/utilize virtual reach back capabilities
- Consider multiple operational periods

- Identify and consider performance decrements
- Balance consequence management and investigative needs

REQUEST SPECIALIZED RESOURCES

Upon consultation between the Incident Commander and the FBI on-scene commander (or Unified Command) and the MMRS leader, the NMRT-WMD and any necessary specialized federal response capabilities will be requested.

EVALUATE AND IMPLEMENT PROTECTIVE ACTIONS

- Consider PPE needs
- Determine potential protective actions
- Consult with the MMRS leader and technical specialist for advice on suggested protective actions (i.e., evacuation or in-place protection)

ADDRESS POTENTIAL FOR SECONDARY ATTACK

- Conduct search for secondary devices
- Clear site support areas (CP), (S), (LZ), (D), (T) for secondary, tertiary devices

DEVELOP AN ON-GOING INTELLIGENCE COLLECTION PLAN

Continual reevaluation of tactical situation, weather conditions, situation status, and resource status is necessary. Consider use of field observers to act as a "directed telescope" to gather and relay situation status to the command.

REAL-TIME INTELLIGENCE

- Synchronize disease surveillance, epidemiological investigation, and criminal intelligence/investigation
- Sample collection (consider multiple agents)
- Monitor disease trends and potentials (consult online surveillance[1] and reporting sources such as ProMED or Outbreak, and epidemiological agencies)
- Integrate veterinary surveillance

DEVELOP A SITE SECURITY/FORCE PROTECTION PLAN

- Develop an incident traffic plan
- Consider impact on the community

While many specialized resources will be mobilized to respond to an NBC incident, it will take time for that assistance to arrive. Many specialized resources (such as military response teams) need to be airlifted to the area requiring local resources to manage the initial phases of an NBC emergency. This initial response phase may range from a few to many hours; response times for federal resources ranging from two to twenty-four hours can be expected.

---

1. www.cdc.gov

Community panic, intense media interest, and the convergence of contaminated persons at local hospitals and urgent care centers can be expected. Rapid assessment of the scope of the incident, activation of the emergency management infrastructure, and designation of CCPs and decontamination points are essential to mitigating potential community panic. Efforts to assess the situation and provide clear, easy-to-follow emergency management instructions to the public are essential.

INITIAL RESPONSE CONCERNS

DOWNWIND POTENTIALS

A large release may result in a lethal plume that may travel for miles. Emergency agencies in neighboring jurisdictions must be advised of the release and included in incident-management activities.

TRAFFIC RESTRICTIONS AND CONGESTION

Roads, freeways, and transit systems may need to be closed to contain the incident. Regardless of the need, panic may cause some persons to self-evacuate. Traffic congestion and gridlock conditions and confusion may result, which will slow response by emergency agencies and specialized resources to affected areas. Detailed traffic management plans will need to be developed.

SELF-TRANSPORT TO MEDICAL PROVIDERS (CONVERGENT CASUALTIES)

Injured and contaminated victims may leave the immediate site of the incident and go to hospitals,[1] urgent care centers, or individual physicians seeking medical care. In most cases, the care provider will not be equipped to decontaminate victims or treat bioterrorist casualties. This can extend the scope of the incident, lead to potential secondary contamination, and strain local medical and emergency response resources. Hospitals impacted by an influx of casualties who have not been decontaminated will have to establish decontamination areas and may not be able to continue providing treatment.

PANIC VICTIMS

In the immediate aftermath of an NBC incident, responders will anticipate a number of people who think they have been exposed to or contaminated by the agent(s) even though there has been no actual exposure. Provisions must be made to manage these persons and provide supportive care as necessary.

SCARCE SUPPLIES

Equipment and supplies needed to manage a bio-event will be in short supply. Sufficient pharmacological supplies may not be available at first or at all. Antidotes and other drugs used to treat victims are usually not stockpiled in sufficient quantities[2]

1. See Appendix K, Fixed Hospital DECON System
2. See Appendix I, Pharmaceutical Push Package Contents

for use in a mass casualty incident. Efforts to secure additional supplies will be an immediate need.

Personnel involved in managing potential NBC incidents must be aware of these concerns. Measures to address these issues must be incorporated into the Incident Action Plan and will be considered and assessed throughout the management of a bio-incident.

SUSPICIOUS OUTBREAK OF DISEASE

In the event of a suspicious outbreak of disease, biological problem-solving activities should be initiated:

- Recognize that attack may present as unusual outbreak of disease
- Distinguish between natural and intentional outbreak
- Consider threat information and background threat level
- Consider potential for hybrid and multivector attack

CONSIDER DIFFERENT CRITERIA

A number of single outbreaks may be considered as an attack when a combination of evidence clusters or illnesses of an intentional release is found. The following factors must be evaluated:

- Population characteristics (who is sick and incidence vs. expected occurrence)
- Epidemic curve (plot temporal patterns of onset, rate of onset)
- Number of cases
- Clinical presentation
- Strain/variant
- Economic motivations (targeting of food supply or animals for economic impact)
- Geographic/spatial characteristics
- Morbidity/mortality rates (unusual rate of attack or disease behavior)
- Antimicrobial resistance patterns (consider importation of new strain or intentional strain manipulation)
- Seasonal distribution (variation from the norm)
- Zoonotic potential (presence or lack of natural host)
- Residual infectivity/toxicity (suspect atypical route of transmission)
- Prevention/therapeutic potential (can the attackers immunize themselves?)
- Route of exposure (suspect atypical route of transmission)

- Weather and climatic factors (downwind/downstream, etc.)
- Incubation period (suspect clustering/common exposure)

DEFINE THE DISEASE/AGENT

- Identify agents employed and mechanism of delivery
- Consider predictability of effect on target population
- Assess vulnerability of target population to exposure
- Assess rapidity of effect/define expected incubation period
- Evaluate human-to-human transmissibility
- Identify appropriate prophylaxsis (antimicrobial agents or vaccines)

REAL-TIME INTELLIGENCE

- Synchronize disease surveillance, epidemiological investigation, and criminal intelligence/investigation
- Sample collection (consider multiple agents)
- Monitor disease trends and potentials (consult online surveillance and reporting sources, such as ProMEand Outbreak,and epidemiological agencies)
- Integrate veterinary surveillance

CONSIDER INCIDENT CONSEQUENCES

- Evaluate current situation
- Consider the range of consequences/alternative scenarios
- Estimate post-attack illness and death
- Determine whether single attack or campaign
- Establish whether multiple sites or venues are impacted
- Identify potential for in depth decontamination (multiple sites)
- Consider downwind/downstream potentials
- Consider secondary and systematic cascading effects
- Review target (response information) folder or develop spontaneous target folder

DEVELOP A COURSE OF ACTION

- Contain the situation
- Consider options/evaluate intervention
- Evaluate prospects (practicality and availability) for prophylaxsis (antimicrobial agents or vaccines)
- Evaluate resource availability and capability

- Define response resources available (<15 minutes, 30 minutes, 1 hour, 2 hours, >2 hours)
- Consider response rehearsal
- Consult technical specialists/utilize virtual reach back capabilities
- Consider multiple operational periods
- Identify and consider performance decrements
- Balance consequence management and investigative needs

EVALUATE AND IMPLEMENT PROTECTIVE ACTIONS

- Consider PPE needs
- Determine potential protective actions
- Consult with the MMRS leader and technical specialists for advice on suggested protective actions (i.e., evacuation or in-place protection)
- Activate critical incident debriefing (MMRS/CISM team)

COMMUNITY AWARENESS AND PUBLIC INFORMATION CONCERNS

Effective management of the impact of a bio-incident on the community requires the coordinated and timely dissemination of accurate information in a manner that minimizes confusion and unwarranted panic. Conflicting information must be avoided and information regarding protective actions, appropriate evacuation measures, self-aid and decontamination must be provided in a timely manner. At the operational area level, the dissemination of such information will be handled by the county EOC in close coordination with all affected individual agencies and jurisdictions.

SYNOPSIS OF BW AGENT CHARACTERISTICS

Biological agents include pathogens, which are living, reproducing, disease-producing organisms; toxins, which are nonliving poisons derived from living organisms; and Endogenous Biological Regulators (EBRs), which are chemical substances produced in the body to regulate bodily functions.

Pathogens include

- Bacteria such as Anthrax, Tularemia, and Bubonic Plague. Bacteria are capable of reproducing outside of living cells.
- Viruses such as Yellow Fever, Smallpox, HIV, Ebola, or Marburg. Viruses are infective agents composed of DNA or RNA that can only reproduce inside living cells.
- Rickettsia such as Q Fever and Rocky Mountain Spotted Fever. These are parasitic microorganisms that naturally transmit disease through the bites of fleas, ticks, etc. These parasites require a living host.

- Additional pathogens include yeast and fungi, as well as genetically engineered pathogens.

Toxins such as ricin, BTX, and saxitoxin are nonliving, poisonous chemical compounds. They are thousands of times more lethal than standard chemical agents. EBRs include hormones, adrenalin, and peptides. The unusual or atypical presence of swarms of insects may be indicative of BW attack with the insects serving as the delivery vector. Unlike victims of exposure to chemical or radiological agents, victims of a BW attack are not themselves contaminated or contagious; however, they may serve as carriers of the disease.

BW FIRST RESPONDER CONCERNS

Treat all incidents involving BW agents as intentional HAZMAT situations. In all cases, safely isolate and deny entry into the Hot Zone and make appropriate notifications. In addition, whenever it is believed that a BW agent has been released, assume that all personnel and property have been contaminated.

Immediately don PPE if available and request specialized resources. Public health officials, along with experts from the CDC, the MMRS, NMRT-WMD, and the USAMRIID are needed to identify the exact nature of the BW agent.

Pending identification of the agent, measures must be taken to prevent an epidemic, including isolation, quarantine,[1] and restriction of personnel movement. These procedures apply to both victims and first responders. Identify the source of contamination and designate zones of operation (i.e., a Hot, Warm, and Cold Zone). Consider weather effects during zone designation. If large numbers of exposures are involved, quarantine may be necessary with all victims being treated on-site. If a small number of persons are exposed, they will be decontaminated and transported to a hospital capable of isolating the patients. Initiate protective actions (i.e., evacuation or in-place protection) and avoid all exposed food and water.

BW AGENT DISSEMINATION

Dissemination of bio-agents can include aerosol or covert techniques, as well as manually or spread achieved by breaking containers holding the agent to achieve release. The most practical method of initiating a bio-attack is through the dispersal of aerosol particles. Biological agents may enter the body through respiration, ingestion, or direct contact with skin or membranes.

Unlike chemical agents, exposure to biological agents may not be immediately apparent, with casualties occurring hours, days, or weeks after exposure. In many

1. For information on quarantine, check with your local department of health and your state's DOH laws and statues. For example, in Florida the statues are in: *s. 381.0011, Florida Statues 1997. Duties and Power of the Department of Health, Rule 64D-3.007, Florida Admin. Code - Quarantine, Requirements and s. 252.311 Legislative intent ("State Emergency Management Act")*.

cases, the first indication of a bio-attack may occur after a number of unusual illnesses begin to appear in local hospital's emergency departments. Without advance warning, first responders may not recognize the existence of a bio attack.

WEATHER EFFECTS

The impact of biological agents is affected by weather conditions. Accordingly, detailed and accurate assessments of weather conditions and forecasts are critical elements in the tactical management of the emergencies. Weather effects to consider include

- **Sunlight**—Ultraviolet light found in sunlight helps kill biological agents.
- **Temperature**—Temperatures above 100° F begin killing off biological agents; freezing temperatures can render biological agents dormant.
- **Temperature Gradient**—Elevation influences temperature. For each ten meters from ground level there is a different temperature known as the temperature gradient. This factor causes biological agents to hold close to the ground.
- **Wind**—Wind aids the dispersal and spread of chemical, biological, and nuclear agents. Wind direction and speed influence the resulting plume and must be considered when setting up zones of operation and making evacuation decisions.
- **Precipitation**—Precipitation can influence agent dispersal and the spread of contaminated areas (i.e., run-off). In biological situations, the quantity of rain can either kill or stimulate the growth of individual agents.
- **Humidity**—Higher humidity levels cause the pores on human skin to open up and aid the absorption of agents.

All weather effects need to be continually assessed and evaluated to ensure an accurate understanding of the current situation.

DECONTAMINATION (DECON) CONSIDERATIONS

Determine if decontamination (DECON) is necessary for the involved agent.[1] Bio incidents may potentially involve civilians, law enforcement, fire service, and medical personnel that have been exposed to potentially lethal agents. Prompt, safe, and effective DECON procedures are essential to protect exposed persons, equipment, and the environment from the harmful effects of these agents.

DECON is the process used to reduce the hazards of bio agents to safe levels and minimize the uncontrolled transfer of contaminants from the hazard site to clean

1. For additional information on DECON standards, refer to local fire department SOP's and the National Fire Protection Associations standards.

areas. DECON will be performed any time contamination with a bio-agent or hazardous material is suspected.

During DECON operations, the safety of emergency response personnel is the first and most important consideration. Proper use of PPE, including Personal Protective Clothing (PPC) and respiratory protection such as Self-Contained Breathing Apparatus (SCBA), reduces hazards to response personnel.

The risk of secondary contamination to rescue personnel and medical personnel, both on scene and at the hospital—specifically victims or first responders, and transport vehicles and equipment—must be adequately assessed and protected against to avoid spreading the incident. Any contamination of the skin must be decontaminated immediately.

Decontamination prevents victims' response personnel and equipment from spreading the contaminant beyond the exclusion (or Hot Zone). A contamination reduction corridor and decon equipment must be set up prior to any entry by response personnel (with the exception of an emergency decon to save a life). Limited access to the exclusion zone through the contamination reduction corridor helps to keep the incident from spreading.

RECOVERY CONCERNS

MANAGING MASS FATALITIES/DECEASED DISPOSITION

A terrorist incident involving biological weapons of mass destruction may yield fatalities. The number of deaths is dependent upon the specific conditions present at an incident. Factors influencing the number of fatalities include the agent released, the dispersal method, the location of release, the number of persons present during the attack, and the response and mitigation measures employed. The most complex situation would involve a mass-casualty situation requiring the establishment of fatality or decedent collection points, as well as the activation of mutual aid protocols to effectively manage a mass-fatality situation.

A mass-fatality situation resulting from a biological terrorist incident is compounded by the presence or risk of biological contaminants. Deceased persons and their personal effects contaminated by agents must be decontaminated before removal from the incident scene, and should be managed by the medical examiners office. However, contaminated bodies or items shall not be transferred to medical examiners personnel prior to decontamination. Additionally, no medical examiners personnel shall conduct operations within a contaminated area unless equipped with proper PPE. Medical examiners personnel shall receive training in the proper use of PPE prior to utilizing it at an incident scene.

The management of a mass fatality situation involving NBC agents may require specialized assistance. The MMRS and the NMRT-WMD are designed to provide technical assistance regarding deceased disposition in WMD situations. Additional technical assistance is available from the United States Public Health Service (USPHS) and military specialists. Military mass-fatality management resources (i.e., graves registration units) may be appropriate in some cases; if required, they are accessed through regular Military Support to Civil Authorities (MSCA) channels.

Medical examiners personnel will coordinate their activities with the MMRS to ensure appropriate mass-fatality management. All requests for specialized mass-fatality assistance will be coordinated with law enforcement and emergency services to ensure that there is no duplication of effort. The medical examiners office is responsible for determining the number of fatalities. Casualty tracking shall be coordinated with the MMRS to ensure an accurate accounting of the number of fatalities and their disposition. The medical examiner shall provide the official death count during any disaster. It is crucial that all involved agencies immediately relay all fatality information to the medical examiner.

SITE DECONTAMINATION AND RESTORATION

Restoration efforts include site decontamination, cleanup and/or removal of contaminated soil, materials, vehicles, etc.

OVERSIGHT

Regulatory oversight is required following HAZMAT releases to ensure that restoration of the site, equipment, and all contaminated items is conducted within current environmental and Occupational Safety and Health Administration (OSHA) statutes and regulations. The area, facilities, and items affected by the release must be held until the oversight agency declares them "fit for re-occupancy" or reuse. All personnel who assist in restoration efforts must be properly trained and equipped per OSHA regulations.

INVESTIGATION

Investigative operations must be closely coordinated with the restoration effort. Coordination of investigations and restoration ensures proper evidence preservation and limits potential health and safety risks to investigators.

SUMMARY

This chapter covers general topics dealing with areas needed for the development of a biological plan. The areas listed in this chapter should be used as a reference. Your plan should be flexible enough so new ones can be easily added and old ones removed. The point for your terrorism plan is that is can be integrated into your comprehensive emergency management system at the local, state and federal levels. While developing your plan it is important to remember that the plan must be flexible enough for use when there is a biological attack. Therefore, you must include the unforeseen events that may relate to this type of an attack or event.

Planning and Action Guidelines

OVERVIEW

This chapter provides details for specific organizations concerning planning and actions during a bioterrorism emergency. Effective actions are dependent on successful pre-incident planning and training, and on the familiarity of responders in various jurisdictions with your Comprehensive Emergency Management Plan (CEMP). This plan must be carefully prepared with reference to the capabilities and needs of those responders, among other information, in order for it to address all likely contingencies in an actual emergency. The details that follow will assist you in creating an effective CEMP or in augmenting your existing plan. This chapter covers

- Passive and active surveillance
- Biological Terrorism Stakeholders Group (BTSC)
- Detecting a bioterrorism incident
- Mass prophylaxis: details and responsibilities
- Obtaining and distributing medicines
- Emergency Support Function #8 (ESF#8) Health and Medical
- Mass fatality management implementation
- Medical Examiner Expansion Program

PASSIVE SURVEILLANCE

Passive surveillance is the ongoing monitoring of certain predefined medical and public-health criteria outside of and before any evidence or suspicion of a bioterrorism event. Listed below are the key persons and agencies in passive surveillance followed by the signs that might indicate bioterrorism pertinent to those persons and agencies. The existence of any of these signs should be immediately reported to the local, county and/or the state DOH.

HEALTH-CARE PROVIDERS AND INFRASTRUCTURAL PERSONNEL

Those working in hospitals, public health facilities, public and private employee health programs, and EMS, among others, should look out for any occurrences of unusual or exotic diseases or of fevers of unknown origin, outbreaks of Influenza-Like Syndrome (ILS) outside of the regionally defined flu season, an increase in the number of ILS patients to levels exceeding the seasonal averages, or other unexpected increases in disease-service demands.

SENTINEL VETERINARIANS AND ANIMAL CONTROL AGENCIES

Veterinarian clinics, municipal and county animal control agencies, and other animal facilities must monitor any exotic animal diseases (especially those that are zoonotic), unusual rises in animal infectious-disease encounters, and unexplained animal deaths (especially those that occur in clusters).

MEDICAL EXAMINERS

Medical examiners should report any unusual increases in the death rate due to unknown pathogens, an unusual or exotic disease, ILS outside of normal flu seasons, or zoonotic diseases and pathogens that are known to have been manufactured as bioterrorism agents.

SENTINEL PHARMACEUTICAL CENTERS

Pharmacists should report any unusual rise in, or cluster of, purchases of antiviral, antibiotic, and antidiarrhea medications, both prescription and over-the-counter.

ACTIVE SURVEILLANCE (EPIDEMIOLOGICAL SERVICES)

When any abnormality is identified through the passive surveillance program, a preliminary investigation will be conducted by the local or county (for simplicity, hereafter referred to as *county*) DOH in concert with the state DOH. When an abnormality (excursion that is statistically above the expected value) occurs, the DOH will expand its observational activities and poll emergency departments, pediatricians, family physicians, internists, and other infection-control practitioners to ascertain the context and possible cause of the nonspecific indicators.

If an abnormal community health problem is confirmed and no specific cause for the increased patient load is ascertained, the county DOH will begin an active investigation. The decision to initiate the investigation will have a low threshold since these initial activities have a modest cost and impact on the community.

During a phase of active surveillance, the services involved in passive surveillance should submit their data to the county DOH as required. An official from this department will collate and organize the data in a visually friendly format in a County Health Indicators report (CHI). The CHI will include raw data generated in spreadsheet format as well as bar graphs to illustrate trends. The responsible person (preferably someone with infection-control experience) will analyze the data, comparing it to previously established baselines and look for unusual spikes in any of the categories. Such anomalies will be reported immediately to the public health and emergency management agencies for the affected jurisdictions.

Designated officials in these agencies will determine if a full epidemiological investigation is needed, and if so, local law enforcement authorities and the FBI will be notified and invited to join the investigation. A full epidemiological investigation will seek additional information from numerous sources, notably from additional surveillance and sampling among the affected elements of the population and from laboratory analyses of materials.

The director of county health is responsible for activating the Biological Response Plan (BRP) when a bioterrorism event is suspected. The decision for activation will be based on alerts, threats, suggestive evidence collected by other agencies and/or components, or as a result of the full epidemiological investigation.

Mass-care decisions will be based on the number and location of victims involved in the incident. Early epidemiological recognition is therefore critical to the success of this plan. Tools used to identify the affected population will vary depending upon whether the release of the bioterrorism agent is announced through a threat or is discovered by one of the elements identified in the early recognition plan.

BIOTERRORISM INDICATIONS

Unusual or unexplained disease patterns or clusters of patients may be the first indicator of a bioterrorism incident. Features that may be indicative of a suspicious outbreak include the following:

- A single, definitively diagnosed or strongly suspected case of an illness due to a potential bioterrorism agent occurring in a patient with no history suggesting another explanation for the illness
- A cluster of patients presenting symptoms of a similar disease with either an unusually high morbidity or mortality rate and without an obvious etiology or explanation
- An unexplainable increase in the incidence of a common syndrome above seasonally expected levels or with higher than expected morbidity and mortality

- An unexplainable and rapid increase in the incidence of a disease (e.g., within hours or days) in a normally healthy population
- An epidemic curve that rises and falls during a short period of time
- Unusual increases in the number of people seeking care, especially with fever, respiratory, or gastrointestinal complaints
- An endemic disease rapidly emerging at an uncharacteristic time or in an unusual pattern
- Any outbreak of disease in which there is a lower attack rate among people who had been indoors, especially in areas with filtered air or closed ventilation systems, compared with people who had been outdoors
- Clusters of patients arriving from a single locale
- Large numbers of rapidly fatal cases
- Any patient presenting a disease that is relatively uncommon and has bioterrorism potential (e.g., pulmonary anthrax, tularemia, plague)

The CDC web site (www.cdc.gov) lists an extensive group of pathogenic agents. There are many pathogens and toxins that are not on the typical DOH list of reportable diseases. Some of these pathogens have the potential to be employed as bioterrorism agents. Therefore, updating these DOH databases remains an important objective.

BIOLOGICAL TERRORISM STAKEHOLDERS GROUP

The Biological Terrorism Stakeholders Group (BTSC) should be composed of senior members from various critical services in communication with the governor's office, county officials, and municipal officials from impacted communities. The services should include but not be limited to the following:

- EMS
- Hospitals
- Emergency management
- Law enforcement
- Fire service

The BTSG should consider the following when recommending policy that will stimulate an operational response:

- MMRS-activation thresholds based on region-wide casualty counts and activities at local medical examiner offices. This might include the activation and operation of OPTCs and AMCs
- A region-wide mass fatality strategy

- Locations/facilities that will be used as OPTCs and AMCs for prophylaxis and casualty care
- Staffing for OPTCs and AMCs
- Other potentially important medical resources, including personnel and materials
- Ensuring sufficiently rapid initiation of prophylaxis, treatment, or quarantine (if the agent is contagious)
- Guidelines for requesting state and federal support

The BTSG should regularly communicate with subject-matter experts and, if possible, organize support groups composed of such subject-matter experts, including personnel with expertise in the following fields:

- Local and state public health (including adult and pediatric infectious diseases)
- Veterinary science
- Laboratory science
- Law enforcement agencies (including other local police departments, state agencies, and the FBI)
- National Guard (representing the adjutant general directly)
- U.S. HHS (CDC, OEP, and the USPHS regional health administrator/regional emergency coordinator)
- U.S. DoD (USAMRIID senior scientific representation)
- Universities with schools of public health (senior science advisors)

Finally, the BTSG should consider the following in their analysis of the evolving situation:

- Region-wide casualty counts and fatalities
- Possible primary and alternate pathogen(s), based on general information being received from medical facilities and hospitals
- Contiguousness of the pathogen
- Prophylaxis/treatment modalities based on the tentatively or positively identified pathogen(s)
- Rapidity with which prophylaxis, treatment, or quarantine (if the agent is contagious) will be initiated

DETECTION

In the case of an unannounced bioterrorism incident, detection occurs when irregularities appear in one or more key local indicators during passive medical and public health surveillance.

EMS

Unusually high overall demand on EMS provides an easily monitored indicator for the detection of unannounced bioterrorism attacks. Particular attention should be paid to increases in the number of patients complaining of ILS or respiratory problems such as those caused by ILS, asthma, or other known or unknown phenomena.

MEDICAL EXAMINERS (ME)

ME records should be monitored for a large increase in overall reported deaths, deaths accepted for investigation/post mortem, deaths associated with ILS, or encounters with unusual deaths. Attention should also be given to higher than normal mortality associated with a common disease.

HOSPITALS AND OTHER HEALTH-CARE INDICATORS

Increases in overall rates in hospital admissions are another important indicator. Attention should also be given to the following indicator groups: emergency department activity; Intensive Care Unit (ICU) admissions; demands on private practitioners, especially if the patients in any of these domains complain of ILS or other respiratory problems; any unusual illness suddenly occurring in a selected population (e.g., outbreak of severe rash affecting adults); higher morbidity associated with a common disease or syndrome or failure to respond to usual therapy; a single case of a disease caused by an uncommon agent (e.g., smallpox, viral hemorrhagic fever, pulmonary or cutaneous anthrax); endemic disease with unexplained increased incidence (e.g., tularemia, plague).

The following list of indicators was created for the World Trade Organization (WTO) meeting in Seattle under a CDC grant (Jeffrey S. Duchin, M.D., Communicable Disease Control, Epidemiology and Immunization Section Public Health—Seattle and King County):

- Unexplained deaths
- Meningitis, encephalitis, acute encephalopathy, or acute delirium
- Unexplained paralysis/paresis or other neurologic symptoms of rapidly progressive onset including ptosis, diplopia, dysphagia, or seizures
- Rash or other skin/mucosal lesions with a history of fever
- History of fever and respiratory disease that includes either dyspnea, tachypnea, hypoxia, abnormal CXR, or any combination of these
- Shock or sepsis syndrome with history of fever or hypothermia
- Diarrheal illness with fever

AGENT SURETY

Once the immediate bioterrorism event has terminated, measures should be taken to ensure that conditions for the resident population return to normal levels of safety. Such measures include sampling the environment in which the event has taken place, designing control measures, and communicating these control measures to the population. Safety measures should be based on the specific agent, which will have been identified in earlier stages of the incident response.

ANTHRAX

Exposure, sampling, and treatment

Anthrax used in a bioterrorism incident could be dispersed as an aerosol, exposing large areas of soil and vegetation to anthrax spores. Although the spores can persist for decades, it is extremely difficult to create a secondary aerosol because of strong adhesive forces between the infectious particle and the surface area on which it rests. Therefore, the danger of direct infection to humans through this dispersal method is relatively minor. Samples should be obtained, however, to fully understand the spatial extent and concentration of spores. During this sampling, personnel will wear protective masks capable of blocking inhalation of the spores. In the later part of 2001, for the first time in U.S. history, the postal service was exposed to anthrax-laced letters. These letters killed postal workers and news reporters, and exposed many others. Local, state, and federal agencies had plans in place for the overwhelming increase in calls related to the postal anthrax scare. Overall, the emergency services, in conjunction with public health officials, were able to mitigate the incidents nationwide.

Animal exposure and treatment

There are no gross physical signs evident to the casual observer that an animal is infected with anthrax. Near the terminal stages of the disease, hemorrhages from the mouth, nose, and anus would be suggestive of anthrax. All exposed animals, whether or not they appear to be infected or have high temperatures, should be isolated and given antibiotics and vaccinations. Infected and exposed animals should be placed in a separate location from nonexposed animals and given long-acting antibiotics immediately. High dosages of antibiotics (50 ml of long-acting penicillin; 300,000 IU/ml) are recommended. Such antibiotic therapy can stop anthrax intoxication if given early. Infected or exposed cattle should be vaccinated as soon as possible using Thraxol® (Miles Laboratories) or Anthrax Spore Vaccine® (Colorado Serum). The vaccine provides protective immunity starting three to five days after vaccination. A booster vaccination should be given according to label directions. Antibiotic therapy can prevent death until the vaccine can provide immunity. There are currently no approved vaccines for domestic animals other than cattle.

Disposal of dead animals

If anthrax is suspected or confirmed in an animal carcass, the animal should not be moved. This will aid in preventing contamination of the surrounding area. Bury the carcass deep in the ground and cover it with anhydrous calcium oxide (quicklime).

Decontaminate the soil with 5% lye and quicklime. Some have advocated incineration; however, a concern of incineration is that during the process, the smoke produced may aerosolize some of the anthrax spores before they are killed and create an airborne threat. It is theoretically possible for flies and biting insects to transmit the disease, but this has not been documented and is unlikely. Transmission to humans under normal circumstances has a low probability.

Vector control

There are no arthropod vectors associated with anthrax. Therefore, no control measures are required.

Decontamination

In areas of heavy contamination, decontamination should be considered. Decontamination methods for all types of bacterial agents include burning the area or spraying the area with a mixture of bleach and water. Spraying water or oil on the area helps prevent secondary aerosol exposure but does not decontaminate the bacteria. Anthrax spores are highly resistant to decontamination. Any commercial hypochlorite (bleach) can be used to produce a decontaminant that will rapidly kill all potential biological threat agents, including *Bacillus anthracis* spores. Chlorine dosages sufficient to kill anthrax spores rapidly would kill other microorganisms even faster. Sodium hypochlorite, formaldehyde, and phenol are also effective sporicidal decontaminates. These chemicals are caustic and corrosive in addition to being toxic and offensive to humans and animals. A new commercial sporicidal product, Exspor, has been found to be less corrosive than hypochlorite bleach, not caustic, and generally harmless to humans; however, inhalation of the aerosolized vapor during decontamination may result in breathing difficulties due to the acidity of the solution. In all cases, steps must be taken to capture the runoff produced by the decontaminate. Failure to capture the runoff may constitute illegal release of a hazardous substance under federal and state laws and regulations. Danger to animals, both domestic and wild, will continue to exist as long as there are significant numbers of spores in the environment.

Building assessment

Building ventilation systems in direct line of any release of anthrax spores may be contaminated and require decontamination. The building must first be assessed to determine if any anthrax spores are present. Then, if spores are found, the level of decontamination must be determined. It is generally believed that re-aerosolization of sufficient spores needed to infect humans is very unlikely. Irrespective of the effectiveness of the decontamination, the public may not accept rehabitation of the building. There is very little danger of anthrax contamination in buildings used to diagnose or treat anthrax patients. Normal hospital infection control techniques will adequately deal with any potential building contamination.

BRUCELLOSIS

Exposure and sampling

Brucellosis used in a bioterrorism incident would likely be dispersed as an aerosol, exposing large areas of soil and vegetation to the organism. An alternative mode of attack would be the deliberate contamination of food products, probably dairy products. Brucellosis is normally transmitted by contact with tissues, blood, urine, vaginal discharges, and especially placentas through breaks in the skin or by ingestion of dairy products. Furthermore, the bacteria are relatively fragile in the environment and will not survive for long periods of time. Therefore, environmental sampling is not required.

Animal exposure and treatment

Animals in the affected area should be tested for the presence of brucellosis. Infected animals exhibit no obvious signs of the infection to the casual observer. Antibiotics can be used to treat infected animals, but the treatment is frequently not effective and the animal remains a carrier of the disease. As a result, infected animals should be destroyed and their carcasses should be disposed of properly.

Disposal of dead animals

Dead or dying animals pose little threat to humans as long as strict sanitation procedures are followed. Gloves (latex or nitrile) should be worn when handling the carcasses and hands should be thoroughly washed and disinfected. As stated above, the primary means of transmission are bodily fluids or consumption of contaminated tissue. Therefore, dead animals must be dealt with prior to rehabitation of an area. Carcasses should be collected and incinerated. If incineration is not feasible, they should be buried and covered with quick lime.

Vector control

There are no arthropod or insect vectors associated with brucellosis. Therefore, no control measures are required.

Decontamination

Due to the frailty of the bacteria, only local decontamination at the site of animal carcasses is required. As previously stated, bodily fluids are the primary contaminants from dead or dying animals. Any disinfectant can be used, but quick lime is probably the most convenient to use.

Building assessment

There are no concerns, beyond those associated with dead or dying animals, in rehabitation of housing. Standard hospital infection control techniques will easily deal with any contamination introduced in treatment and housing of infected humans.

Public information

An aggressive public-information program is essential to ensure that no raw or undercooked meat from potentially infected animals or any unpasteurized dairy products are consumed. This is especially important for the consumption of any wild animals because of the decreased ability to monitor the health of wildlife.

PLAGUE

Exposure and sampling

Plague used in a biological attack would likely be dispersed as an aerosol. An aerosol attack would also infect rodents and they would subsequently become carriers of the disease. An alternate attack mechanism would be to release large numbers of infected fleas to establish a reservoir in rodents and other hosts. Classic environmental sampling of media (air, water, and soil) is not required. Capture and sampling of rodents and fleas should be carried out to determine if the attack has created a reservoir for the disease.

Animal exposure

Animals infected with bubonic plague often show an obvious *bubo*, or swollen lymph node, often draining pus. Carrier animals or those dying from acute pneumonic plague may show no overt signs of infection. Therefore, it is important to test mammalian species. Although over 200 mammalian species have been reported to have the natural form of the disease, rodents, particularly rats, are the natural reservoir for the disease.

Disposal of dead animals

There is a direct danger to humans from dead and dying animals. Infected fleas on the carcasses can bite and subsequently infect humans or other animals. Carcasses should be handled with strict sanitary precautions (gloves, hand washing, etc.) and incinerated as soon as possible. Fleas should be prevented from leaving the host animal and infesting a bystander. This can be accomplished by using an insecticide to kill the fleas.

Vector control

If the disease is found to be present in the local rodent/flea populations, aggressive rodent and flea control measures will be undertaken. This, coupled with an aggressive public health monitoring program, is essential to keeping the subsequent disease incidence under control.

Decontamination

Basic sanitation is the only concern in buildings. That, coupled with the aggressive rodent control measures, will minimize the likelihood of subsequent infections.

Public information

As stated earlier, a forceful public-information program stressing the need for sanitation and rodent/flea control measures will help to keep the subsequent infections to a minimum.

SMALLPOX

Exposure and sampling

Smallpox will not have any significant environmental consequences.

Animal exposure/disposal of dead animals/vector control

There are no animal hosts for smallpox.

Decontamination

Natural weathering in the environment will virtually eliminate any hazard after one or two days.

Building assessment

Standard infection control measures will reduce the likelihood of further spreading of the disease due to contamination in buildings.

TULAREMIA

Exposure and sampling

Tularemia used in a bioterrorism incident would likely be dispersed as an aerosol, exposing both animals and humans to the agent. Both humans and animals can acquire the infection from inhalation. The organism can remain viable for weeks or months at low temperatures in water, soil, carcasses and hides, and for years in frozen meat. Classical environmental sampling of media (air, soil and water) is probably not required.

Although the infection can be acquired by ingestion of contaminated soil and water by animals and humans, the most common mode is through contact via skin abrasions and mucous membranes. To determine if tularemia has been established in the affected area, collection and testing of wild rabbits will yield an understanding of the extent of the disease in the environment. Arthropods, especially ticks, are another indicator of the extent of the disease in the animal population.

Human or animal exposure

Visible evidence of the infection often depends on what clinical form of tularemia the animal possesses. In ulceroglandular tularemia there is often a necrotic ulceration at the site of inoculation, often on the extremities, due to the bite of an arthropod and swelling of regional lymph nodes. Infection is not visibly evident to the causal observer in the intestinal, pneumonic, and typhodial forms of tularemia. If tularemia is found to be present in the wildlife population, domestic animals should be tested in the affected area for the disease.

Disposal of dead animals

There is a direct danger to humans from dead and dying animals. Infected arthropods, especially ticks, that remain on the carcasses can bite and subsequently infect humans or other animals. Carcasses should be handled with strict sanitary precautions (gloves, hand washing, etc.) and buried or incinerated as soon as possible. Ticks should be prevented from leaving the host animal and infesting bystanders. This can be accomplished by using an insecticide to kill the ticks.

Vector control

Aggressive control measures should be taken to eradicate ticks and other arthropods if the disease is found to be present in the wildlife population.

Decontamination

Decontamination of affected areas is not required, although heat and disinfectants kill the bacteria rather easily. Natural aging will be sufficient to eliminate any contamination in the environment.

Building assessment

There is no need to decontaminate buildings used for treatment or housing of infected patients beyond normal hospital infection control measures. There is no evidence to support acquisition of the disease through inhalation of dust.

Public information

An aggressive public information campaign should be mounted to fully inform the public of the dangers associated with handling and consumption of infected carcasses and symptoms associated with arthropod bites from infected animals.

NOTIFICATION

It is recommended that the DOH serve as the repository and clearinghouse for surveillance data. Health-care providers, MEs, sentinel veterinary clinics, animal control agencies, and sentinel pharmaceutical centers should immediately notify the county DOH when they encounter any bioterrorism indicators (see Passive Surveillance and Detection sections earlier in this chapter).

DIAGNOSIS

Diagnosis includes the identification of a disease caused by an agent listed as a possible agent of bioterrorism. Preliminary medical diagnosis of the infectious outbreak will occur locally, in cooperation with local hospitals, health clinics, and doctor's offices. Although a preliminary diagnosis may be made locally, local public health officials should send biological samples through their chain of custody to the state specialty laboratory and the CDC for final confirmation. Veterinary diagnosis also should be obtained in cases where animals are diseased or act as vectors.

Appropriate biological samples (i.e., sputum, blood, or urine) should be taken for laboratory identification, and these samples should be sent through appropriate channels set up by the FBI, your state's DOH, and the CDC to a designated local laboratory that has the ability to identify bioterrorism agents and has biosafety level 2 (BL2) and BL3 capabilities.

When a patient is diagnosed with such a disease or when a laboratory makes a presumptive identification of any of these agents, the public health department should be contacted to initiate an epidemiological investigation and aid in determining the scope of the event. A formal protocol should be in place to ensure that appropriate agencies are notified promptly.

According to the CDC, the critical path to determining the use of a biological agent will include six primary steps:

1. *Identifying an outbreak.* An outbreak is determined by examining public health surveillance data, population at risk, and signs and symptoms.
2. *Verifying the diagnosis.* The diagnosis is verified by examining signs and symptoms and laboratory diagnosis
3. *Establishing a case definition.* An effective case definition begins with simple objective criteria that are stated broadly at first, gradually focused, and narrowly stated later, allowing individuals who meet the definition to be easily identified.
4. *Identifying and counting cases.* Everyone exposed and potentially exposed must be accounted for, allowing for incubation time as well.
5. *Describing epidemiology.* A complete description is made by collecting demographics, including names, ages, genders, workplaces, and residences, by identifying the time of illness onset and of potential exposure, and by tracing the outbreak curve.
6. *Developing a hypothesis.* Everything must be considered regarding how and why the outbreak occurred, the point of dissemination, the mode of spread, the population at risk, why some individuals became ill and others do not, and whether the outbreak poses an environmental hazard.

LABORATORY DETECTION

States differ in the quality of their of health laboratory. You should check with you state DOH for the location of a BL3 laboratory and other labs. The labs should work with your county DOH on the detection and laboratory confirmation of an outbreak and specifically on bioterrorism agents.

Screening times depends on the type of agent and the level of the lab. Screening and confirmation times for a high-quality lab are shown in Figure 10–1.

Bacteria	Screening (hours)	Confirmation (hours)
Anthrax (*Bacillus anthracis*)	4	48
Tularemia (*Francisella tularensis*)	4	72
Brucellosis (*Brucella spp.*)	36–48	72
Plague (*Yersinia pestis*)	4	96
Virus		
Smallpox (*Variola virus*)	Laboratory work performed at CDC	

Figure 10–1 Screening and confirmation times

MASS PROPHYLAXIS

OVERVIEW

The mass prevention/prophylaxis plan is integrally related to the mass care plan. Successful implementation of this plan will reduce demands on the care system, while unsuccessful prevention/prophylaxis will increase demand on the care system. For this reason, the implementation of the mass prevention/prophylaxis campaign will likely be concurrent with implementation of ESF#6, mass care.

As the scope of an event increases, separate resources should be used to provide medication, which in turn moves a large segment of the affected population away from the patient care system. The intent of this separation is to reduce the burden on the mass care system. The goal of the prevention component of the plan is to stop the spread of disease associated with a bioterrorism incident. Timely use of preventive measures will reduce the need for acute care and forestall the possibility of communicating illness to those who are not initially exposed.

The first two stages of this plan are shared with the mass care strategy. The third stage, however, involves a process that is completely separate from the mass care plan.

DETAIL AND RESPONSIBILITIES

Stage I: developing public health crisis, fewer than 100 patients

- Strategy: assess needs
- General action: utilize existing response structure

Mass prevention and prophylaxis actions

Determination of need for prophylaxis. In the mass prevention/prophylaxis plan, the focus includes not only those people directly exposed to an agent but also those at risk of contracting a disease through secondary exposure. For highly contagious agents, this could include anyone who comes in contact with an exposed individual, including health-care workers. The county DOH will be the lead agency in determining the population that could be affected in this manner.

Delivery of prophylaxis/immunization. Some biological agents can be effectively prophylaxed. Preventive medicine is the most effective treatment for those agents. This section describes how the city/county should provide such medication to an affected population.

For smaller events involving fewer than 100 people, the hospital system can be used to dispense prophylactic medication with stores on hand or readily accessible through the pharmaceutical distribution system. These hospitals should have to establish separate and distinct areas within their hospitals to provide this service, especially for those agents that are highly contagious.

Consideration of forward movement of the CDC stockpile. If an event results in the exposure of more than 100 people, large-scale distribution of preventive medication will become necessary. The Logistics Section outlines how these medications will be obtained and sorted from the CDC stockpile. In the early stages of a bioterrorism event, the EOC must consider requesting the forward deployment of the CDC stockpile.

Issuance public health recommendations. When logistical considerations have been suitably fulfilled, the public can be informed according to the demands posed by the situation.

Stage II: public health disaster, 101 to 1,000 patients

- Strategy: establish alternate response structure
- General action: augment resources within the region

Mass prevention and prophylaxis actions

Request for CDC stockpile. When the EOC determines that a stage II incident is developing, the NPS 12-hour push package must be requested immediately. Each NPS 12-hour push package can provide a full course of treatment and, as applicable, prophylaxis for a determined number of persons as indicated following:

Chemical exposure

- Therapeutic treatment for 2,000 persons

Anthrax

- Treatment for symptomatic anthrax for 6,125 persons
- Postexposure prophylaxis to prevent anthrax for 45,425 persons

Plague

- Treatment for symptomatic plague for 4,125 persons
- Postexposure prophylaxis to prevent plague for 357,500 persons

Tularemia

- Treatment for symptomatic tularemia for 3,000 persons
- Postexposure prophylaxis to prevent tularemia for 178,700 persons

Targeted population prophylaxis. County DOHs will need to establish strict triage guidelines to evaluate patient needs. As events exceed 100 victims, DOHs must determine which patients can be helped by direct care and which patients would best be served by prevention strategies. The criteria for this evaluation will depend upon the agent but must include method of exposure, time since exposure, and extent of symptoms. Nonsymptomatic victims with a strong likelihood of recent exposure will be the primary population referred to the prevention/prophylaxis system.

There must also be PPE available for the general public to ensure at least a minimum level of protection when they go for triage, treatment, and/or prophylaxis medications.

Establishment of points of distribution (PODs). The most resource-efficient method for large-scale events is bringing people to the medication by directing them to distribution centers placed throughout the community. The three goals of PODs are:

- Provide medication to the community
- Provide basic information and fact sheets
- Direct victims away from hospitals

A bioterrorism event may result in a transportation-system gridlock caused by a panicked public. In this case, transportation infrastructure will need to be secured and security personnel will need to be assigned to delivery trucks to ensure that medication reaches the PODs.

In addition to the medication itself, information fact sheets regarding precautions necessary to prevent the spread of disease must be available at PODs.

During an event of biological terrorism, hospitals may become overwhelmed by people (either as actual victims or the worried well). PODs will diffuse and redirect those people away from hospitals. Off-site clinics associated with the hospital system may be integrated into the POD network and act, at the hospital's discretion, as ancillary PODs. Additional PODs such as employee health services at large companies, may be developed as necessary.

Local pharmacies may be asked to augment their staff with pharmacists from other locations. The EOC (ESF#5, Information and Planning) in conjunction with the DOH could be asked to make up any staffing shortfall with qualified personnel. In addition, security personnel must be placed at each POD to ensure orderly distribution of medication. The DOH must consider the availability of security personnel when determining the number of PODs to establish.

Stage III: catastrophic public health event, more than 1,000 patients

- Strategy: utilize all available federal and state resources
- General action: integrate resources into response structure

Mass prevention/prophylaxis actions

Direct delivery. Direct delivery of medication to an affected population may need to be considered for home-bound individuals and for highly contagious agents.

Home-bound individuals

The POD concept relies on the ability of the affected population to reach a POD. This may not be practical for homebound or physically challenged individuals. For these people, the city must consider direct delivery of prophylactic medication. This process presents several challenges.

First, the city must confirm the location and verify the status of homebound individuals. This may be facilitated by databases on in-home nursing care providers and assisted living organization. However, verifying the status of people not served by these care organizations will be difficult.

The second challenge presented by direct delivery is the delivery process itself. PODs limit the number of delivery points so that a relatively small number of delivery trucks can serve a large population. For direct delivery, the reverse is true; many trucks may be needed to serve a small part of the affected population. This plan assumes private delivery fleets should be capable of accomplishing direct delivery. These resources would be inadequate if direct delivery were needed for thousands of victims.

Additionally, security resources will be taxed to a much greater extent in a direct-delivery process. Affected people within a neighborhood may converge on delivery personnel in an attempt to obtain medication without going to a POD. If direct delivery is attempted, the EOC may need to provide security personnel to each delivery unit to ensure medication reaches the targeted direct-delivery population.

Highly contagious agents

In the case of a disease that is highly communicable, the best course of action may be to isolate individuals from one another and avoid mass gatherings. In order to prevent a person-to-person spread of the disease, officials could choose to instruct people to remain home and would distribute medications by teams sent to every residence in the affected areas.

The ability to carry out such a plan in public emergency conditions is speculative and untested.

Quarantine. Quarantine may be the only method available to stop or slow the spread of a disease associated with a bioterrorism incident. Quarantine is medically recommended in two of the five target biological agents, *Yersinia pestis* and smallpox virus. County health authorities may recommend quarantine for other disease agents as deemed appropriate under the prevailing conditions and circumstances. Check with your state's DOH on quarantine regulations and laws. For example, in Florida the section is s.381.0011, Florida Statutes 1997, Duties and Power of the Department of Health and Rule 64D-3.007, Florida Admin.code-Quarantine Requirements.

Record Keeping. Record keeping is a vital adjunct of a successful medical care and prevention/prophylaxis operation. It is the principal means by which the inoculated population can be identified and tracked. This will help in determining who is and is not at risk and it should facilitate follow-up if that is required.

In a threatened or announced release it is anticipated that patients should be processed at the scene. The *tag-and-release* method and appropriate care, prophylaxis, or treatment should be directed by the county DOH guided by laboratory analysis of the sample. In the year 2000, police/fire/HAZMAT/EMS have responded to over 300 anthrax threats nationwide. The tag-and-release procedure ensures victims subjected to a real or false threat are given appropriate treatment and follow up and processed efficiently. In the autumn of 2001, confirmed anthrax threats using the United States to mail letters filled with anthrax spores, killing people in Washington DC and Florida and exposing people in Florida and northeastern states.

Patients who enter the triage centers or hospitals as walk-ins or through EMS transport will be triaged using the emergency medical-tag forms developed by the CDC (included in the National Pharmaceutical Stockpile). All patients identified as probable WMD/biological victims will be given medical forms for tracking purposes. The number on the medical-tag form will run concurrently with the hospital's medical record number. It will be shared with the patient-tracking system. This will allow for an accurate accounting of those who are affected by an agent.

The large number of people to be served and the potentially large number of sites involved make it likely that records should be hand scribed. The forms should be machine readable, so that data-entry bottlenecks can be minimized and monitoring of progress of the operation can be computer tracked.

OTHER DETAILS ON MASS PROPHYLAXIS

Medical prophylaxis is the distribution and medical application of appropriate antibiotics, vaccines, and other medications to prevent disease and death in exposed victims. Preventive medicine, in the form of prophylaxis, is the most effective treatment available in a BW attack. Preventing illness eliminates the need for acute care and forestalls the possibility of communicating the illness to others. Some biological agents can be effectively treated in this manner.

During a declared bio-emergency, the Emergency Management Agnecy (EMA), in concert with the state DOH and the CDC, should assess the need to activate phase I of the MREP mass care strategy, which focuses on medical prophylaxis of a fixed population. Area hospitals, neighborhood health clinics, urgent care centers, employee health centers, private medical doctors, and the Red Cross should become the primary sites for medication distribution. These facilities will need to forego their normal autonomy and begin functioning as a unified body under the city's ICS.

The speed with which medical prophylaxis can be effectively implemented is critical to the success of the plan. Application of medical prophylaxis traditionally required identification of the population at risk, which may not be simple in an intentional BW attack. Consequently, treatment must be applied to a much greater number of people than those actually exposed to the agent, perhaps even to an entire city or county population.

After a confirmed BW incident (when the agent has been identified and medication susceptibility determined), personnel, supplies, and pharmaceuticals will flow into the local, county, or regional area from other local, state, and federal resources. Representatives from the local emergency-management agencies should be designated as the formal POCs that should coordinate and manage the arrival, inventory, storage, and distribution of these critical items to central receiving areas. Local law-enforcement agencies should be tasked to move the supplies from the central receiving area to the points of distribution. Additional storage sites located away from the central receiving area, in addition to mobile refrigeration units, may also be used, depending on the size of the incident.

Two distribution paths for prophylaxis/immunization will be established simultaneously. First, mass prophylaxis centers (fixed and mobile) will be set up as the primary means of distribution. Second, a system of home delivery will be initiated to accommodate those unable to travel to the distribution center.

Designated sites for medication distribution should execute the internal disaster plan once the decision to initiate mass prophylaxis is made based on the advice and direction of the emergency management. Health officials should work with each of these institutions to ensure their plan specifically includes a mass prophylaxis policy.

Additional medication-distribution centers should be set up by county officials at predesignated locations to ensure appropriate access to all individuals requiring prophylaxis within a specified geographical location. Quickly implementing these additional distribution sites should help prevent existing clinics from becoming overwhelmed. In underserved or heavily populated areas, alternative methods should be used to distribute medications.

Minors

Emergency management and the county's DOH will determine if parental permission is required to dose children or to seek a waiver because of the State of Emergency.

Mobile distribution vehicles

These vehicles (commuter buses, General Service Administration trucks, school buses, etc.) should be dispatched to retirement communities, group homes, extended-care facilities, and homes of individuals with special needs (including homebound individuals). The EMA should identify these groups of individuals in conjunction with local DOHs, home-care provider agencies, hospitals, and other city human-service agencies.

Schools and commercial facilities

Schools, businesses (with more than 100 employees) and large office buildings (including government offices) may be used to dispense medications. A supply of medications should be dispatched to these facilities and distributed by either internal student or employee health services or by EMA-designated individuals (such as the National Guard).

High-rise residential units

A supply of medications should be distributed directly to the office manager of high-rise residential complexes with with large volume of units to be dispersed to their residents.[1]

Neighborhood canvassing

The U.S. Postal Service should deliver medication and instruction packets to affected areas. The role of neighborhood canvassing will increase in the event that the BW agent is highly contagious and individuals are instructed to remain home to prevent person-to-person spread. Under this circumstance, the services of the police, fire service, EMS, National Guard, military (both active and reserve components), community health personnel, and volunteers may be called upon to assist the U.S. Postal Service in allocating medication to ensure rapid distribution of prophylaxis.

---

1. The number of units that determines a residential complex to be considered a high-rise differs from state to state and county to county. Plan according to your communities needs.

Nondomiciled outreach

The homeless population in the city should be provided medication and information packets through a coordinated effort spearheaded by the EMA.

Personnel for distributing medications

The personnel needed to provide immunization and medication prophylaxis should be drawn from existing staff at the primary distribution points (i.e., hospitals, neighborhood health centers, and freestanding urgent-care centers). Auxiliary personnel to staff these and other designated distribution centers (including door-to-door care) should be recruited from the following sources: Red Cross volunteers; EMTs; National Guard and reservists; retired physicians; nurses; dentists; veterinarians; medical, dental, nursing, veterinary, physician-assistant, and public health students; home health-care agencies; public health departments; religious organizations (i.e., churches, synagogues, etc.); and state and federal supporting agencies.

All participating volunteers will be credentialed through the EMA by a simple application process. All of these individuals will have an information packet that includes an identification card, two HEPA filter masks, two pairs of surgical gloves, and a standard operating policy that is brief and easy to understand for use in the event of a BW incident.

Emergency responders and health-care providers will be supplied with emergency prophylaxis therapy at the direction of the DOH and under advisement from the CDC. This policy should be implemented to assure their continued presence during the response. The medication should be distributed to them through their place of employment or affiliation. Should shortages develop during the early phases of the incident, the medications that are issued should be limited to a one- to two-day course of treatment, pending agent identification. Medications for emergency responders should be provided from a stockpile separate from the general public resources to assure availability.

Treatment facilities and information

Treatment facilities that agree to participate in your planning process should be supplied with prepackaged medications and multilingual information pamphlets from the EMA after the BW emergency declaration is made. Treatment centers should be expected to operate at full capacity using their internal emergency-operations plans. Facilities should execute their plans for augmenting personnel to ensure continuous operations in a standalone mode until such time that relief can be provided through mutual aid or federal assets. The EMA, with the assistance of the county DOH, should direct in the planning and implementation phases of these treatment centers to identify, credential, and train supplemental health-care personnel.

An Internet-based Web application should release timely information and locations of treatment centers and hours. EMA should have the responsibility to maintain the accuracy of this site. This address should be made available to the public in frequent press releases. The Web site should include a search link in which the public can enter street addresses and cities. The application will then cross-reference the address against a database of treatment sites and provide the closest location. The application also can be expanded to include a graphical map display as well as transportation directions or instructions relative to a specific address.

The EMA should establish and operate (with support of other agencies) a 24-hour hotline for use by the public to acquire up-to-the-minute information and to determine the location of their closest treatment site. The telephone number for the 24-hour hotline should be 800-123-HELP or some other easy-to-remember phone number. Press releases should clearly state that those members of the public who do not have Internet access should call the 1-800 number. Hotline operators should have direct access to the identical web-based database at their workstations. On the other hand, those who do have Internet access should be encouraged to use the Web interface to reduce telephone traffic in order to free as many HELP lines as possible.

All avenues of communication with the public including press releases,Web communications, and telephone hotlines should have multilingual capability. TTY capability also should be available at the 24-hour hotline center, with trained operators on duty at all times.

AUGMENTATION ACTIVITIES

Increase of Local Stockpiles of Pharmaceuticals and PPE

Local emergency services should look into the purchase of a cache of pharmaceuticals and PPE to immediately equip and provide prophylaxis to the health-care and public-safety workers and the families of public-safety workers. A maintenance plan will need to be developed for this cache.

Record Keeping

An abbreviated mass casualty medical record that provides a level of confidentiality, access to patient information (name, allergies, concurrent health problems, etc.), and ongoing documentation of care at a local or national level is in existence. The ability of our current information technology to accommodate the influx of hundreds to thousands of patient entries in a short span of time is questionable. Ways of record keeping must be designed by local emergency management and is still under development.

Statewide Mass Immunization Plan

A plan should be written at the state level and should be implemented into the state CEMP in the near future.

MEDICAL RESPONSE EXPANSION PROGRAM

Successful implementation of the MREP requires a number of assumptions:

- A large-scale biological terrorist incident could produce thousands to hundreds of thousands of casualties and/or fatalities.
- During a biological terrorism event, actual infected casualties and the worried well seeking aid will overwhelm EMS, outpatient clinics, and hospitals.
- Hospital resources should be redirected to care for the most seriously ill. Elective admissions should temporarily cease while critical medical/surgical and 911 functions will continue.
- Establishing a system to rapidly expand outpatient and inpatient acute care facilities is necessary to provide rapid treatment to a large population of severely ill BW patients.
- A simple system that rapidly integrates medical resources and provides massive casualty management will be needed.
- Emergency officials should communicate to the medical community in advance (preplanning activities) and at the time the event is recognized, to assure health-care workers that their safety has been planned for and that prophylaxis should be provided. It will be necessary to have accurate and timely dissemination of information to medical professionals in order to decrease the fear factor and persuade them to care for BW patients.
- During a large-scale biological terrorism event the standard of care will change to provide effective care to all those affected. In a mass casualty situation, health-care workers will provide care to as many victims as possible, but individualized treatment plans may be rare or non-existent. A decentralized team approach to providing basic medical care may be the most effective use of resources. Advanced life-saving technology and treatment options will be either unavailable or impracticable due to lack of specially trained medical personnel.
- The expanded inpatient facilities (AMCs) will be most efficient if directed to victims of BW disease only. Victims of BW illness who also require acute or critical medical treatment of urgent conditions such as heart attack or traumatic injuries should receive care at the existing medical facility where more diverse resources are available. The AMC should be an extension of a nearby medical facility (hospital) and transparent to the public. Ideally, the general public would seek initial care from either the OPTC or the emergency department of their local hospital.
- The type of agent used and resulting illness will determine the composition of the AMC. The number of casualties expected to survive versus expire will dictate the allocation of medical staff.

- OPTCs and AMCs will function more efficiently and require less dedicated, specialized resources if located adjacent or very close to a hospital in the affected region.
- The EMA is responsible for establishing, maintaining, and overseeing the system-wide operation of the OPTCs and AMCs, including credentialing of personnel in concert with sponsoring hospitals.
- The bioterrorism emergency response system for your region must be self-sufficient for at least twenty-four to forty-eight hours until state and federal resources become available.
- Medical and support personnel as well as supplies required to establish and sustain the AMC together with some facets of the OPTC will need to be drawn from both local and outside resources. Requirements should be identified through state and federal emergency management plans, including the FRP. Local and regional memoranda of agreement may exist with some agencies but must be coordinated for all appropriate agencies.

As the need for medical services increases, the EMA should implement a Medical Response Expansion Program (MREP) and a mass casualty incident emergency response, which focuses on expanding the community's existing outpatient and inpatient treatment capabilities. Outpatient centers will need to implement a triage policy that directs its staff to care for the sickest patients first while also increasing staffing to help manage the increased patient load. Hospitals will need to activate their internal disaster plans and begin redirecting their limited resources to care for the most seriously ill. Inpatients stable enough for discharge or transfer to another suitable facility should be identified and processed as quickly as possible. In addition, hospitals will need to begin maximizing their use of internal bed space by admitting patients to outpatient surgical areas, treatment rooms, and other short-term treatment and holding areas (e.g., radiology, endoscopy suites, waiting rooms).

Subsequently, the EMA should implement phase II of the MREP. During this phase, designated mass care facilities capable of offering OPTC and AMC services are mobilized to provide care to BW victims. These facilities should be quickly established in structures (such as military tents) or buildings of sufficient size that are located close to existing hospitals. From this location, the OPTC and AMC should be able to more easily share hospital resources and services, including food preparation, laundry services, the pharmacy, and laboratories. Examples of suitable buildings include schools, community centers, hospital cafeterias, and hotel conference rooms. These structures are advantageous because they contain adequate floor space for patient care, bathrooms, kitchens, refrigeration, laundry service, electricity, and generator backup. Ideally, the building selected should be large enough to allow all patient services to be provided on a single floor. This type of layout should minimize the need for additional support personnel.

The identity of the agent and its infectious characteristics, combined with real-time epidemiological information, will influence the number of mass care facilities activated. These factors will also determine whether emphasis should be focused on outpatient or inpatient care. For example, biological agents that are incapacitating but have a low-mortality rate (e.g., Q-fever, brucellosis, VEE) will require a greater emphasis on outpatient treatment. Biological agents known to have a high-mortality rate (anthrax, plague, and tularemia) will necessitate more inpatient medical care.

The level of care provided at the OPTC and AMC will include agent-specific therapy (i.e., antibiotic or antiviral medication); noninvasive respiratory care (i.e., bronchodilators and oxygen therapy); intravenous hydration; pain management; treatment of nausea, vomiting, diarrhea, and anxiety; and minor exacerbation's of the underlying disease. OPTCs should also provide mass distribution of medications along with minor wound and trauma management (i.e., wound irrigation, dressings, and orthopedic splinting). A temporary morgue should be set up at the OPTC and AMC to manage the deceased.

The OPTC and AMC will not have the capability to provide advanced airway management (i.e., intubation and ventilator support), advanced cardiac life support, pediatric advanced life support, advanced trauma life support, or neonatal advanced life support. Patients requiring these advanced levels of critical care support, including labor and delivery services, should be transferred to the closest hospital when bed space and EMS transportation permit. Otherwise, these patients should be provided supportive care at the AMC while awaiting final disposition. OPTCs and AMCs may be established in areas distant from hospitals also to create a medical presence in the greater community and permit local residents to travel shorter distances. Dispersed AMCs may be particularly prudent if the bioterrorism agent is contagious. NDMS Level I DMAT should be used to accomplish this mission.

LOGISTICS FOR THE NATIONAL PHARMACEUTICAL STOCKPILE PROGRAM

OVERVIEW

The key factor for logistical success lies in the ability to utilize existing systems and resources within the infrastructure to gain a proactive advantage in the mitigation of the event. By utilizing the collateral support system, the city can proceed with its overall strategy while awaiting assistance from state and federal resources.

The immediate needs for this strategy include

- Additional staffing
- Obtaining and distributing medicines

- Providing infrastructure security
- Providing adequate transportation resources (ESF#1)
- Obtaining necessary equipment
- Establishing staging areas

It can be anticipated that once a biological agent has been determined and the response has been activated, the EOC should also be activated for addressing the medical needs of the population. Many local capacities can withstand eight to twenty-four hours of operations. Federal resources should not be relied upon for twenty-four to forty-eight hours.

Logistically, the personnel requested through ESF#6 would need to be internally processed, fed, and sheltered. The immediate need for medicines will require the organization of several logisticians to move medicines from the points of arrival (POAs) to the PODs. These logisticians will then need to facilitate the dispersal of the medicines at these distribution sites. Equipment needs would include temporary shelters, transportation, food, water, temporary utilities, and communications augmentation. Additionally, supporting all ancillary functions, such as alternate care facilities needs for additional toilets, food, blankets, beds, will also need to be addressed.

The overall biological plan should be determined through the BTSG and the DOH's biological response committee through the EOC. Field logistics will be coordinated through the logistics command function of the ICS.

DETAIL AND RESPONSIBILITIES

Staffing

Regional call-out

The assistance of local health-care providers would be requested through a media broadcast. Contact would also be made with local professional organizations for their assistance in gaining medical personnel. In addition, entire units from agencies such as DoD and DMAT may be requested. The request for and utilization of DoD assets and support personnel should be considered early on. The request for those assets should be specific and follow established guidelines.

Qualified medical personnel will be sought through a variety of sources:

- State medical association
- State nursing association
- State association of pharmacists
- State dental society

- State university system
- Veterinary medical association
- National Guard
- In-state Department of Defense medical personnel

While the county DOH should pursue all means available, ultimately plans for such a far-reaching contingency can only be done on a statewide scale. Working with other DOHs, health practitioners, and organizations that represent health practitioners, it should make an effort to organize and train these skilled resources to serve as emergency workers in a state pool. As such they could be called upon to respond to help local communities that face dire health emergencies and a critical shortage of professional health-care workers.

The scope of duties for medical emergency workers can include but not be limited to the following:

- Medical and surgical field teams
- Triage, general emergency, and mobile hospitals
- Nursing service
- First aid and ambulance service
- Sanitation
- Mortuary and laboratory service
- Medial-related radiological monitoring
- Precautionary measures for biological and chemical incidents
- Identification of sick and injured
- Critical incident stress debriefing teams

Staff needs

Mass care or prevention interventions will require large numbers of qualified medical personnel to administer immunizations and symptomatic treatment. In addition, nonmedical personnel should be required to staff phone lines, sort pharmaceuticals, and provide support for all basic services, both within the hospital system and throughout the city. The logistics and personnel section of the EOC should coordinate integration of these personnel into the response system through central processing, the provision of prophylactic medication, and the provision of PPE.

Central processing

Primary consideration must be given to providing a singular point of entry for incoming response personnel. This single entry point, otherwise known as a *staging area* or *base camp*, will provide clear direction to incoming personnel. It will also serve as a point for redistributing these medical assets.

To "capture" the medical health-care workers that arrive as voluntary or as a result of requests, you will need to incorporate them into the ICS system so that they may be woven into the overall strategy. Overhead command teams (for example, in Florida the "Fire-Rescue Disaster Response Plan") can accomplish this internalized processing. Check with your state fire chiefs association for information on this system. These teams can provide processing and perform all the functions of an ICS plans section and will report to the plans command. This includes the validation of the medical licensing of these health-care workers.

Entire units or groups of requested personnel, such as a federal DMAT team, will also be captured by the plans section and instructed to report to identified staging area.

Prophylaxis

In order to have enough qualified trained personnel available to assist with a WMD or biological response, it is imperative that responders do not fall victim to the agent. Therefore, as soon as the biological agent is identified, all response personnel shall be given the appropriate preventive intervention.

The possibility of a shortfall in vaccines or antibiotics raises the question of which groups should get first priority for prophylaxis or immunization. Under adverse conditions such as a bioterrorism event, first priority must go to those delivering health care and guarding public safety. In addition, the families of these workers must be provided with preventive interventions to mitigate the increased risk of exposure they face when potentially contaminated workers return home.

Personal Protective Equipment (PPE)

In addition to prophylactic treatment, response personnel must be provided with PPE suitable for the hazards associated with the agent. This may include HEPA masks, gloves, and glasses. These must be immediately available in order for the health-care workers to carry out the mass care and prevention strategy. A cache of PPE necessary to equip large number of workers should be purchased and stored within the equipment stockpile of the MMRS or similar system.[1] An adequate cache of PPE must be available for the distribution to the health-care workers in order to develop a minimum level of protection in and around the symptomatic and exposed population.

---

1. The amount of equipment needed will differ from state to state and county to county. Plan according to your communities needs.

Obtaining and Distributing Medicines

National Pharmaceutical Stockpile Program

The CDC has established a National Pharmaceutical Stockpile (NPS).[1] The NPS is a national repository of antibiotics, chemical antidotes, life-support medications, IV administration and airway maintenance supplies, and medical/surgical items. It is designed to resupply state and local public health agencies in the event of a biological and/or chemical terrorism incident anywhere and anytime within the United States. The stockpile is segregated into several packages. First, there are eight separate yet identical prepackaged, prepositioned caches referred to as 12-hour push packages that are fully stocked, positioned in environmentally controlled and secured warehouses, and ready for immediate deployment to reach any affected area within twelve hours of the federal decision to release the assets.

If the incident requires a larger or multiphased response, follow-on Vendor Managed Inventory supplies (known as *VMI packages*) will be shipped to arrive within twenty-four to thirty-six hours. The follow-on VMI packages will be comprised of pharmaceuticals and supplies that can be tailored to provide pharmaceuticals, supplies and/or products specific for the confirmed agent or combination of agents. The VMI portion of the NPS is being contractually held at vendor warehouses throughout the United States. These vendors are either manufacturers or prime vendors for each of the items in the VMI formulary.

The stockpile can be accessed by one of three methods:

1. The local EOC can request this stockpile through the governor (via the state Emergency Management), who can contact the CDC.
2. The local EOC can go through the state EM, who can then contact FEMA, who contacts the CDC.
3. The county DOH can go through the state DOH.

The decision to deploy the NPS assets rest with the director of the CDC. The director of the CDC should consult with the surgeon general, the secretary of the DHHS, FEMA, and the FBI prior to making any decisions.

Logistical equipment considerations for the distribution of the NPS to the PODs. Each push package is packed in fifty air cargo containers and fills the space of a Boeing 747 or similar type of aircraft. The push package needs 5,000 square feet of floor space to be properly staged and managed (approximately 50 square feet of space is needed to properly store and manage one air cargo container).

---

1. These specifications for the National Pharmaceutical Stockpile are directly quoted from the CDC Planning and Information Sheets dated 5/00.

The best trucks for moving this load could handle either a 40 or 55 ft. flat bed tractor trailer. You would need five of the 40 ft. or four of the 55 ft. trailers to move one entire push package. To move an individual air cargo container requires a truck that has a cargo base greater than 62" long, 88" wide, and 79.3" tall and can handle 2,000 to 6,000 gross pounds.

The NPS shipment will also contain the following to assist in the logistical operation:

- two dolly jacks to move air cargo containers within the secured staging area
- sixteen tablet counting machines
- eight electric power strips
- eight heavy-duty 50-ft. power cords
- eight safety-cap removal devices
- eight cotton-removal devices
- a supply of computer labels and a compact disc providing dosage bag label wording in a range of languages common in the United States.
- a sample medical tag form

Logistical personnel considerations for the distribution of the NPS to the PODs. The following work model is a reference for developing a timeline for a given workforce dedicated to breaking down this cache. For a work team of thirty-two persons, with one person filling the bags and one person labeling, this cache can be broken down in twelve hours. These crews should be rotated every two to three hours.

NOTE

Included in the NPS delivery will be advisors, including pharmacists, public-health experts, and emergency-response specialists that can be utilized to prepare this package for distribution.

Other logistical considerations to ensure the transition from air pallets to the PODs. Two computers (one as a back-up) with a laser printer will be needed to print the dosage labels and enter other pertinent data.

The entire shipment will require security from pick-up through distribution. For this aspect, the most suitable security elements would be local law enforcement, state police, and the National Guard.

Some of the antibiotics or vaccines will require refrigeration. For this challenge, some of the solutions will include the following:

- PODs (most likely pharmacies) have some refrigeration capacity.
- Some refrigerated trucks will be used deliver the medicines. They can be left at the PODs and secured, maintaining an exterior cold box to store the medicines.
- There are commercial refrigerated warehouses, notably grocer store warehouses, sea ports, and airport facilities, that may be considered if the medicines are needed for an extended period or are of great bulk.
- Existing pharmaceutical distribution centers have refrigeration capacity.

Local/county-wide security

Transportation corridors and essential public services must be secured during a bioterrorism event in order for the mass care and prevention strategies to be successful. Security needs will depend upon the size of an incident. Public safety may require that the scene of a small release be secured, that entry into hospitals be restricted (including closing roadway to general traffic), or that more extreme measures be imposed to limit citizen movement and disease spread.

Securing transportation arteries will allow for quick distribution of medicines to PODs and efficient movement of an affected population to triage centers and hospital care. For large-scale events, local law enforcement can request assistance from state and federal law enforcement personnel to provide this security.

Transportation

Medication Transportation

The primary transportation needs will be for the movement of medicines and equipment between POAs to PODs. Patient transportation will not be as urgent of an issue as in a chemical event unless there is a need for extensive quarantine (for example, in Florida see s.381.0011, Florida Statutes 1997, Duties and Powers of the Department of Health and Rule 64D-3.007, Florida Administration Code—Quarantine Requirements)

For medicines, whether it is from pharmacy to pharmacy, or pharmacy to PODs or administration, there are fire department units and police units located throughout the city. These units can move blocks of medicines between the pharmacies and alternate care facilities. There are also the established package carriers (U.S. Postal Service, Federal Express, United Parcel Service, etc.), which can deliver medicines to their designated location.

Patient Transportation

Should the situation overwhelm the existing capacity of emergency medical transportation assets, assistance should be coordinated through ESF#7. If necessary, ESF#7 is also capable of arranging assistance in moving emergency medical supplies and equipment.

Equipment

Most equipment needs will be in support of existing systems and to augment the surge of patients with their need for additional shelters. There will be a pressing need for temporary shelters, toilets, food, water, essential clothing, and bedding. These issues can be addressed through several existing systems:

- EOC ESF#6, Human Services
- EOC ESF#7, Resource Support
- Local agencies' cache equipment
- National Guard

The majority of the medical equipment will be arriving via the NPS. Any additional needs will be requested as determined through the action mass care plan as developed at the EOC.

The need to maintain a cache of biological PPE is mandatory. The PPE include HEPA filter masks for airborne agent protection and protective gloves for contact protection.

Staging Areas

The following points of entry will be used to receive and to inventory these resources before dispatching them to a location determined by the plans section of the ICS structure.

Air

The best local, private, and/or international airport in your area should serve as the primary air support for assets arriving near the impacted area. Military air bases[1] can be utilized, if necessary. An on-site incident liaison officer should direct their activities at the receiving airfield(s).

Ground

All goods shipped via ground transportation should be directed by the EOC to a suitable secure location. An on-site incident liaison officer will direct their activities at these sites.

Large open areas, such as a university campus or a shopping mall, would be a suitable location for a personnel in-processing staging area or center.

Sea

If your city or county is located on or near the water, the local port can be the primary reception center for any goods arriving by ship. This includes the support of

1. Note that if a biological incident occurs, many if not all military installations will go into lock-down mode.

U.S. government hospital ships. An on-site incident liaison officer will direct the personnel and supplies.

Augmentation Activities

If the situation stresses the capabilities of the local medical personnel, independent, retired, or out-of area medical personnel should be organized to assist in the event. Other resources include public and private health nurses, Red Cross, and local university health-related students. The county DOH should develop a working group to form a resource list of medical assets for use in a biological event.

EMS RESOURCES

The EOS will designate EMS resources in coordination and cooperation with local and regional fire and EMS agencies to provide stand-by and transport services in and around clusters of hospitals, clinics, OPTCs, and AMCs. These EMS resources should not, if possible, be used for 911 services. Rather, these units should be dedicated exclusively to providing rapid intervention and transport services for the mass casualty incident emergency response plan.

EM, in cooperation with command, will designate and define EMS transport zones (groupings of hospitals, OPTCs, and AMCs) known as clusters. Each cluster should have at least one designated EMS unit with more available as the need rises. If the demand for EMS services is great in the city, mutual aid will be requested through established agreements within and outside the regional operating area. Commercial ambulance providers and federal assistance through FEMA may also be considered.

The operation and activity of cluster EMS units should be coordinated through ESF#8 (Health and Medical) in EM's EOC.

HOSPITAL PLAN

Hospitals should make every attempt to discharge stable patients and transfer suitable non-BW patients to other comparable outlying facilities in order to make bed space available for critically ill victims. This transport may be accomplished through the hospital's usual transport mechanism or, if ambulance resources are not readily available, through ESF#8 in the EM's EOC.

Hospitals should close to elective, noncritical admissions while continuing to provide services to acutely ill BW and non-BW patients. Hospitals may consider the transport of inpatients, either BW or non-BW, to facilities out of the region in coordination with the NDMS. This should be coordinated through ESF#8 in the EOC, in which DHHS has a senior leadership seat.

EM, through ESF#8, will ensure that adequate medical transportation and logistical support are provided to each of these centers to initiate and sustain operations.

See appendix E for a sample of area hospital listings and important information to be included in your plan.

EMERGENCY SUPPORT FUNCTION #8 (ESF#8), HEALTH AND MEDICAL

Area hospitals, clinics, and private medical doctors should forego their normal autonomy and function as a unified body under the city's ICS during a BW emergency declaration as detailed in its entirety in the command-and-control strategy of this plan.

OPTCs and AMCs will report to the local incident management command system through ESF#8 in the EOC and should be linked to a sponsoring hospital's medical command center (for staffing and supplies in the early stages of an incident). ESF#8 in the EOC should serve as the medical command-and-control element.

A medical facility chief, either a senior administrator or senior clinical officer, should manage each hospital, OPTC, and AMC.

Hospitals, clinics, OPTCs, and AMCs will provide situation reports (SITREPS) directly to ESF#8. The DHHS and DMATs should submit their reports to ESF#8 through the DHHS management support team in an effort to streamline communication and affect a rational span of control where possible. SITREPS should be transmitted to ESF#8 either by fax or by secure email and should include the following:

- General status of activities and operations
- Current patient count
- Cumulative patient count
- Logistics or staffing needs

Hospitals, clinics, OPTCs, and AMCs will make requests for resources such as material, human resources, and pharmaceuticals directly to ESF#8. This will allow ESF#8 to coordinate the distribution of assets throughout the region based on resource availability.

ESF#8 should be led by a section chief who reports directly to the EOC's director. The ESF#8 section chief will coordinate with other ESF section chiefs to ensure that activities are cohesive and all objectives and requests are met. The section chief should be a member of EM or the DOH.

The JIC will receive and handle all media inquiries. The JIC leader may, however, authorize a facility chief to release information once it is cleared and verified to be consistent with JIC information. In any interaction with the media, the goal is to ensure that information is accurate, timely, and consistent.

MASS FATALITY MANAGEMENT IMPLEMENTATION

During an event, a *mass casualty incident emergency response* should be implemented. With regard to mass fatality management for a bioterrorism incident, it is assumed that

- Hospital and municipal mortuaries have limited resources for conducting mass fatality operations, including those for administration, autopsies, and storage of corpses.
- Mutual-aid resources and federal and state assets will be needed to support local ME activities.
- Most victims will have been identified prior to expiring.

The BTSG will consider the following fatality issues in its analysis of the evolving situation:

- Region-wide casualty counts
- Region-wide fatality counts
- Possible primary and alternate pathogen(s) based on general information being received from medical facilities and hospitals
- Contagiousness of the pathogen

The BTSG should consider the following when recommending policy that will stimulate an operational response:

- Locations/facilities that will be used as alternative mortuary facilities
- Staffing for AMFs
- Need for ME resources including personnel and materials
- Request for regional, state, and federal support

Once the decision to implement the mass fatality management plan has been made, the office of the chief ME is responsible for the following:

- Identifying and designing a morgue area, as required
- Maintaining security of bodies and personal effects
- Covering, tagging, and protecting bodies prior to transport to the morgue
- Coordinating the removal of bodies from other areas to the morgue area
- Performing tasks to ensure accurate certification of death
- Contacting social and religious concerns will be considered during final disposition.

ESF#15[1] (Volunteers and Donations) will be activated for contact of the religious community. Members of ESF#15 include the Salvation Army, Interfaith Council Disaster Relief Committee, Lutheran Social Services, among others.

MEDICAL EXAMINER EXPANSION PROGRAM

Local MEs will establish and sustain operations of main ME facilities and AMFs in the early stages of an incident of bioterrorism. While the burden on already-taxed ME offices is recognized, this is necessary in order to process large numbers of anticipated fatalities as well as for creating the infrastructure necessary for the integration of mutual-aid, state, and federal assets, funeral director teams, and disaster mortuary teams.

The Medical Examiner Expansion Program (MEEP) will expand based on the recommendations of EM as the crisis evolves and ME facilities become overwhelmed. This will include internal expansion and augmentation of mortuary facilities and capabilities as well as the establishment and integration of AMFs. ME augmentation is further accomplished by the request for and integration of mutual-aid assets. Requirements will be communicated and coordinated with the DHHS through ESF#8 in the state EOC and with FEMA to set up, staff, and operate fatality-management activities. Federal resources that may be requested at the direction of the EOC include the following:

- Disaster Mortuary Teams (DMORTs), which can augment staff in overwhelmed hospitals, municipal mortuaries, and ME offices. They may also establish free-standing ME operations, albeit in direct coordination and cooperation with the local ME office, and as part of the overall MEEP.
- Other medical assets, which may include other assets tasked under ESF#8 (through DHHS) including DoD mortuary assets, Veterans Administration, and others.

ESF#8 (Health and Medical) will coordinate with all ME offices in the region to acquire and field mortuary transport vehicles to support transport activities. Based on the numbers of fatalities, a large number of transport vehicles may be needed. Communications with all transport assets will be coordinated through their primary dispatch centers. Where this is unavailable, such as with some private funeral directors vehicles, ARES (an amateur radio station licensed to an emergency management agency) personnel or other communications personnel (e.g., National Guard) will be assigned to accompany the vehicles. Other resources that may be accessed for this task include ME transport vehicles from outlying areas, National Guard and DoD assets, General Service Administration vehicles, and commercial vehicles. Funeral directors associations will also be asked to contribute vehicles and personnel.

---

1. ESF#15 is within the state of Florida's response plan. Your state may by different

The ME office having jurisdiction, as well as other governmental ME offices as needed, will provide resources such as carry devices, stretchers, and PPE, including gloves and masks. Funeral directors and funeral director associations will also be asked to contribute resources.

Ideally, deceased persons should be transported to a central location[1] that will be the point of access for entry into a tracking system for case management and where the body will be stored, usually in refrigeration at 34 to 37 °F, until processing is possible. Accompanying the tagged body may be documents indicating the medical history of the patient (if transported from a medical-care facility) or scene of death information (if the body is found at some other location). Bodies will be identified, clothing and property will be inventoried, evidence or biological specimens will be collected, and the cause and manner of death will be determined.

Hospitals, ME offices, and AMFs will track victims, including at the very least the victim's name, address, gender, date of birth, cause of death (actual or assumed), social security number, ME of record, and ME facilities (including where the body is currently as well as where it has been or is going). Tracking of fatalities is critical to ensure that the location and cause of death is properly recorded, to ensure that law enforcement officials have access to bodies and records for follow-on investigation, and to ensure that families and friends of the deceased can be notified as to the death as well as the location of the body.

The ME having jurisdiction will receive all death notifications at its central office or communications center. The chief ME will establish a uniform cause of death and may require that each attending physician certify the cause of death. Surviving family will be made aware that the identification process might take a minimum of three or four days, particularly, if casualty numbers are extremely high.

An autopsy of each deceased person may not be performed for events that result in fatalities in numbers within the range of several thousand but can be accomplished if desirable with supplemental resources and staff for fatalities that number several hundred. DHHS and DMORTs may be used to augment local ME activities. Since a terrorist event implies that the deaths are homicidal in nature, the decision by the chief ME to limit post-mortem procedures to external examinations will be coordinated with law-enforcement agencies of the appropriate jurisdiction.

Disposition of bodies will be according to the number of victims, the pathogen involved, and facilities available. Burial is the preferred method if it can be accomplished in a timely manner and there is no threat to the public health. In extreme, catastrophic incidents, where overwhelming numbers of fatalities exist, it may be necessary to perform mass burials. In general, there is a very limited capability to

1. Many ports are equipped with refrigeration warehouses.

perform cremation en masse. However, for certain biological agents, this may be the safest permanent disposition. It may be necessary to incinerate bodies en masse if the agent is transmissible and body counts are very high. The decision to deny release of the body to the next-of-kin should be supported by governmental officials in order to gain public acceptance over the anticipated objection of family members, who should be provided a death certificate signed by the chief ME.

A representative from Red Cross will oversee the establishment of a Family Assistance Center (FAC) staffed by disaster and crisis counselors trained to assist grieving family members. It is critical that an FAC be located away from the incident scene and that security be provided by local law enforcement so as to shield the families from the media.

A public information officer designated by the chief ME will, at regular intervals, provide briefings to family members.

ENVIRONMENTAL CLEAN UP

OVERVIEW

The first step in an effective environmental clean-up strategy is to identify the health risks associated with a particular site. This will be done by obtaining environmental samples, transporting samples to appropriate testing facilities, and analyzing the samples to determine the nature and level of contamination. Once the nature and level of contamination is established, appropriate action will be taken to decontaminate the site by cleaning the area or by removal and disposal of possible infection vectors such as animals. Reentry to an area will be allowed only after decontamination is complete.

DETAIL AND RESPONSIBILITIES

Environmental health and engineering services will be responsible for identifying the environmental hazards, developing a decontamination plan, and establishing criteria for re-entry into an area or building. Environmental Health (EH) will consult with state and federal agencies for additional expert advice on identification and mitigation of environmental hazards. EH will coordinate decontamination of an area and make a recommendation to the DOH director regarding when (or if) a building or area can be reoccupied. The final decision to allow reentry will be made by the DOH director. Until that decision is made, law enforcement will be responsible for the security of a building or area to preclude entry or property damage.

EH and the county DOH recognize that a chemical, biological, or other terrorism act falls under federal jurisdiction and will cooperate with local and federal law enforcement in investigating and prosecuting any related criminal case.

AUGMENTATION ACTIVITIES

Decontamination Resources

Environmental health and engineering services possess the expertise to evaluate an environment; however, they do not have the staffing or tools to decontaminate an area or building. Decontamination efforts would be coordinated between the county and state DOHs, the EPA, and a civilian contracted company. There is currently a shortage of civilian contractors who have indicated a capability to handle an emergency clean up after a biological release.

This is an untested field. Decontamination and reentry criteria will have to be coordinated by local, state, and federal public health and environmental agencies.

SUMMARY

This chapter deals with coordination and communication, which will determine if an effective response is possible. Effective actions depend on proper preplanning and response by agencies that know how to communicate during an event. Agencies that are vital to a biological incident may not use lights and sirens but must be included in the preplanning stage. These agencies should also understand the role everyone plays during a biological situation. Groups of important players include veterinarians, medical examiners, pharmacists and health-care workers. It has been shown over the years that agencies that have participated in preplanning exercises and have learned to communicate, respond more effectively during an event. The local and county agencies should know the state and the state should understand the federal level. Those who will provide the record keeping during a biological event will play an important part of the response yet how often do we train them and include them in our exercises? In the anthrax events after the September 11, 2001 attack, we are finding out that an aggressive public information campaign will help the public understand both the responder and public role that must be followed after an incident. This chapter reinforces the need to communicate and coordinate as often as necessary to learn each role and understand the chain of events that will occur.

APPENDIXES

Bioplan References

Appendix A	Check List for Agencies
Appendix B	Public Safety Precautions/Actions
Appendix C	Public Health Anthrax Threat Advisory
Appendix D	Complete Agent Descriptions
Appendix E	Area Hospital Listings
Appendix F	County Public Health Clinics/Centers
Appendix G	An Anthrax Threat Field Guide
Appendix H	Pharmaceutical Needs for the Five Types of Biological Agents
Appendix I	Pharmaceutical Push Package Contents
Appendix J	Biological Agent Signs and Symptoms
Appendix K	Israel's Fixed Hospital Decontamination System
Appendix L	Sample Domestic Preparedness Training Courses: Public Health Focus
Appendix M	Biological Incident Preparedness Training
Appendix N	Sample School Emergency Plan
Appendix O	Internet Resources for Terrorism/Disaster Planning

Check List for Agencies

THE EMERGENCY OPERATIONS CENTER

- Serve as the interagency coordinator.
- Coordinate interagency response to any biological event.
- Notify the SO, FR and FBI of any reported illness caused by a biological agent of concern.
- Oversee the coordination of the arrival and distribution of supplies and personnel.
- Coordinate the request for all mutual-aid, federal, and state assistance during a biological incident.
- Coordinate and integrate personnel with credentials.
- Coordinate the city's efforts to provide mass medication and mass treatment.

COUNTY DEPARTMENT OF HEALTH, BIOLOGICAL RESPONSE COMMITTEE

- Evaluate the hazard of the biological agent and assist in determining the need for evacuation as well as victim and environmental decontamination.
- Develop the strategic medical plan for all medical/clinical matters related to biological terrorism agents. Medical control will be consulted prior to the issuance of any orders.
- Advise EOC on the appropriate laboratories for specific biological agents.
- Identify all potentially exposed individuals and determine the need for acute medical care, including but not limited to the need for prophylaxis and/or vaccination.
- Collect and collate all surveillance reports for the surveillance sentinel system and analyze the data for aberrancies. Aberrancies are reported to the EOC, medical control, and other agency heads, as necessary.

- Ensure that the appropriate clinical personnel are trained to recognize reportable diseases.
- Notify EOC and medical control of any diseases caused by the biological agents of concern.
- Notify department of health.
- Monitor waterborne illness.
- Monitor influenza-like illness.
- Maintain contact with the CDC and alert EOC of any health threats to the city and county.
- Review all unexplained deaths due to potentially infectious agents in otherwise healthy persons aged 1 to 49 years old.
- Develop and update the mass care/casualty management strategy (action plan).
- Activate clinics that can be used as ancillary PODs.
- Support school mass medication programs.

COUNTY MEDICAL EXAMINER'S OFFICE

- Notify county's DOH and medical control of any unusual deaths, including but not limited to deaths due to uncommon infectious diseases.
- Manage the fatalities that result from a biological incident.

DEPARTMENT OF HEALTH (DOH)

- Perform laboratory analysis of samples.
- Assist the county DOH with the overall plan and implementation of its strategies.
- Coordinate the distribution of pharmaceuticals, from reception points to distribution centers.
- Oversee the receipt and distributions of all requested medicines.

FIRE AND RESCUE

- Perform a comprehensive hazard analysis of the site including but not limited to oxygen levels, the presence of carbon monoxide or other toxic gases, and radioactive substances.
- Obtain a biological sample in coordination with investigators from the JSO and FBI.
- Package biological materials.
- Contain or mitigate the release of any biological agents.
- Perform decontamination of highly contaminated individuals and entry team members as advised by the county DOH.

- Assist the county DOH in evaluating and caring for patients.
- Monitor all entry team members.
- Monitor daily call volume of EMS runs.
- Monitor EMS activity for unusual variations in call type or call volume and report these to the county DOH.
- Notify medical control and the county DOH of any unusual patient contacts, including but not limited to unusual deaths or patients exhibiting the signs of uncommon infectious or communicable disease.
- Ensure that field personnel are trained to the awareness level in weapons of mass destruction.
- Ensure that field personnel are trained in procedures for reporting unusual deaths or uncommon infectious diseases.
- Assist with the coordination of the logistics operation.

SHERIFF'S OFFICE

- Secure the building/area.
- Defuse or otherwise render safe any explosive devices.
- Assist in the evacuation of at-risk individuals.
- Perform crime scene investigation in coordination with the FBI.
- Assist in the transport samples to the county DOH or state lab(s) for testing.
- Notify the DOH of any calls for a suspicious package involving biological materials.
- Ensure that field personnel are trained to the awareness level in weapons of mass destruction.
- Notify the JFR, EOC, and FBI of any threatened use of biological agents.
- Determine the credibility of any threatened use of biological agents.
- Provide security and escorts for supplies and personnel.
- Provide security at PODs, ACFs and CCPs.
- Staff PODs.
- Augment security, if needed, at hospitals.

INTERNATIONAL AIRPORT

- Provide assistance and security at airports.
- Identify hangar space and other facilities to be used for logistics operations.

HOTEL ASSOCIATION

- Provide facilities for billeting responding personnel.

UNITED STATES POSTAL SERVICE

- Provide direct delivery in support of mass medication operations.
- Provide logistical support for mass medication and mass treatment operations.

UNITED STATES DEPARTMENT OF DEFENSE

- Provide personnel and material to support mass medication and mass treatment operations.
- Provide military field hospitals, hospital ships, etc.
- Provide technical support in agent identification.
- Provide security and logistical support as requested.

NATIONAL GUARD

- Provide logistical support for the mass care/casualty management.
- Provide personnel and material to support mass medication and mass treatment operations.
- Provide technical support in agent identification.
- Provide security and logistical support as requested.

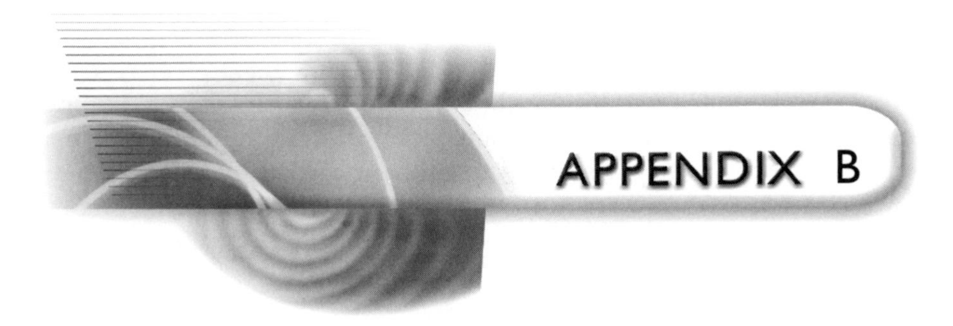

Public Safety Precautions/Actions

GENERAL FACTORS

INDICATORS OF BIOLOGICAL AGENT

Biological agents have the potential to be more lethal than chemical agents and are primarily deployed through aerosol spray or introduction into a water system. There are only two documented biological weapons (BW) attacks in the United States. One occurred in 1984 when followers of the Rajineesh Bagwhan produced and dispensed salmonella bacteria in Oregon. In that case, the assailants spread the agent via restaurant salad bars. The second occurred in the latter part of 2001 when anthrax laced letters were mailed through the United States postal system. At the time of the printing of this book, the perpetrator of this crime was unknown.

The following are indications of a biological terrorism event:

- A single, definitively diagnosed or strongly suspected case of an illness due to a potential bioterrorist agent occurring in a patient with no history suggesting another explanation for the illness.
- A cluster of patients presenting with a similar disease with either unusually high morbidity or mortality, without an obvious etiology or explanation.
- An unexplained increase in the incidence of a common syndrome, above seasonally expected levels, or with higher than expected morbidity and mortality.

BIOLOGICAL (BW) FIRST RESPONDER CONCERNS

Treat all incidents involving biological agents as intentional hazardous materials situations. In all cases, safely isolate and deny entry and make appropriate notifications. In addition, whenever it is believed that a biological agent has been released, assume that all personnel and property has been contaminated. The most practical method of initiating a biological attack is through the dispersal of aerosol particles. Biological agents may be able to enter the body through the respiratory tract, ingestion or direct contact with skin or membranes. Unlike chemical agents, exposure to biological agents may not be immediately apparent with casualties occurring hours, days, or weeks after exposure. In a silent release scenario, the first indication of a biological attack may occur after a number of unusual illnesses begin to appear in local hospital emergency departments. Without advance warning, first responders may not recognize the existence of a biological attack. Additionally, first responders should

- Immediately don PPE if available
- Immediately request specialized resources

  These resources include public health officials from the county DOH, and state DOH. Experts such as the CDC, the MMRS, NMRT—WMD, and the USAMRI are also needed to identify the exact nature of the biological agent.

- Take measures to prevent an epidemic (pending identification of the agent)

  These measures include isolation, quarantine, and restriction of personnel movement based on local, state, and federal laws and regulations. These procedures apply to both victims and first responding personnel. Identify the source of contamination and designate zones of operation (i.e., hot, warm and cold zones). Consider weather effects during zone designation. If large numbers of exposures are involved, quarantine may be necessary with all victims being treated on-site. If a small number of persons are exposed, they should be decontaminated and transported to a hospital capable of isolating the patients.

- Initiate protective actions (i.e., evacuation or in-place protection); avoid all exposed food and water
- Consider the impact of weather conditions

  The impacts of biological agents are affected by weather conditions. Accordingly, detailed and accurate assessments of weather conditions and forecasts are critical elements in the tactical management of biological emergencies. Weather effects to consider include

  - Sunlight

    Ultraviolet light found in sunlight helps kill biological agents.

  - Temperature

    Temperatures above 100° F begin killing off biological agents. Freezing temperatures can render biological agents dormant.

  - Temperature Gradient

    Elevation influences temperature. For each ten meters from ground level, there is a different temperature known as the temperature gradient. This factor causes biological agents to hold close to ground.

- Wind

  Wind aids the dispersal and spread of biological agents. Wind direction and speed influence the resulting plume and must be considered when setting up zones of operation and making evacuation decisions.

- Precipitation

  Precipitation can influence agent dispersal and the spread of contaminated areas (e.g., run-off). In biological situations, the quantity of rain can either kill or stimulate the growth of individual agents.

- Humidity

  Higher humidity levels cause the pores on human skin to open up aiding the absorption of agents.

INITIAL ACTIONS BY FIRST RESPONDERS

In cases of actual release of biological agents, first responding units must immediately take steps to protect themselves. First responders suspecting a biological release must (1) remain calm, (2) don PPE, (3) from a safe vantage point, reassure victims that assistance is on the way, and (4) wait for properly equipped help at a safe location (upwind, uphill, upstream). Essentially this involves safely isolating and denying entry and making notifications. The following checklist summarizes the essential ingredients of an initial biological notification:

- Observed biological indicators
- Wind direction and weather conditions at scene
- Plume direction (direction of cloud or vapor travel)
- Orientation of victims (direction, position, and pattern)
- Number of apparent victims
- Type of injuries and/or symptoms presented
- Witness statements or observations
- Nature of biological agents (if known) from detection equipment or monitors
- Exact location of reporting unit
- Suggested safe access route and staging area

INITIAL ACTIONS BY DISPATCH PERSONNEL

Dispatch personnel play a key role in mobilizing proper response and support to a biological incident. Public-safety dispatchers (both law enforcement and fire service) are vital elements in recognizing and assessing biological events. Dispatchers must be cognizant of potential target locations and the indicators of possible criminal or terrorist activity involving biological agents. Dispatchers must know the indicators, signs and symptoms of exposure to biological agents and recognize unusual trends or patterns of activity indicative of a possible biological incident. Dispatchers must also be able to discern and solicit critical information regarding threats and biological indicators encountered by field personnel.

DECONTAMINATION

Biological incidents may potentially involve civilians, law enforcement, fire service, and medical personnel that have been exposed to potentially lethal agents. Prompt, safe, and effective decontamination procedures are essential to protect exposed persons, equipment, and the environment from the harmful effects of these agents.

Decontamination is the process used to reduce the hazards of biological agents to safe levels. Decontamination minimizes the uncontrolled transfer of contaminants from the hazard site to clean areas. Decontamination should be accomplished any time contamination with a biological agent or hazardous material is suspected.

During decontamination operations, the safety of emergency response personnel is the first and most important consideration. Proper use of PPE such as SCBA, reduces hazards to response personnel.

The risk of secondary contamination to rescue personnel, medical personnel on the scene and at the hospital, other persons and to transport vehicles and equipment must be adequately assessed and protected against to avoid spreading the incident. Any contamination of the skin must be decontaminated immediately.

The MMRS, as well as supporting hazardous materials teams, must establish standard decontamination procedures for a range of biological incidents. These procedures should include provisions for selecting and establishing a decontamination site as well as specific operational protocols. All personnel assigned to these teams shall be thoroughly trained to safely and effectively carry out their responsibilities. Specific decontamination protocols must retain the flexibility to respond to a range of hazards or conditions at the incident scene or decontamination site. Factors that can affect the decontamination process include

- Prevention of further contamination
- Minimizing contact with potential contaminants to keep the incident from escalating
- Physical and chemical properties of the agent
- The very properties that make the agent hazardous make it difficult to decontaminate
- Amount and location of contamination
- The greater the area contaminated, the more involved the decontamination process

- Contact time and temperature
- The longer a contaminant is in contact with an object, the greater the probability and extent of contamination
- Level of protection and work function
- Decontamination requirements may vary somewhat according to the particular type of protective clothing

INFECTION CONTROL PRACTICES FOR PATIENT MANAGEMENT

The management of patients following suspected or confirmed bioterrorism events must be well organized. Strong leadership and effective communication are paramount.

ISOLATION PRECAUTIONS

Agents of bioterrorism are generally not transmitted from person to person; re-aerosolization of these agents is unlikely. All patients in health-care settings, including symptomatic patients with suspected or confirmed bioterrorism-related illnesses, should be managed utilizing standard precautions. Standard precautions are designed to reduce transmission from both recognized and unrecognized sources of infection in health-care facilities, and are recommended for all patients receiving care, regardless of their diagnosis or presumed infection status. For certain diseases or syndromes (e.g., smallpox and pneumonic plague), additional precautions may be needed to reduce the likelihood for transmission.

Standard precautions prevent direct contact with all body fluids (including blood), secretions, excretions, nonintact skin (including rashes), and mucous membranes. Standard precautions routinely practiced by health-care providers include

- Hand washing

  Hands are washed after touching blood, body fluids, excretions, secretions, or items contaminated with such body fluids, whether or not gloves are worn. Hands are washed immediately after gloves are removed, between patient contacts, and as appropriate to avoid transfer of microorganisms to other patients and the environment. Either plain or antimicrobial soaps may be used according to facility policy.

- Gloves

  Clean, nonsterile gloves are worn when touching blood, body fluids, excretions, secretions, or items contaminated with such body fluids. Clean gloves are put on just before touching mucous membranes and nonintact skin. Gloves are changed between tasks and between procedures on the same patient if contact occurs with contaminated material. Hands are washed promptly after removing gloves and before leaving a patient care area.

- Masks and eye protection or face shields

  A mask and eye protection (or face shield) are worn to protect mucous membranes of the eyes, nose, and mouth while performing procedures and patient care activities that may cause splashes of blood, body fluids, excretions, or secretions.

- Gowns

  A gown is worn to protect skin and prevent soiling of clothing during procedures and patient-care activities that are likely to generate splashes or sprays of blood, body fluids, excretions, or secretions. Selection of gowns and gown materials should be suitable for the activity and amount of body fluid likely to be encountered. Soiled gowns are removed promptly and hands are washed to avoid transfer of microorganisms to other patients and environments.

PATIENT PLACEMENT

In small-scale events, routine facility patient placement and infection-control practices should be followed. However, when the number of patients arriving at a health-care facility are too large to allow routine triage and isolation strategies (if required), it will be necessary to apply practical alternatives. These may include grouping patients who present with similar syndromes into a designated section of a clinic or emergency department, a designated ward or floor of a facility, or setting up a response center at a separate building.

Designated patient grouping sites should be chosen in advance by the Incident Commander, in consultation with facility engineering staff, based on patterns of airflow and ventilation, availability of adequate plumbing and waste disposal, and capacity to safely hold potentially large numbers of patients. The triage site should have controlled entry to minimize the possibility for transmission to other patients at the facility and to staff members not directly involved in managing the outbreak. At the same time, reasonable access to vital diagnostic services, e.g., radiography departments should be maintained.

PATIENT TRANSPORT

Most infections associated with bioterrorism agents cannot be transmitted from patient to patient. In general, the transport and movement of patients with bioterrorism-related infections should be limited to movement that is essential to provide patient care, thus reducing the opportunities for transmission of microorganisms within health-care facilities.

CLEANING, DISINFECTION, AND STERILIZATION OF EQUIPMENT AND ENVIRONMENT

Principles of standard precautions should be generally applied for the management of patient-care equipment and environmental control. Each facility should have adequate procedures in place for the routine care, cleaning, and disinfection of environmental surfaces, beds, bedrails, bedside equipment, and other frequently touched surfaces and equipment, and should ensure that these procedures are being followed.

Facility-approved germicidal cleaning agents should be available in patient-care areas to use for cleaning spills of contaminated material and disinfecting noncritical equipment.

Used patient-care equipment soiled or potentially contaminated with blood, body fluids, secretions, or excretions should be handled in a manner that prevents exposures to skin and mucous membranes, avoids contamination of clothing, and minimizes the likelihood of transfer of microbes to other patients and environments.

Policies should be in place to ensure that reusable equipment is not used for the care of another patient until it has been appropriately cleaned and reprocessed, and to ensure that single-use patient items are appropriately discarded.

Sterilization is required for all instruments or equipment that enter normally sterile tissues or through which blood flows.

Rooms and bedside equipment of patients with bioterrorism-related infections should be cleaned using the same procedures that are used for all patients as a component of standard precautions, unless the infecting microorganism and the amount of environmental contamination indicates special cleaning. In addition to adequate cleaning, thorough disinfection of bedside equipment and environmental surfaces may be indicated for certain organisms that can survive in the inanimate environment for extended periods of time. The methods and frequency of cleaning and the products used are determined by facility policy.

Patient linen should be handled in accordance with standard precautions. Although linen may be contaminated, the risk of disease transmission is negligible if it is handled, transported, and laundered in a manner that avoids transfer of microorganisms to other patients, personnel and environments. Facility policy and local and state regulations should determine the methods for handling, transporting, and laundering soiled linen.

Contaminated waste should be sorted and discarded in accordance with federal, state and local regulations. Policies for the prevention of occupational injury and exposure to blood borne pathogens in accordance with standard precautions and universal precautions should be in place within each health-care facility.

DISCHARGE MANAGEMENT

Ideally, patients with bioterrorism-related infections will not be discharged from the facility until they are deemed noninfectious. However, consideration should be given to developing home-care instructions in the event that large numbers of persons exposed may preclude admission of all infected patients. Depending on the exposure and illness, home care instructions may include recommendations for the use of appropriate barrier precautions, hand washing, waste management, and cleaning and disinfection of the environment and patient-care items.

POST-MORTEM CARE

Pathology departments and clinical laboratories should be informed of a potentially infectious outbreak prior to submitting any specimens for examination or disposal. All autopsies should be performed carefully using all PPE and standards of practice in accordance with standard precautions, including the use of masks and eye protection whenever the generation of aerosols or splatter of body fluids is anticipated.

POSTEXPOSURE MANAGEMENT

DECONTAMINATION OF PATIENTS AND ENVIRONMENT

The need for decontamination depends on the suspected exposure and in most cases will not be necessary. The goal of decontamination after a potential exposure to a bioterrorism agent is to reduce the extent of external contamination of the patient and contain the contamination to prevent further spread. Decontamination should only be considered in instances of gross contamination. Decisions regarding the need for decontamination should be made in consultation with state and local DOHs. Decontamination of exposed individuals prior to receiving them in the health-care facility may be necessary to ensure the safety of patients and staff while providing care.

Depending on the agent, the likelihood for re-aerosolization, or a risk associated with cutaneous exposure, clothing of exposed persons may need to be removed. After removal of contaminated clothing, patients should be instructed (or assisted, if necessary) to immediately shower with soap and water. Potentially harmful practices, such as bathing patients with bleach solutions, are unnecessary and should be avoided. Clean water, saline solution, or commercial ophthalmic solutions are recommended for rinsing eyes. If indicated, after removal at the decontamination site, patient clothing should be handled only by personnel wearing appropriate personal protective equipment, and placed in an impervious bag to prevent further environmental contamination.

Be aware that the FBI may require collection of exposed clothing and other potential evidence for submission to FBI or DoD laboratories to assist in exposure investigations.

PROPHYLAXIS AND POSTEXPOSURE IMMUNIZATION

Recommendations for prophylaxis are subject to change. However, up-to-date recommendations should be obtained in consultation with local and state DOHs and CDC. Facilities should ensure that policies are in place to identify and manage health-care workers exposed to infectious patients. In general, maintenance of accurate occupational health records will facilitate identification, contact, assessment, and delivery of postexposure care to potentially exposed health-care workers.

TRIAGE AND MANAGEMENT OF LARGE SCALE EXPOSURES AND SUSPECTED EXPOSURES

Each health-care facility, with the involvement of the Incident Commander, administration, building engineering staff, emergency department, laboratory directors, and nursing directors; should clarify in advance how they will best be able to deliver care in the event of a large-scale exposure. Facility needs will vary with the size of the regional population served and the proximity to other health-care facilities and external assistance. Triage and management planning for large-scale events should include

- Establishing networks of communication and lines of authority required to coordinate on-site care
- Planning for cancellation of nonemergency services and procedures
- Identifying sources able to supply available vaccines, immune globulin, antibiotics, and botulinum anti-toxin (with assistance from local and state DOHs)
- Planning for the efficient evaluation and discharge of patients
- Developing discharge instructions for patients determined to be noncontagious or in need of additional on-site care, including details regarding if and when they should return for care, or if they should seek medical follow-up

PSYCHOLOGICAL ASPECTS OF BIOTERRORISM

Following a bioterrorism-related event, fear and panic can be expected from both patients, health-care providers, and the general public. Psychological responses following a bioterrorism event may include horror, anger, panic, unrealistic concerns about infection, fear of contagion, paranoia, social isolation, or demoralization.

Consider the following to address patient and general public fears:

- Minimize panic by clearly explaining risks, offering careful but rapid medical evaluation and treatment, and avoiding unnecessary isolation or quarantine.
- Treat anxiety in unexposed persons who are experiencing somatic symptoms. Do this with reassurance (or diazepam-like anxiolytics as indicated for acute relief of those who do not respond to reassurance).

Consider the following to address health-care worker fears:

- Provide bioterrorism-readiness education, including frank discussions of potential risks and plans for protecting health-care providers.
- Invite active, voluntary involvement in the bioterrorism readiness planning process
- Encourage participation in disaster drills.
- Fearful or anxious health-care workers may benefit from their usual sources of social support, or by being asked to fulfill a useful role (e.g., as a volunteer at the triage site).

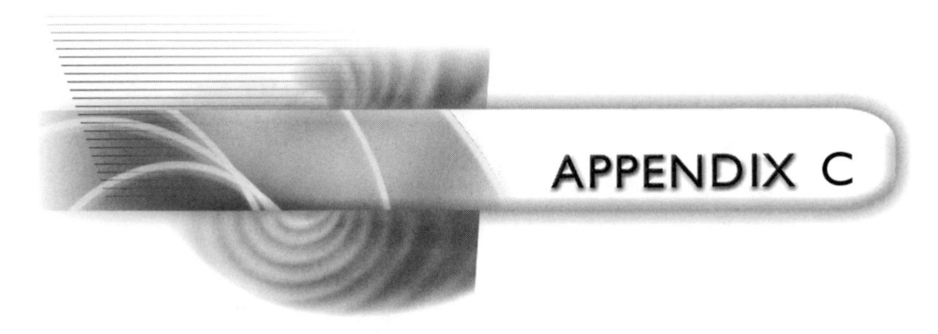

Public Health Anthrax Threat Advisory

RECIPIENTS OF POTENTIAL ANTHRAX THREATS

This advisory is to provide potential recipients of letters or packages containing an anthrax threat with useful information and guidance to help them deal more effectively with such incidents, should they occur. If you have any questions or need further assistance, please call the county DOH or the CDC.

BACKGROUND

Recently, throughout the United States there have been numerous threats of exposure to anthrax through letters and phone calls. During threatening phone calls, the person answering the call has usually been told that anthrax was somewhere in the building or the ventilation system. Typically, threats by letter have advised the reader that anthrax was present in the envelope. Some of these envelopes contained nothing other than a threatening note, while others were found to contain a variety of nontoxic substances (e.g., detergents, baby powder). Until the fall of 2001 all threats had been hoaxes and many of the threats in 2001 remain to be hoaxes, but reactions to these threats (including quarantine, evacuation, decontamination, and medicating victims) have disrupted normal routines.

Anthrax is a disease caused by a bacterium (*Bacillus anthracis*) that is spread as microscopic spores, and is harmful only if inhaled, ingested, or introduced into an open cut or wound. If inhaled, the disease may begin with mild flulike symptoms (e.g., muscle aches, fever, fatigue, and slight cough) usually one to six days after exposure, which may then progress after two to four days to more severe symptoms (e.g., high fever, shortness of breath). Anthrax is treatable with currently available antibiotics.

Anthrax is not contagious (i.e., not spread from person-to-person); therefore, potentially exposed individuals should not be isolated or quarantined.

IMMEDIATE ACTIONS

If you receive a letter or package containing an anthrax threat

- Close the envelope or package and gently put it down. Do not blow into the envelope or examine the contents further (leave it alone for the appropriate authorities to handle).
- Walk promptly out of the room and advise others in the *immediate* area (sharing enclosed airspaces) to leave at once. Close the door. No one should be permitted to reenter. (Note: those *not* in the immediate area of the opened package have very little risk of exposure.)
- The person(s) who directly handled the contents of the package should immediately wash their hands and arms with soap and water.
- Call 911 and report the incident.

EVACUATION

Evacuation is defined as the orderly movement of individuals to a safe distance from a hazard. During an evacuation the following should occur:

- People should immediately remove themselves from risk of exposure, *however,* nobody should leave the grounds. The extent of a building evacuation will depend on the circumstances of the situation and the comfort level of the individual(s) in charge. When in doubt, moving individuals outside or to an adjacent building is an acceptable option.
- While waiting for authorities, make a list of the names, addresses, and phone numbers of all persons in the immediate area of the incident to provide to the official in charge.
- If possible, have the facility manager turn off the ventilation to the involved area(s).
- Upon arrival, the authorities will provide assistance with appropriate evacuation procedures, collect pertinent information surrounding the incident and exposed individuals, and facilitate decontamination activities (if necessary).
- The authorities will remove the package from the facility and decontaminate the affected area. Instructions will be provided as to when the affected area may be reoccupied

DECONTAMINATION

Decontamination is defined as the removal of potentially harmful substances from the skin and clothes. If decontamination is necessary, consider the following:

- Decontamination should be done for person(s) who directly handled or who may have come into contact with the package contents.

- Potentially contaminated individuals should proceed calmly (this is not an emergency procedure) to the nearest showering facility, when instructed to do so by authorities.
- Remove clothing and personal effects and place in a sealed, airtight triple plastic bag (i.e., trash bags). Plastic bags with personal effects should be clearly labeled with the owner's name, address, phone number, and an inventory of the bag's contents.
- Individuals should shower with soap and water, focusing on exposed skin surfaces such as hair, face, neck, and arms. Bleach solutions are NOT necessary and should not be used to decontaminate individuals.
- For those individuals undergoing decontamination, your personal effects may be held by the authorities or returned to you. In the event that your personal effects are returned to you, the clothing should be laundered as normal, but as a separate load.

THE NEXT STEPS

Medical treatment and follow-up will include the following:

- The authorities will have the contents of the package tested in a laboratory within forty-eight hours to ensure that you have not been exposed to any harmful substances. In the unlikely event that you have been potentially exposed to a harmful substance you will be contacted immediately and given further instructions for appropriate medical follow-up.
- Under most circumstances, those people potentially exposed to the contents of a threat letter do NOT need any further medical evaluation by a physician or treatment (e.g., medications). However, if you develop even a mild *fever* (above 100°) before laboratory results are available you should call the county DOH or 911 and inform them about the incident and your potential exposure. Individuals who wish to consult with their local physician should have their physician contact the DOH for appropriate medical information

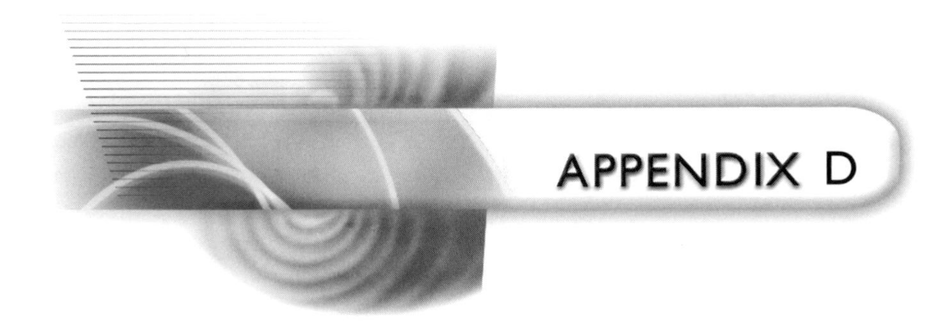

Complete Agent Descriptions

ANTHRAX

DESCRIPTION OF AGENT

Anthrax is a highly lethal infection caused by the Gram-positive bacterium, *Bacillus anthracis*. In naturally acquired cases, organisms usually gain entrance through skin wounds (causing a localized infection), but may be inhaled or ingested. Intentional release by belligerents or terrorist groups would presumably involve the aerosol route, as the spore form of the bacillus is quite stable and possesses characteristics ideal for the generation of aerosols.

SIGNS AND SYMPTOMS

The incubation period for inhalational anthrax is one to six days. Fever, malaise, fatigue, cough, and mild chest discomfort are rapidly followed by severe respiratory distress with dyspnea, diaphoresis, stridor, and cyanosis. Shock and death occur within twenty-four to thirty-six hours of the onset of severe symptoms. In cases of cutaneous anthrax, a papule develops, then vesicles, finally developing into a black eschar surrounded by moderate to severe edema. The lesions are usually painless. Without treatment, cutaneous anthrax may progress to septicemia and death, with a case-fatality rate of twenty percent. With treatment, fatalities are rare.

DIAGNOSIS

Physical findings are typically nonspecific in inhalational cases, with initial complaints of malaise, fever, headache, and possibly substemal chest pain. A widened mediastinum is sometimes seen on an x-ray late in the course of the illness, and correlates with a pathologic finding of hemorrhagic mediastinitis, the "classic" presentation of inhalational anthrax. The bacterium may be detected by a Gram stain of blood and by a blood culture late in the course of the illness.

TREATMENT

Although usually ineffective in inhalational cases once symptoms are present, antibiotic treatment with high-dose penicillin, ciprofloxacin, or doxycycline should nonetheless be administered. Although typically sensitive T2: HPBYLA to penicillin, resistant isolates are readily produced in the laboratory. For this reason, in the case of an intentional release, and in the absence of antibiotic sensitivity data, treatment should be initiated with 117V ciprofloxacin (400 mg q 8–12 hrs) or IV doxycycline (200 mg initially, followed by 100 mg q 12 hrs). Supportive therapy may be necessary.

PROPHYLAXIS

A licensed vaccine is available for use in those at risk of exposure. Vaccination is undertaken at zero, two, and fourweeks (initial series), followed by booster doses at six, twelve, and eighteen months, and then yearly. Oral ciprofloxacin (500 mg po bid) or doxycycline (100 mg po bid) are useful in cases of known or imminent exposure. Following confirmed exposure, all unimmunized individuals should receive three 0.5 ml SQ doses of vaccine over thirty days, while those vaccinated with less than three doses prior to exposure should receive an immediate 0.5 ml booster. Anyone vaccinated with the initial three-dose series in the previous six months does not require boosters. All exposed personnel should continue antibiotic therapy for four weeks. If the vaccine is unavailable, antibiotics may be continued beyond four weeks and should be withdrawn only under medical supervision.

DECONTAMINATION AND ISOLATION

Drainage and secretion precautions should be practiced. Anthrax is not known to be transmitted via the aerosol route from person to person. Following invasive procedures or autopsy, instruments and surfaces should be thoroughly disinfected with a sporicidal agent (high-level disinfectants such as iodine or 0.5 percent sodium hypochlorite).

OUTBREAK CONTROL

Although anthrax spores may survive in the environment for many years, secondary aerosolization of such spores (such as by pedestrian movement or vehicular traffic) generally presents no problem for humans. The carcasses of animals dying in such an environment should be burned, and animals subsequently introduced into such an environment should be vaccinated. Meat, hides, and carcasses of animals in affected areas should not be consumed or handled by untrained and/or unvaccinated personnel.

BRUCELLOSIS

DESCRIPTION OF AGENT

Human brucellosis is an infection caused by one of four species of Gram-negative coccobacilli of the genus *Brucella. B. abortus* is normally a pathogen of cattle, while *B. melitensis, B. suis, and B. canis* are pathogens of goats, pigs, and dogs, respectively. Organisms are acquired by humans via the oral route through the ingestion of unpasteurized milk and cheese, via inhalation of aerosols generated on farms and in slaughterhouses, or via inoculation of skin lesions in persons with close animal contact. Intentional exposure by belligerents would likely involve aerosolization, but could involve contamination of foodstuffs.

SIGNS AND SYMPTOMS

The incubation period is quite variable, with symptoms often requiring months to appear. This marked variability would appear to somewhat temper the use of brucellae as weapons. Symptoms of acute and subacute brucellosis are nonspecific and consist of irregular fever, headache, profound weakness and fatigue, chills and sweating, and generalized arthralgias and myalgias. Depression and mental status changes are noteworthy. Osteoarticular complications, particularly involving the axial skeleton (sacroiliitis, vertebral osteomyelitis) are common. Fatalities are uncommon, even in the absence of therapy.

DIAGNOSIS

Naturally occurring cases may often be suspected based on a history of close animal contact or consumption of implicated foodstuffs. Brucellae may be isolated from standard blood cultures, but require a prolonged period of incubation; cultures should thus be maintained for six weeks if brucellosis is suspected. Bone marrow cultures yield the diagnosis in a higher percentage of cases than do peripheral blood cultures. A serum agglutination test is available and often helpful.

TREATMENT

Doxycycline (100 mg po bid) plus rifampin (600-900 mg po qd) administered for six weeks is the regimen of choice for uncomplicated brucellosis. Doxycycline + streptomycin, TMP/SMX + gentamicin, and ofloxacin + rifampin are acceptable alternative regimens.

PROPHYLAXIS

Avoidance of nonpasteurized milk products and appropriate veterinary vaccination practices are sufficient to prevent most naturally occurring brucellosis. Persons inadvertently exposed to veterinary vaccine strains of brucella have been successfully prophylaxed with doxycycline + rifampin for ten days. No human brucellosis vaccine is available in the western world.

DECONTAMINATION AND ISOLATION

Drainage and secretion precautions should be practiced in patients who have open skin lesions; otherwise no evidence of person-to-person transmission of brucellosis exists. Animal remains should be handled utilizing universal precautions and disposed of properly. Surfaces contaminated with brucella aerosols may be decontaminated by standard means (0.5 percent hypochlorite).

OUTBREAK CONTROL

In the event of an intentional release of brucella organisms, it is possible that livestock will become infected. Thus, animal products in such an environment should be pasteurized, boiled, or thoroughly cooked prior to consumption. Proper treatment of water, by boiling or iodination, would also be important in an area subjected to intentional contamination with brucella aerosols.

PLAGUE

DESCRIPTION OF AGENT

Plague is an infectious disease caused by the Gram-negative, bipolarstaining bacterium, *Yersinia pestis*. Naturally occurring plague is most often acquired by the bite of a flea that had previously fed on infected rodents. In such cases, plague classically presents as a localized abscess with secondary formation of very large, fluctuant regional lymph nodes known as buboes (bubonic plague). Plague may also be transmitted via aerosols and by inhalation of sputum droplets from coughing patients. In such instances, a primary pneumonic form may develop and, in the absence of prompt therapy, progress rapidly to death within two to three days. Intentional release by belligerents or terrorist groups would presumably involve aerosolization, but could also involve the release of infected fleas. Plague may be considered a lethal agent.

SIGNS AND SYMPTOMS

Pneumonic plague has an incubation period of two to three days and begins with high fever, chills, headache, hemoptysis, and toxemia, progressing rapidly to dyspnea, stridor, and cyanosis. Death results from respiratory failure, circulatory collapse, and bleeding diatheses. Bubonic plague has an incubation period of two to ten days, and presents with malaise, high fever, and tender lymph nodes (buboes). Bubonic plague may progress spontaneously to the septicemic form, with spread to the CNS, lungs, and elsewhere.

DIAGNOSIS

To facilitate prompt therapy, plague must be suspected clinically. A presumptive diagnosis may also be made by a Gram or Wayson stain of lymph node aspirates, sputum, or CSF. The plague bacillus may be readily cultured from aspirates of buboes or from the blood of septicemic patients.

TREATMENT

Early administration of antibiotics is quite effective, but must be started within twenty-four hours of onset of symptoms in pneumonic plague. The treatment of choice is streptomycin (30 mg/kg/day IN4 in two divided doses x 10 days) or gentamicin (2 mg/kg, then 1.0–1.5 mg/kg q 8 hrs x 10 days). Intravenous doxycycline (200 mg, then 100 mg q 12 hrs x 10–14 days) is also effective; chloramphenicol should be added in cases of plague meningitis. Supportive therapy for pneumonic and septicemic forms is typically required.

PROPHYLAXIS

A licensed, killed vaccine is available. The primary vaccination series consists of a 1.0 ml IM dose initially, followed by 0.2 ml doses at one to three months, and three to six months. Booster doses are given at six, twelve, and eighteen months and then every one to two years. As this vaccine appears to offer no protection against aerosol exposure in animal experiments, victims of a suspected attack with aerosolized plague, or respiratory contacts of coughing patients, should be given doxycycline (100 mg po bid x 7 days or the duration of exposure, whichever is longer).

DECONTAMINATION AND ISOLATION

Drainage and secretion precautions should be employed in managing patients with bubonic plague; such precautions should be maintained until the patient has received antibiotic therapy for forty-eight hours and has demonstrated a favorable response to such therapy. Care must be taken when handling or aspirating buboes to avoid aerosolizing infectious material. Strict isolation is necessary for patients with pneumonic plague.

OUTBREAK CONTROL

In the event of the intentional release of plague into an area, it is possible that local fleas and rodents could become infected, thereby initiating a cycle of enzootic and endemic disease. Such a possibility would appear more likely in the face of a breakdown in public-health measures (such as vector and rodent control), which might accompany armed conflict. Care should be taken to rid patients and contacts of fleas utilizing a suitable insecticide. Flea and rodent control measures should be instituted in areas where plague cases have been reported.

TULAREMIA

DESCRIPTION OF AGENT

Tularemia is an infection caused by the Gram-negative coccobacillus, *Francisella tularensis*. Two biogroups are known. Biogroup *tularensis*, also known as type A, is the more virulent form and is endemic in much of North America. Naturally acquired tularemia is contracted through the bites of certain insects (notably ticks and deerflies) or via contact with infected rabbits, muskrats, and squirrels. Intentional release by belligerents would presumably involve aerosolization of living organisms. Although naturally acquired tularemia has a case fatality rate of approximately 5 percent, the pneumonic form of the disease, which would predominate in the setting of intentional release, would likely have a greater mortality rate.

SIGNS AND SYMPTOMS

Naturally acquired tularemia frequently has an ulceroglandular presentation, although a significant minority of cases involve the typhoidal or pneumonic forms. The incubation period averages three to five days, but varies widely. Use of tularemia as a weapon would likely lead to a preponderance of pneumonic and typhoidal cases, and large aerosolized innocula would be expected to shorten the incubation period. Ulceroglandular disease involves a necrotic, tender ulcer at the site of inoculation, accompanied by tender, enlarged regional lymph nodes. Fever, chills, headache, and malaise often accompany these findings. Typhoidal and pneumonic forms often involve significant cough, abdominal pain, substernal discomfort, and prostration in addition to prolonged fever, chills, and headache.

DIAGNOSIS

Prompt diagnosis relies on clinical suspicion. Routine laboratory tests are rarely helpful, and *F. tularensis* does not typically grow in standard blood cultures, although special media are available for the culturing (under BL-3 containment conditions) of blood, sputum, lymph node material, and wound exudates if the diagnosis is suspected. Serology is available to confirm the diagnosis in suspected cases.

TREATMENT

Streptomycin (7.5–15 mg/kg im q 12 hrs for 7–14 days) is the drug of choice for all forms of tularemia. Gentamicin (3–5 mg/kg/d q 8–12 hrs for 7–14 days) is an acceptable alternative. Relapses are more common with tetracycline therapy (500 mg po q 6 hrs for 14 days), although this alternative may be employed in patients who cannot tolerate aminoglycosides.

PROPHYLAXIS

A live, attenuated vaccine is available as an investigational product through USAMREED (Fort Detrick, MD 21702). It may be given to those individuals, such as laboratory workers, at high risk of exposure. A single dose is administered by scarification. Intramuscular streptomycin will prevent disease following documented exposure, but is not recommended following tick bites or animal contact.

DECONTAMINATION AND ISOLATION

Tularemia is not transmitted person-to-person via the aerosol route, and infected persons should be managed with secretion and drainage precautions. Heat and common disinfectants (such as 0.5 percent hypochlorite) will readily kill *F. tularensis* organisms.

OUTBREAK CONTROL

Following intentional release of *F. tularensis* in a given area, it is possible that local fauna, especially rabbits and squirrels, will acquire the disease, setting up an enzootic mammal-arthropod cycle. Persons entering such an area should avoid skinning and eating meat from such animals. Water supplies and grain in such areas might likewise become contaminated and should be boiled or cooked before consumption. Organisms contaminating soils are unlikely to survive for significant periods of time and present little hazard.

SMALLPOX

DESCRIPTION OF AGENT

Smallpox is an infection caused by the Variola virus, a member of the chordopoxvirus family. Naturally occurring smallpox has been eradicated from the globe, with the last case occurring in Somalia in 1977. Repositories of the virus are known to exist in only two laboratories worldwide. Monkeypox, cowpox, and vaccinia are closely related viruses that might lend themselves to genetic manipulation and the subsequent production of smallpox-like disease.

SIGNS AND SYMPTOMS

The incubation period of smallpox is about twelve days. Clinical manifestations begin acutely with a prodromal period involving malaise, fevers, rigors, vomiting, headache, and backache. After two to four days, skin lesions appear and progress uniformly from macules to papules to vesicles and pustules. Lesions progress centrifugally and scab in one to two weeks. In unvaccinated individuals, variola major, the classical form of the disease, is fatal in approximately 30 percent of cases.

DIAGNOSIS

In its full-blown form as typically seen in unimmunized individuals, smallpox is readily diagnosed on clinical grounds. Differentiation from other vesicular exanthems such as varicella and erythema multiforme might be difficult, however, in cases of variola minor or in disease modified by prior vaccination. Electron microscopy can readily differentiate variola virus from varicella, but not from vaccinia and monkeypox when performed on lesion scrapings. The virus can be grown in chorioallantoic membrane culture.

TREATMENT

Supportive care is the mainstay of smallpox therapy. No specific antiviral therapy exists.

PROPHYLAXIS

A licensed, live vaccinia virus vaccine is available and is administered via a bifurcated needle using a multiple puncture technique (scarification). Given the eradication of smallpox, the vaccine would only be indicated in laboratory settings or where biological warfare was a distinct possibility. Vaccination is probably protective for at least three years. Exposed persons may be managed with prompt vaccination. Vaccinia Immune Globulin (VIG), given IN4 at a dose of 0.6 ml/kg, may prove a useful adjunct to vaccination, although its precise role is unclear.

DECONTAMINATION

Given the extreme public-health implications of smallpox reintroduction, patients should be placed in strict isolation pending review by national health authorities. All material used in patient care or in contact with smallpox patients should be autoclaved, boiled, or burned.

OUTBREAK CONTROL

Smallpox has considerable potential for person-to-person spread. Thus, all contacts of infectious cases should be quarantined for sixteen to seventeen days following exposure and given prophylaxis as indicated. Animals are not susceptible to smallpox.

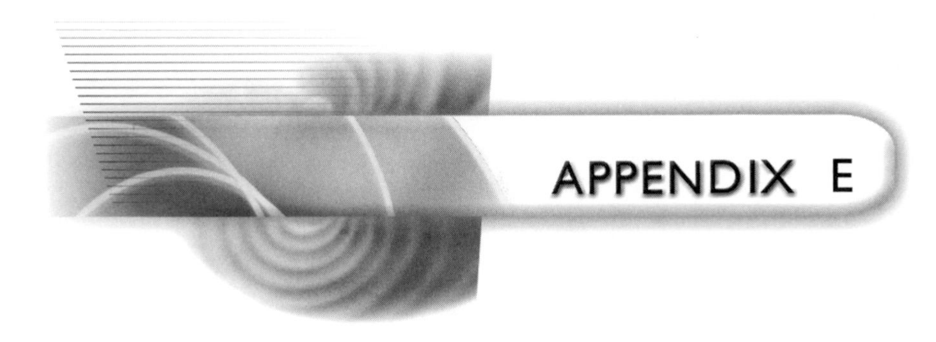

Area Hospital Listings

EXAMPLES

When developing your plan insert the proper names of local hospitals and level 1 trauma centers, as well as the number of beds available at those locations.

BEDS	
ABC Medical Center	579
XYZ Medical Center	52
Columbia Medical Center	224
123 Medical Center	107
Memorial Medical Center	3443
Methodist Hospital	144
University Medical Center (Level 1 Trauma Center)	685

APPENDIX F

County Public Health Clinics/Centers

AREA CODE 123

ABC Health Clinic
1824 N. Pearel St.
123-4567

Family Health Center
123 S. 6th Ave.
234-5678

Comprehensive Care Center
1812 Boulevard
798-450

123 Family Health Center
8264 Old Plank Road
785-4170

Community Health Services
1 University Blvd.
960-3245

Dental Clinic – Children's
515 North 6th St.
530-3571

Mobile Dental Unit – Children's
123-45672

South City Family Health Center
678 S. University Blvd.
123-4922

Taylor's Family Health Center
1220 37th Ave. North
573-4922

West Family Health Center
123 Acme St.
3567-6761

County Women's Health Center
4456 Acme St
813-4567

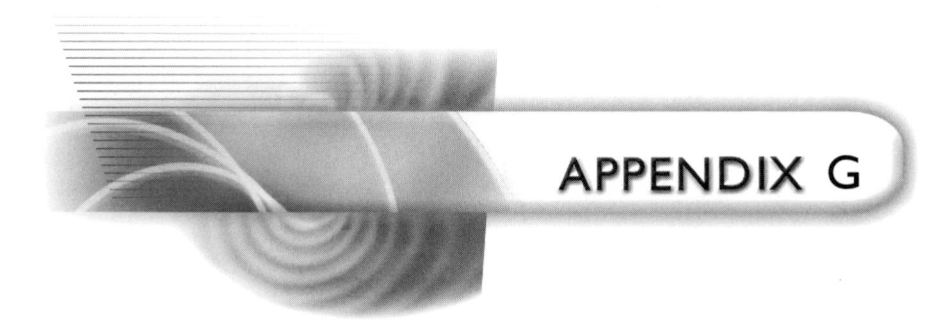

An Anthrax Threat Field Guide

INDICATORS OF POSSIBLE CHEMICAL WARFARE (CW) AGENT USAGE

UNUSUAL DEAD OR DYING ANIMALS

- Lack of insects in the air

UNEXPLAINED CASUALTIES

- Multiple victims
- Serious illnesses
- Nausea, disorientation, difficulty breathing, convulsions
- Definite casualty patterns

UNUSUAL LIQUID, SPRAY OR VAPOR

- Droplets, oily film
- Unexplained odor
- Low-lying clouds or fog unrelated to weather

SUSPICIOUS DEVICES/PACKAGES

- Unusual metal debris
- Abandoned spray devices
- Unexplained munitions

INDICATORS OF POSSIBLE BIOLOGICAL AGENT (BW) USAGE

UNUSUAL DEAD OR DYING ANIMALS

- Sick or dying animals

UNUSUAL CASUALTIES

- Unusual illness for region or area
- Definite pattern inconsistent with natural disease

UNUSUAL LIQUID, SPRAY OR VAPOR

- Spraying and suspicious devices or packages

IF RELEASE IS SUSPECTED

- Remain calm
- Don PPE
- Establish control of scene
- From a safe vantage point, reassure victims assistance is on the way
- Request properly equipped help at a safe location (upwind, uphill, upstream)
- Safely isolate and deny entry to affected area

NOTIFICATION ESSENTIALS

- Observed NBC indicators
- Wind direction and weather conditions at scene
- Plume direction (direction of cloud or vapor travel)
- Orientation of victims (direction, position, pattern)
- Number of apparent victims
- Type of injuries, symptoms presented
- Witness statements or observations
- Nature of NBC agents (if known) from detection equipment or monitors
- Suggested safe access route and staging area

INCIDENT OBJECTIVES

- Secure a perimeter and designate zones of operation (hot, warm, cold)
- Control and identify agent
- Rescue, decontamination, triage, treat, and transport affected persons
- Move uninvolved crowds or persons to safe zones
- Stabilize the incident
- Avoid secondary contamination
- Secure evidence and crime scene
- Protect against secondary attack

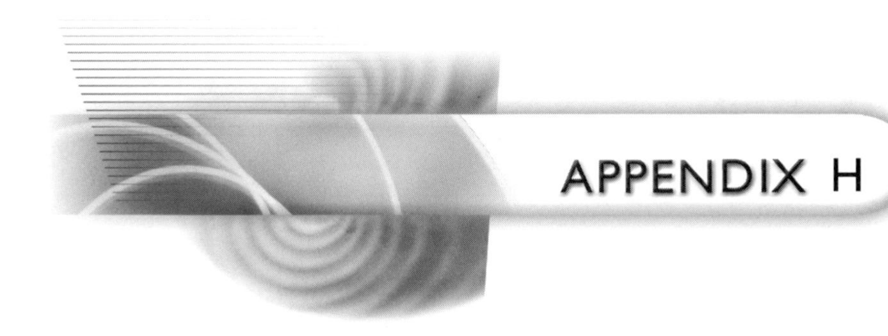

Pharmaceutical Needs for the Five Types of Biological Agents

INHALATION ANTHRAX

	Initial Therapy	Optimal Therapy	Duration
Adults, including pregnant women and immunocompromised	Ciprofloxacin, 400 mg IV q 12 hours[a]	Penicillin G[b], 4 million units IV q 4 hours or Doxycycline[c], 200 mg IV, then 100 mg IV q 12 hours	60 days[d]
Children (the high mortality rate from the infection outweighs any risk from Cipro or Doxycycline)	Ciprofloxacin, 20–30 mg/kg per day IV divided into two daily doses, not to exceed 1 g/day	Age <12 y: Penicillin G, 50,000 U/kg IV q 6 hours Age ≥12 y: Penicillin G, 4 million units IV q 4 hours or Weight ≤45 kg: Doxycycline[b], 2.5 mg/kg IV q 12 hours Weight >45 kg: Doxycycline[b], 100 mg IV q 12 hours	60 days[d]

a. Ofloxacin, 400 mg IV q 12 hours, or Levofloxacin, 500 mg IV q 24 hours, could be substituted for Cipro

b. Amoxicillin, 500 mg IV q 8 hours, is an alternative regimen

c. Tetracycline could be substituted for Doxycycline

d. Oral antibiotics should be substituted for IV antibiotics as soon as clinical condition permits, with prompt reinstitution of IV therapy if condition worsens or symptoms recur. Treatment may be shortened to 30 to 45 days IF the patient receives all three doses of the anthrax vaccine (see Anthrax Vaccine). Treatment must be initiated early and during initial symptoms: highly fatal even with early treatment

POSTEXPOSURE ANTIBIOTIC PROPHYLAXIS

	Initial Therapy[a]	Optimal if strain is susceptible	Duration
Adults, including pregnant women and immunocompromised	Ciprofloxacin[b], 500 mg po bid or Doxycycline, 100 mg po bid	Penicillin V Potassium, 30mg/kg po in four divided doses or Amoxicillin, 500 mg po q 8 hours	60 days[c]
Children	Ciprofloxacin[b], 20–30 mg/kg per day po divided into two doses, not to exceed 1 g/day or Doxycycline, 5 mg/kg/day po divided into two doses	Amoxicillin (Wt. ≥ 20 kg), 500 mg po q 8 hours Amoxicillin (Wt. < 20 kg), 40 mg/kg divided into three doses q 8 hours	60 days[c]

a. In the event of mass casualties, this regimen may need to be substituted for IV therapy for inhalation anthrax because of the logistics associated with treating large numbers with IV therapy

b. Levofloxacin, 500 mg po QD or Ofloxacin, 400 mg po bid may be substituted for Cipro

c. Treatment may be shortened to 30 to 45 days if appropriate doses of vaccine are given

Vaccine prophylaxis	Dose	Indications
Anthrax vaccine: • inactivated cell-free vaccine	0.5 ml SC at 0, 2, and 4 weeks postexposure	Indicated for all patients being treated for anthrax and for all patients who have been exposed to anthrax and are receiving prophylactic antibiotics (including children)

PRECAUTIONS

- Standard precautions

PLAGUE

	Drug of choice	Alternative
Pneumonic or bubonic plague	Streptomycin[a], 30 mg/kg/day IM in two divided doses x 10 days	Doxycycline, 200 mg, then 100 mg IV q 12 hours x 10–14 days
Plague meningitis	Chloramphenicol, 1 gm IV QID x 10–14 days, in addition to above	

a. Gentamicin may be substituted if streptomycin is not available

Treatment is highly effective when started within 24 hours of initial symptoms

PROPHYLAXIS (VACCINE)

	Primary series (3 doses)	Boosters
Plague vaccine: • licensed, killed, whole cell vaccine • Provides protection for bubonic plague, but NOT pneumonic plague (aerosolized *Y. pestis*)	Initial dose: 1.0 ml IM, then 0.2 ml IM at 1 and 6 months	0.2 ml IM at 6 month-intervals x 3 following the third dose of the primary series, then one dose every 1 to 2 years

PROPHYLAXIS (POSTEXPOSURE ANTIBIOTICS)

Indications	Drug of choice	Hypothesized, but unproven in humans
After face-to-face contact with pneumonic plague or after contamination with confirmed or suspected intentional exposure (BW attack)	Doxycycline, 100 mg po bid x 7days or the duration of the exposure, whichever is longer	Ciprofloxacin has been shown to be effective prophylaxis in exposed mice

PRECAUTIONS

- Standard precautions for bubonic plague
- Droplet precautions for pneumonic plague until patient on antibiotics for more than forty-eight hours and has a favorable response to treatment

TULAREMIA

Treatment	Drug of choice	Alternative
For all forms of tularemia (ulceroglandular, glandular, typhoidal, oculoglandular, pharyngeal, and pneumonic)	Streptomycin[a], 1 gm IM q 12 hours x 10–14 days	Tetracycline or Chloramphenicol are also effective, but associated with significant relapse rates

a. Gentamicin 3 to 5 mg/kg/day IV divided TID x 10 to14 days may be substituted for streptomycin

PROPHYLAXIS (VACCINE)

Tularemia vaccine:

- Live, attenuated vaccine is available as an Investigational, New Drug (IND)
- Given by scarification

PROPHYLAXIS (POSTEXPOSURE ANTIBIOTICS)

Indications	Drug of choice
Postexposure	Tetracycline 500 mg po qid x 14 days

PRECAUTIONS

- Standard precautions

SMALLPOX

TREATMENT

- Supportive care; currently no effective chemotherapy

POSTEXPOSURE PROPHYLAXIS

	Route/timing	Contraindications
Vaccination or revaccination with smallpox vaccine Give concurrently with Vaccinia Immune globulin (VIG), if possible	Intradermal inoculation with a bifurcated needle (scarification); effective if given up to 7 days postexposure	Contraindicated with pregnancy, immunosuppression, hx or evidence of eczema, or close contact with person(s) with one of these conditions[a]
(VIG)	Within one week of exposure	None

a. Most authorities state that, with exception of significant immune compromise, there are no absolute contraindications. However, pregnant women and those with eczema should receive concomitant VIG.

ISOLATION AND QUARANTINE PRECAUTIONS

- Droplet and airborne precautions for a minimum of seventeen days following exposure of all persons in direct contact with the index case, especially the unvaccinated
- Patients should be considered contagious until all scabs separate

BRUCELLOSIS

TREATMENT

	Drug of choice	Alternative
Acute brucellosis (adults)	Doxycycline, 200 mg po QD, plus Rifampin, 600-900 mg/day for a minimum of 6 weeks	Ofloxacin, 400 mg/day po, plus Rifampin
Brucellosis with complications (endocarditis or meningitis)	Combination therapy with rifampin, a tetracycline, and an aminoglycoside	

VACCINE

- No human vaccine available

PRECAUTIONS

- Wound precautions only

Pharmaceutical Push Package Contents

Current contents of CDC Pharmaceutical Push Package

Antibiotics	
Ciprofloxacin PO, 500mg	Erythromycin IV 500 mg
Ciprofloxacin IV, 400mg	PO Meds = 997,140 pts for 5 days
Doxycycline PO, 100mg	IV Meds = 17,435 pts for 5 days
Doxycycline IV, 100mg	

Controlled Substances
MS, 10 mg
Lorazepam 2 mg Tubex
Diazepam Auto Injectors, 10 mg
Diazepam for Injection, 10 mg 2 cc

Dressings and Bandages
Sterile Dressings 4×4
Sterile Dressings 8×10
Conforming Gauze 4" × 5 yd

Airway Supplies		
ET Tubes Size 3	Stylets peds	Laryngoscope Blades Mac 3
ET Tubes Size 4	Suction Cath 8 fr	Laryngoscope Blades Miller 2
ET Tubes Size 5	Suction Cath 14 fr	Laryngoscope Handle
ET Tubes Size 6	Suction Tracheal 14 fr	Gloves (unsterile)
ET Tubes Size 7	Portable Suction	Tape 1"
ET Tubes Size 8	OP Airway 60 mm	Nasal Cannula
NG Tubes	OP Airway 90 mm	O2 Tubing
Stylets 10 fr	BVM Adult	NR Masks Adult
Stylets 8 fr	BVM Peds	NR Masks Peds

Biological Agent Signs and Symptoms

Other Emergency Medications	
Dopamine, 400mg	Epi Pen Junior Autoinjector
Epi 1: 10,000 30 cc multidose	Mark 1 Kits
Methylprednisolone Inj., 125 mg	Atropine, 0.4 mg/cc (20 cc multidose)
Albuterol Inhaler	Pralidoxime 1 gm powder vial.
Epi Pen Autoinjector	

Fluids and IV Administration		
Heparin lock flush kits	Intravenous Set, Butterfly, 12", 21G, 3/4"	Alcohol Pads
Tubex Devices	IV Admin Set, Adult, 78", w/clamp, Vented	Swabsticks, Betadine
Syringe 10cc, 20 ga 1 1/2" needle	IV Admin Set, Adult,78", w/clamp, Unvented, Y-site	IV Site Covering (Tegaderm)
Intravenous Catheter & Needle Unit, 18G x 2"	IV Admin Set, MINI, w/clamp, Vented	Tourniquet
Intravenous Catheter & Needle Unit, 18G x 1 1/4"	IV Admin Set, MINI, w/clamp, Unvented	Double Antibiotic Oint-ment 0.3 gm packets
Intravenous Catheter & Needle Unit, 20G x 1 1/4"	NaCl 0.9%, 1000 ml	
Intravenous Catheter & Needle Unit, 24G x 1"	NaCl 0.9%, 100cc Piggyback mix	

Agent	Symptoms	Incubation	Communicable	Treatment for Symptomatic Patients	Exposed but not Symptomatic Patients
Plague[a] **Bubonic** **Pneumonic**	High fever; chills; headache; SOB; swollen, tender, inflamed lymph nodes; vomiting blood	2–6 days 1–6 days	Contact w/ pus suppurating buboes; pneumatic highly contagious	Doxycycline, 200 mg IV, then 100 mg IV q 12 hrs ×14 days	Doxycycline, 100 mg bid ×7 days Tetracycline 15–30 mg/kg daily ×7 days
Anthrax[b]	Fever; malaise; fatigue; cough w/ chest discomfort; itching skin; lesion becomes papular, then vesiculated, then depressed black eschar	2–7 days	Not generally communicable; contact with pus	Usually not effective after symptoms appear. Doxycycline, 200 mg IV, then 100 mg IV q 12 hours	Doxycycline, 100 mg bid ×4 weeks
Brucellosis[c]	Fever, headache, weakness, profuse sweating, chills, general aching, weight loss	5–60 days	Not generally communicable	Doxycycline, 200 mg with Rifampin, 600-900 mg daily ×6 wks	No treatment
Smallpox[d]	Sudden onset fever, malaise, headache, severe backache, prostration, abdominal pain, rash after 2–4 days	7–17 days	Lesion contact; airborne	Quarantine Symptomatic Treatment	Booster vaccine
Tularemia	Indolent ulcer (particularly on hands), enlarged painful lymph nodes, abdominal pain, diarrhea, vomiting	2–10 days	Not generally communicable	Streptomycin, 1 IM q 12 hours ×10–14 days	Doxycycline, 100 mg po q 12 hours ×14 days

a. Vaccine of killed bacteria is available; it is protective for about three months and prevents Bubonic infection (about 80 percent) but is ineffective against pneumonic. Vaccine not currently available in quantity. Vaccine not effective administered post exposure.

b. Cell-free vaccine of killed bacteria is available; prevents cutneous infection, not effective against inhalation. Vaccine given in six doses over 18 months; not useful in emergency response settings.

c. Strain 19 Brucella vaccine available; requires three-month lead time for response. Gaps for protection across brucella types.

d. Most older patients vaccinated; variable among younger. If vaccinated, booster is prophylactic in exposure. Live vaccine still in investigational status. Not available in quantity.

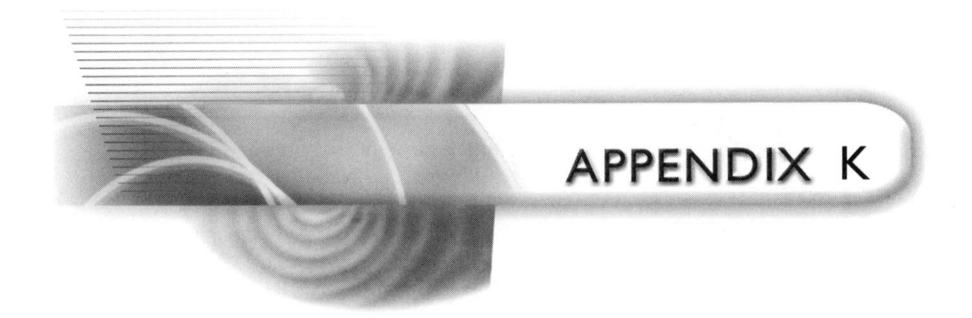

APPENDIX K

Israel's Fixed Hospital Decontamination System

Figure K–1 Folded tent and piping of fixed DECON system against hospital grounds exterior fence. When open, the system serves as gross DECON tented walkway.

Figure K–2 Looking at sidewalk system from the hospital grounds. Looking toward the street.

Figure K–3 As victims leave the tented area, they are guided to a covered area with DECON hoses and sprinklers. At this point all victims are still off hospital grounds.

Figure K–4 Victims being guided to covered area with DECON hoses.

Figure K–5 Victims being guided to covered area with DECON hoses.

Figure K–6 Victims being guided to covered area with DECON hoses.

Figure K–7 Victims being guided to covered area with DECON hoses.

Figure K–8 Drains for DECON system. In the US you must work with local and state environment departments for control of runoff.

Figure K–9 Piping system for street DECON system.

Figure K–10 DECON shower heads at emergency room entrance. ER entrance is opposite this wall. The room to the left houses the police. One hundred self-contained patient DECON kits are stored behind the doctor.

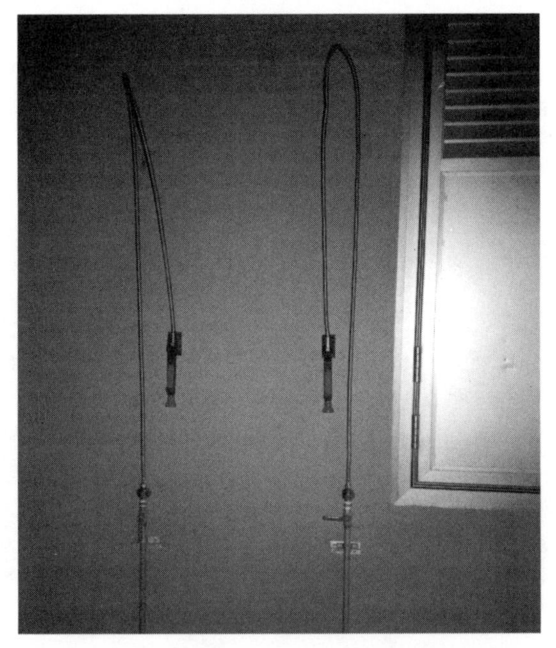

Figure K–11 Close-up of DECON showers.

Figure K–12 Close-up of DECON showers. Decontaminated victims walk through street area outside the secure hospital grounds, where gross DECON is performed, prior to entering the hospital grounds.

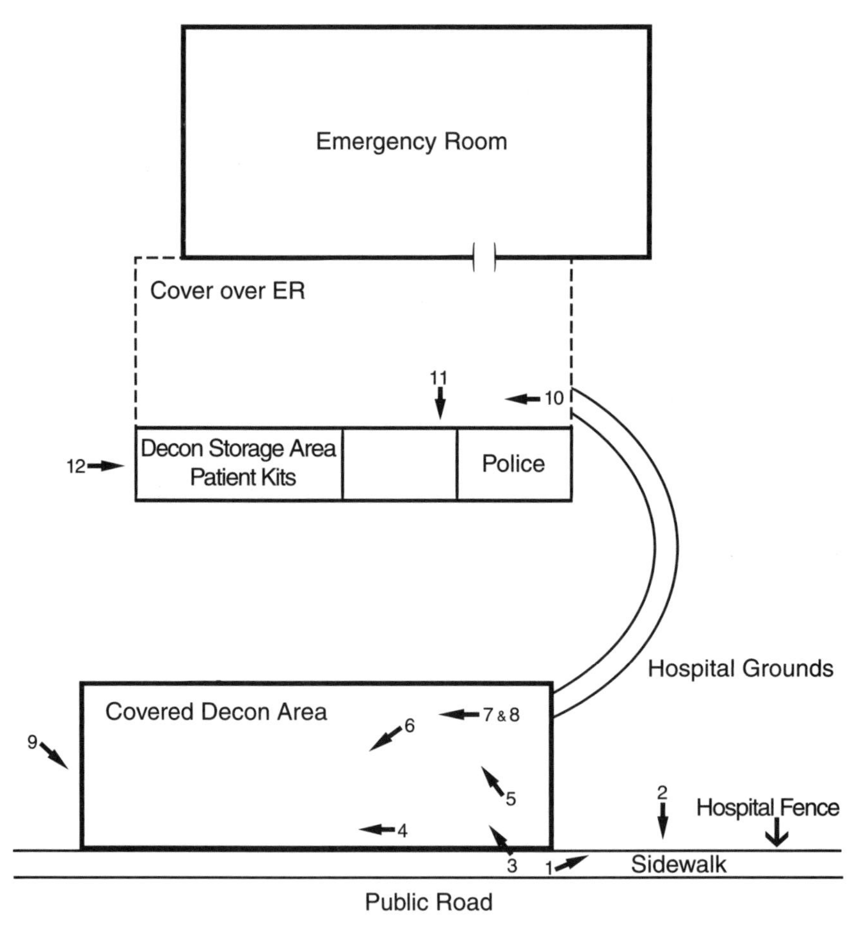

Figure K–13 Location from which photos in Figures K-1 through K-13 were shot.

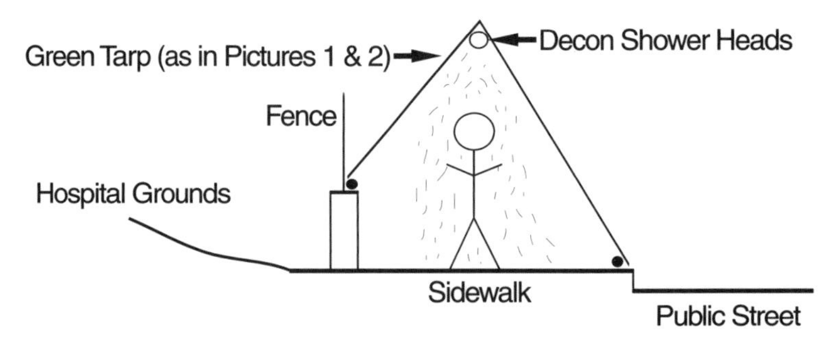

Figure K–14 DECON tent shower

Sample Domestic Preparedness Training Courses: Public Health Focus

Hour	Domestic Preparedness Coordinator	Domestic Preparedness Specialist	Epi & Environ. Health Specialist	Pre-hospital Specialist	Hospital Specialist	Other Personnel (Individual Courses)
8	Introduction to Emergency Mgmt. - ESF8	Introduction to Emergency Mgmt. - ESF8	Introduction to Emergency Mgmt. - ESF8	Introduction to Emergency Mgmt. - ESF8	Introduction to Emergency Mgmt. - ESF8	Introduction to Emergency Mgmt. - ESF8
36/4	Incident Command System	Incident Command System	Incident Command System	Incident Command System	Incident Command System for Hospitals	Incident Command System
8/40	Hazardous Material Awareness	Hazardous Material I	Hazardous Material Awareness			Hazardous Material Awareness
40		Hazardous Material Resp. Ops II				
40		Hazardous Material Resp. Tech III				
40	Emerg. Med. Svcs Ops. & Planning (EMS-WMD)	Emerg. Med. Svcs Ops. & Planning (EMS-WMD)	Emerg. Med. Svcs Ops. & Planning (EMS-WMD)			Emerg. Med. Svcs Ops. & Planning (EMS-WMD)
TBA	Lab Analysis & Sample Acquisition	Lab Analysis & Sample Acquisition	Lab Analysis & Sample Acquisition			Lab Analysis & Sample Acquisition
40		Live Agent Course				Live Agent Course
TBA	Field Exercise	Field Exercise	Field Exercise	Field Exercise	Field Exercise	
12						Food & Waterborne Disease
12						Epideminol-ogy

Coursety of Florida Deptartment of Health—Emergency Operations, August 2000, Bl Dart/excell/K:Matrix - wmd certificate course 1.

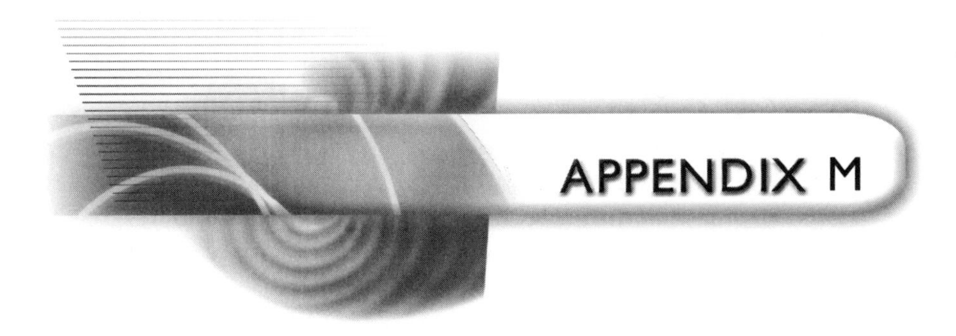

Biological Incident Preparedness Training

There is a need for biological agent awareness, biological agent incident identification, and PPE training to all first responders and appropriate health-care workers within the system.

- Augment the previous round of WMD training to include biological-only awareness training for first responders and health-care workers.
- Identify and purchase the appropriate levels of PPE for the first responders and any other identified high-risk group for this type of event.

METROPOLITAN MEDICAL RESPONSE SYSTEM (MMRS)

TRAINING REQUIREMENTS

I. INTRODUCTION

A The MMRS is intended to provide a level of technical expertise and proficiency not normally found in current civilian-emergency response organizations. To be prepared to respond to incidents involving NBC agents, MMRS personnel must be skilled in the identification of such agents. Additionally, MMRS members must be capable of self-protection, treatment, decontamination, and operation in an NBC environment.

B MMRS personnel must also be skilled in providing technical advice and information to the IC in order to protect supporting emergency-response personnel.

C MMRS personnel must also possess the knowledge required to provide instruction and assistance to the jurisdiction's supporting institutions and agencies to ensure that once casualties are removed from the incident scene, those providing care can do so effectively and without the threat of becoming secondary casualties.

II. TRAINING GUIDELINES

A Initial training will be commensurate with the member's specific assignment in the MMRS.

B Technical NBC training will be obtained from all available existing sources until appropriate training materials and programs are developed for the MMRS.

C Cross-training to enhance the operational capabilities of team members is a desired goal and will constitute an element of annual team member refresher training.

D Once team members meet initial required training objectives, training will be provided to supporting emergency-response personnel and institutions.

III. TRAINING PHILOSOPHY

A Training specified for MMRS members will ensure that each team member is sufficiently technically proficient to safely and effectively deal with the consequences of an NBC incident in addition to demonstrated proficiency in the areas of HAZMAT response, emergency medical treatment, incident management, and appropriate law enforcement areas.

B Personnel assigned to the MMRS will be required to meet all prerequisites mandated by the federal government and respective states for minimum qualifications for HAZMAT, EMTs, incident management, and law enforcement officers.

C Emergency medical and law enforcement personnel assigned to the team must be trained to HAZMAT awareness and operational levels according to their state and federal standards, and each MMRS member must receive appropriate training for the PPE that will be used.

IV. TRAINING PROGRAMS

A Check with the DOH Web site *http://www.mmrs.hss.gov* for an updated list of training programs currently available for all MMRS members.The recommended approach is to utilize existing programs to facilitate training until such time that MMRS-specific certification training programs may be developed and made available. Ideally, NBC modules will be prepared to augment existing HAZMAT programs.

B At a minimum, all training shall be in accordance with nationally recognized standards for HAZMAT emergency responders and will satisfy the OSHA requirements for emergency response to HAZMAT incidents. Where respective state or local jurisdictions have more stringent standards than those proscribed by OSHA, the more stringent standards shall be met.

V. HAZARDOUS MATERIALS TRAINING

A By definition, under OSHA, NBC agents are considered to be hazardous materials. The OSHA requirements for HAZMAT emergency response can be found in 29 CFR 1910.120 (q). The training requirements are task specific. All personnel assigned to the MMRS will receive an appropriate level of HAZMAT training.

B The NFPA has published standards that closely mirror OSHA standards but go into specific training requirements. OSHA standards are more general than NFPA ones and list minimum training hours. Experience has shown that training to comply with the NFPA will require a much greater time commitment than necessary to meet minimum OSHA standards; however, NFPA standards must be met if the MMRS members are to be proficient.

1. The following NFPA standards cover HAZMAT emergency response:

   NFPA 471—HAZMAT Response Standards

   NFPA 472—HAZMAT Training

   NFPA 473—HAZMAT Emergency Medical Services (EMS)

   Currently, the NFPA 472 standards are under revision and will include some specifics for HAZMAT EMS, if adopted.

2. The Department of Transportation (DOT), under the Hazardous Materials Transportation Act (HMTA), has published a document that identifies the

required training and recommended training for each type of emergency responder. This document, titled *Guidelines for Public Sector Hazardous Materials Responders*, also includes other recommended training proposed by the Author Group. The Author Group is coordinated under FEMA.

3. HAZARDOUS MATERIALS TRAINING REQUIREMENTS

Hazardous materials training levels are listed below as Levels 1 through 5. Each numbered level currently has a recognized training course or program. State and department equivalents are acceptable.

Level 1

OSHA—HAZMAT First Responder Awareness.
TASK—For individuals who may discover a HAZMAT incident and take no action other than notification of appropriate authorities.
NFPA—472 awareness, 473 EMS Level I.
RECOMMENDED FOR: All MMRS Team members.

Level 2

OSHA—HAZMAT First Responder Operations.
TASK—For individuals who may respond to a HAZMAT incident and operate in a defensive fashion. This includes working defensively in the Hot Zone and working in the Warm Zone (DECON).
NFPA—472 Operations, 473 EMS Levels I & II.
RECOMMENDED FOR: All MMRS Team members.

Level 3

OSHA—HAZMAT Technician.
TASK—For individuals who may respond to a HAZMAT incident and operate in an offensive fashion.
NFPA—472 Technician.
RECOMMENDED FOR: All fire service team members.

Level 4

OSHA—HAZMAT Specialist.
TASK—Same as technician, but with additional training.
NFPA—472 Technician.
RECOMMENDED FOR: Fire service HAZMAT technicians.

Level 5

OSHA—HAZMAT Incident Management.
TASK—IC, members of the general staff, and safety officers.
NFPA—472 HAZMAT Incident Management.
RECOMMENDED FOR: All MMRS command and general staff members.

VI. OTHER SELECTED EMERGENCY RESPONSE TRAINING

Other selected training levels are listed below as Levels 6 through 13.

Level 6

Confined Space Training
RECOMMENDED FOR: All fire service personnel.

Level 7

Self-Contained Breathing Apparatus Training
RECOMMENDED FOR: All MMRS Team members except medical director and hospital liaison.

Level 8

Safety Officer
RECOMMENDED FOR: All MMRS command and general staff team.

Level 9

Firefighter Levels I and II, NFPA or equivalent
RECOMMENDED FOR: All fire service personnel.

Level 10

Emergency Medical Technician – Basic
RECOMMENDED FOR: All fire service personnel.

Level 11

Emergency Medical Technician – Paramedic
RECOMMENDED FOR: All MMRS medical specialists.

Level 12

EMS HAZMAT Training
RECOMMENDED FOR: All MMRS EMTs.

Level 13

Recognized NBC Agent Training
RECOMMENDED FOR: All MMRS team members.

VII. NBC AGENT TRAINING

All MMRS members will receive training in the following technical subject areas related to NBC agents.

A Nuclear training

1. Basic nuclear radiation fundamentals
2. Self-protection

3. First aid
4. Detection and monitoring
5. Decontamination
6. Emergency medical care

B Biological training

1. Biological agents
2. Self-protection
3. First aid
4. Detection and monitoring
5. Decontamination
6. Emergency medical care

C Chemical agent taining

1. Chemical agents
2. Self-protection
3. First aid
4. Detection and monitoring
5. Decontamination
6. Emergency medical care

VIII. TRAINING DOCUMENTATION

A All MMRS members will provide copies of current training certifications to the Plans Section chief or personnel in charge of training.

B Records of team member training subsequent to documentation of initial certifications will be provided to the Plans Section chief for inclusion in the team member's MMRS membership file.

IX. TRAINING COURSES AND COURSE MATERIALS

A Some available training information and points of contacts are listed below.

B Organizations seeking training must determine the qualifications of potential training based upon careful review of the organizational status and need. Determination of qualifications should include a detailed examination of OSHA and NFPA requirements.

X. MMRS MEMBER'S PROFICIENCY TRAINING AND EXERCISES

A It is essential that the MMRS conduct team training and exercises to develop and maintain team proficiency.

B Training shall be conducted by functional elements in accordance with the training schedule below. Functional element training will include operational checks on team communication, monitoring, detection, and PPE.

C A minimum of one full team training exercise will be conducted annually. It is recommended that this exercise be integrated with jurisdictional HAZMAT or disaster exercises or with local hospital mass-casualty drills.

XI. MMRS SYSTEM HOSPITAL PROVIDER TRAINING

A Local and area hospitals will play an important role in responding to an incident involving WMD or major accidental HAZMAT incidents. A biological plan and/or MMRS Plan shall include awareness, operations; and hospital provider (WMD) training for local hospital providers.

B Hospital preparedness training should include concepts of outside triage and mass decontamination facilities, "lockdown" concepts, and providing definitive care while utilizing appropriate PPE for hospital staff.

C The DoD Domestic Preparedness Program will provide baseline information on treatment protocols. The MMRS Medical Director will implement a program of WMD awareness that can be used during hospital staff meetings, grand rounds or in-service training sessions.

D The intent is to have this program become an integral part of the hospitals orientation and annual refresher programs.

XII. NON-TEAM MEMBER TRAINING NEEDS

Training in WMD-Awareness, Operations and Technician-levels will not be limited to MMRS members or like medical response system. The following groups of responders will need training:

- Initial Responders
  1. Fire service
  2. Law enforcement
  3. HAZMAT
  4. EMS, to include first responders, EMTs, paramedics, vehicle drivers and any other personnel who might provide care for victims of a NBC incident.
  5. Emergency dispatchers
  6. Jail staff
- Medical community (hospitals, urgent care, stadium first aid, etc.) to include Emergency Room staff, security, facilities maintenance, etc.
- EM to include local government officials, DOHs, etc.

XIII. MMRS SYSTEM PERSONNEL TRAINING REQUIREMENTS

A All regional MMRS system components and jurisdictions shall be invited to sponsor student trainers for participation in an instructor training series on NBC incident preparedness, response and management. The DoD will provide this training in addition to a limited amount of training equipment.

B For those personnel who will be caring for victims of a terrorist attack involving WMD, training above and beyond the awareness program will be provided. This will include training specific to decontamination, diagnosis, and treatment with specific antidotes. This category of personnel shall include all of those engaged in the following activities:

1. Patient decontamination
2. Patient triage
3. Patient initial on scene treatment
4. Patient transportation and ongoing medical care during transport
5. Providing definitive patient care

METROPOLITAN MEDICAL RESPONE SYSTEM TRAINING AND EXERCISE SCHEDULE

SUBJECT	FREQUENCY	MINIMUM HOURS
HAZMAT Refresher Personal Protective Equipment Detection and Monitoring NBC Refresher	Annually Annually Annually Annually	8
Communication Team Activation Team External Communication Team Internal Communication	Quarterly Quarterly Quarterly Quarterly	N/A
Full-scale Exercise SARA HAZMAT Jurisdictional Disaster Hospital Mass Casualty	Annually Annually Annually Annually	8

METROPOLITAN RESPONE SYSTEM TRAINING REQUIREMENTS

POSITION	TRAINING COURSES
MMRS PROGRAM DIRECTOR	HAZMAT First Responder Awareness HAZMAT First Responder Operations HAZMAT Incident Management Firefighter Levels I & II, or equivalent Emergency Medical Technician – Basic NBC Agent Training
TASK FORCE LEADER	HAZMAT First Responder Awareness HAZMAT First Responder Operations HAZMAT Technician HAZMAT Incident Management Self-Contained Breathing Apparatus Safety Officer Firefighter Levels I & II, or equivalent Emergency Medical Technician – Basic EMS HAZMAT Training NBC Agent Training
LAW ENFORCEMENT LIAISON	HAZMAT First Responder Awareness HAZMAT First Responder Operations HAZMAT Incident Management Self-Contained Breathing Apparatus Safety Officer NBC Agent Training

POSITION	TRAINING COURSES
SAFETY OFFICER	HAZMAT First Responder Awareness HAZMAT First Responder Operations HAZMAT Technician HAZMAT Incident Management Confined Space Training Self-Contained Breathing Apparatus Safety Officer Firefighter Levels I & II Emergency Medical Technician - Basic EMS HAZMAT Training NBC Agent Training
MEDICAL DIRECTOR	HAZMAT First Responder Awareness HAZMAT First Responder Operations HAZMAT Incident Management Safety Officer EMS HAZMAT Training NBC Agent Training
HOSPITAL LIAISON	HAZMAT First Responder Awareness HAZMAT First Responder Operations HAZMAT Incident Management Safety Officer EMS HAZMAT Training NBC Agent Training
OPERATIONS SECTION CHIEF	HAZMAT First Responder Awareness HAZMAT First Responder Operations HAZMAT Technician HAZMAT Incident Management Confined Space Training Self-Contained Breathing Apparatus Safety Officer Firefighter Levels I & II Emergency Medical Technician - Basic EMS HAZMAT Training NBC Agent Training
FIELD MEDICAL GROUP SUPERVISOR	HAZMAT First Responder Awareness HAZMAT First Responder Operations HAZMAT Technician Confined Space Training Self-Contained Breathing Apparatus Firefighter Levels I & II Emergency Medical Technician - Basic Emergency Medical Technician – Paramedic EMS HAZMAT Training NBC Agent Training

POSITION	**TRAINING COURSES**
MEDICAL SPECIALISTS (FIELD MEDICAL & TEAM MEDICAL)	HAZMAT First Responder Awareness HAZMAT First Responder Operations HAZMAT Technician Confined Space Training Self-Contained Breathing Apparatus Firefighter Levels I & II Emergency Medical Technician – Paramedic EMS HAZMAT Training NBC Agent Training
FIELD HAZMAT GROUP SUPERVISOR	HAZMAT First Responder Awareness HAZMAT First Responder Operations HAZMAT Technician HAZMAT Specialist Confined Space Training Self-Contained Breathing Apparatus Firefighter Levels I & II Emergency Medical Technician – Basic EMS HAZMAT Training NBC Agent Training
FIELD HAZMAT SPECIALIST	HAZMAT First Responder Awareness HAZMAT First Responder Operations HAZMAT Technician HAZMAT Specialist Confined Space Training Self-Contained Breathing Apparatus Firefighter Levels I & II Emergency Medical Technician – Basic EMS HAZMAT Training NBC Agent Training
PLANS SECTION CHIEF	HAZMAT First Responder Awareness HAZMAT First Responder Operations HAZMAT Technician HAZMAT Incident Management Confined Space Training Self-Contained Breathing Apparatus Safety Officer Firefighter Levels I & II Emergency Medical Technician - Basic EMS HAZMAT Training NBC Agent Training

POSITION	TRAINING COURSES
PLANS GROUP SUPERVISOR	HAZMAT First Responder Awareness HAZMAT First Responder Operations HAZMAT Technician Confined Space Training Self-Contained Breathing Apparatus Firefighter Levels I & II Emergency Medical Technician - Basic EMS HAZMAT Training NBC Agent Training
INFORMATION SPECIALISTS (Intelligence, HAZMAT, and Medical)	HAZMAT First Responder Awareness HAZMAT First Responder Operations Self-Contained Breathing Apparatus EMS HAZMAT Training NBC Agent Training
LOGISTICS SECTION CHIEF	HAZMAT First Responder Awareness HAZMAT First Responder Operations HAZMAT Technician HAZMAT Incident Management Confined Space Training Self-Contained Breathing Apparatus Safety Officer Firefighter Levels I & II Emergency Medical Technician - Basic EMS HAZMAT Training NBC Agent Training
LOGISTICS GROUP SUPERVISOR	HAZMAT First Responder Awareness HAZMAT First Responder Operations HAZMAT Technician Confined Space Training Self-Contained Breathing Apparatus Firefighter Levels I & II Emergency Medical Technician - Basic EMS HAZMAT Training NBC Agent Training
LOGISTICIANS	HAZMAT First Responder Awareness HAZMAT First Responder Operations HAZMAT Technician Confined Space Training Self-Contained Breathing Apparatus Firefighter Levels I & II Emergency Medical Technician - Basic EMS HAZMAT Training NBC Agent Training

POSITION	TRAINING COURSES
COMMUNICATIONS GROUP SUPERVISOR	HAZMAT First Responder Awareness HAZMAT First Responder Operations HAZMAT Technician Confined Space Training Self-Contained Breathing Apparatus Firefighter Levels I & II Emergency Medical Technician - Basic EMS HAZMAT Training NBC Agent Training
COMMUNICATION SPECIALIST	HAZMAT First Responder Awareness HAZMAT First Responder Operations HAZMAT Technician Confined Space Training Self-Contained Breathing Apparatus Firefighter Levels I & II Emergency Medical Technician - Basic EMS HAZMAT Training NBC Agent Training
TEAM MEDICAL SEC-TION CHIEF	HAZMAT First Responder Awareness HAZMAT First Responder Operations HAZMAT Incident Management Self-Contained Breathing Apparatus Safety Officer EMS HAZMAT Training NBC Agent Training

SOURCES FOR TRAINING METROPOLITAN MEDICAL RESPONE SYSTEM

NBC agent training can also be obtained from various sources including, but not limited to the following: DoD, federal and state governments, and commercial vendors including colleges and universities.

PLANNING AND TRAINING

Figure M-1 shows suggested planning and training sequence for a biological incident.

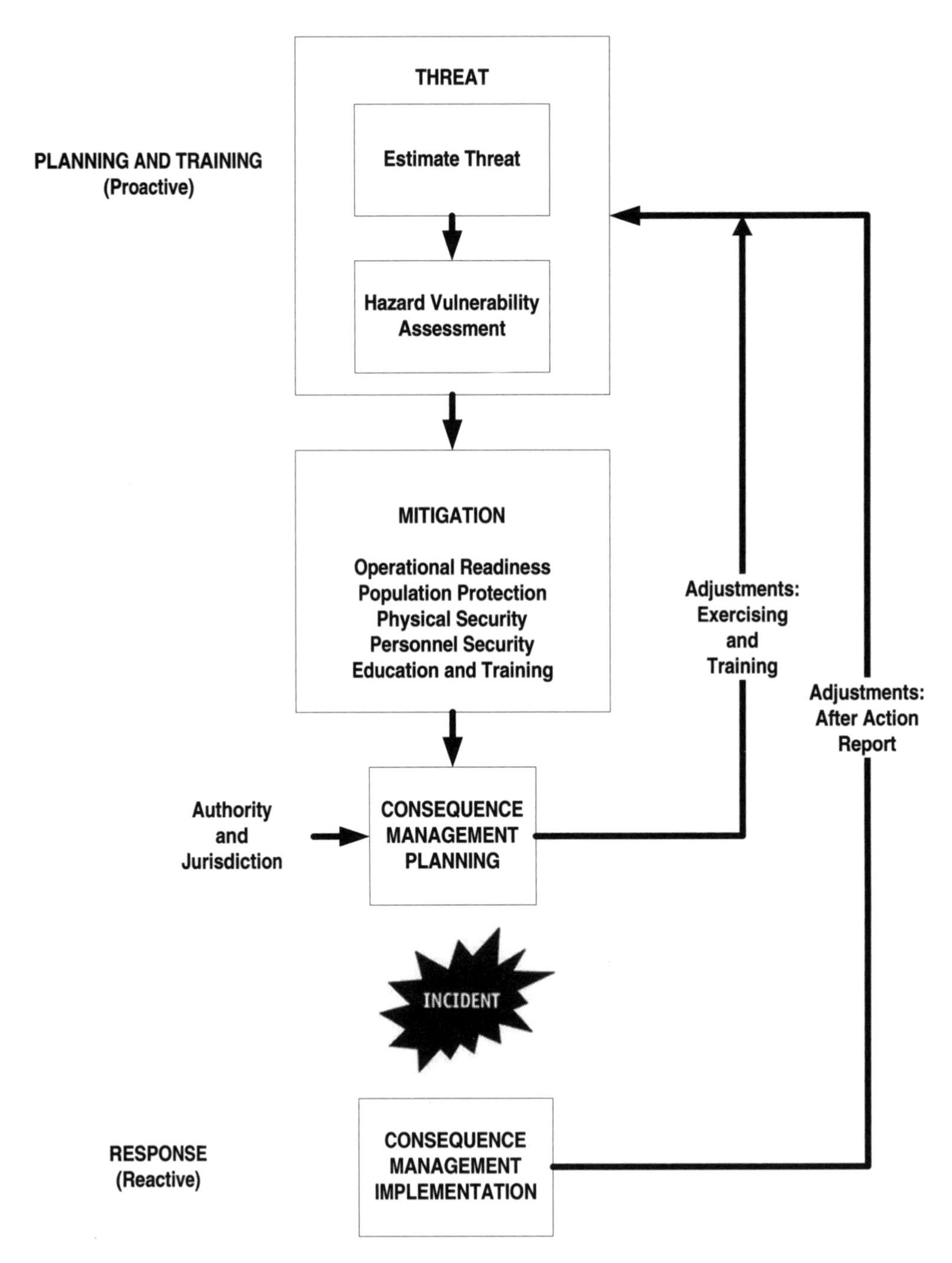

Figure M–1 Incident planning and training

Sample School Emergency Plan

The following is a sample EOP for a preschool or school. This plan was developed and implemented first in Burbank, California by Fire Marshal Darryl Forbes and the local school district. It was modified for the state of Washington by Kitsap County Emergency Manager Phyllis Mann. It is an all-hazards emergency plan and can be adopted and modified by any school. For more information contact the Kitsap County Emergency Management Department. The plan has eight teams and is a great place for schools to start in developing or modifying their emergency plan. Add anannex for specific threats such as a WMD event, gun on campus, earthquake, etc.

EMERGENCYOPERATIONS CENTER (EOC)

PERSONNEL

RESPONSIBILITIES

Location of EOC to be determined by the Site Administrator (SA) in charge of disaster operation based on nature and impact of disaster.

SA SHALL

Declare an emergency

Implement plan and coordinate operations

Account for all students and staff

Establish communications with the district EOC

Assess situation and request needed resources

Control all internal communications

Approve and release any press statements or other external communications

Keep log of communications, decisions, actions

Post current status information; update regularly

Determine when emergency operations cease

Prepare a report to superintendent on disaster operations

Other

EQUIPMENT/SUPPLIES

Building emergency plan

Map of school site

Map of area

Radios, bullhorn

Battery-operated radio

Emergency lighting

FORMS

Message

Resource request

Log of emergency operations

STUDENT/STAFF SUPERVISION TEAM

PERSONNEL

RESPONSIBILITIES

Immediately after the imminent threat subsides,

- Determine as well as possible the extent of any injuries
- Assess the ability of everyone to evacuate
- Determine need to call for medical assistance
- Check your pre-established next-room teacher or buddy
- Evacuate classroom using assigned evacuation route or, if needed use alternate route
- Lead class to assembly area; find assigned class location
- Take roll and report class status to EOC; account for ALL students or report missing students and staff to EOC
- Stay with class to supervise, inform, play, and reassure students throughout the duration of the emergency.

EQUIPMENT/SUPPLIES

- Class roll
- Class-status form
- Missing-student form
- Classroom emergency kit
- Games, books, cards

FIRST AID TEAM

PERSONNEL

RESPONSIBILITIES

Administer first aid

Record information about injuries and first aid administered

Determine need for medical assistance

Assure availability of necessary first aid supplies and equipment

EQUIPMENT/SUPPLIES

First aid supplies

Emergency cards

Health cards

Stretchers

Blankets

Water

Other

FIRE/UTILITY TEAM

PERSONNEL

RESPONSIBILITIES: FIRE

Check utilities according to pre-assigned area of responsibility

Take whatever action is necessary to minimize danger and further damage

Assess what services are still available—water, electricity, telephone, sewer lines, heating/ventilation system, bathrooms, etc.

Report assessment to EOC

Survey and report on extent of apparent structural and site damage to EOC

Other

EQUIPMENT/SUPPLIES

Fire extinguisher

Hard hat

Vest

Gloves

Axe

OTHER

Utility shut-off tools

FORMS

Utilities damage and shut-off survey checklist

Building/Site damage survey checklist

Resource-needs assessment

SEARCH AND RESCUE TEAM

PERSONNEL

RESPONSIBILITIES

Assign specific areas to each team

Follow an orderly, pre-established sweep pattern

Check each classroom, storage room, auditorium, bathroom outdoor area, etc.

Check each area three ways: visually, vocally, and physically

Report location of victims to first aid team as soon as feasible

Record location of victims on checklist

Look for obvious problems as you go—structural damage, hazardous materials spills, and fires

Report imminent danger to search-and-rescue team leader as soon as feasible

Record nature and location of observed problem on checklist

Mark areas searched upon completion to avoid duplication or oversight

ALWAYS STAY WITH YOUR TEAM PARTNER

Other

EQUIPMENT/SUPPLIES

Hard hat, vest

Sturdy shoes/Boots

Leather gloves

Dust mask

Master keys

Two-way radio/Crowbar/Saw

Fire extinguisher

Blankets

Duct tape or other method of marking area searched

Clipboard and checklist

FORMS

Victim located message

Imminent danger location message

SUPPORT/SECURITY TEAM

PERSONNEL

RESPONSIBILITIES

Immediately following evacuation, secure all site buildings; lock doors and gates

Cordon off any areas with apparent structural damage or other danger

Station team members at obvious school access points to direct parents, fire, rescue, police, medical aid personnel to appropriate area

Be prepared to receive neighbors who wish to volunteer

Set up sanitation facilities

Monitor sanitation and properly care for waste until it can be disposed of completely

Gather all food and water supplies for distribution as needed

Set up areas for sheltering, sleeping, and eating

EQUIPMENT/SUPPLIES

Master keys

Two-way radios

Barricades, rope, tape

Prewritten place cards and signs

Site diagrams with each important area clearly indicated

Written instruction to parents

Volunteer job descriptions

Toilet facilities: black plastic, duct tape, wire, toilet paper, toilet seat

STUDENT/PARENT REUNION TEAM

PERSONNEL

RESPONSIBILITIES

Establish reunion points at each school assess location

As soon as parents arrive, begin to process release of students to their parents or other authorized guardian

Check student identification card to assure adult is authorized to take child

Ask for identification; be sure to confirm that each student recognizes the adult as a legitimate, authorized adult

Other

EQUIPMENT/SUPPLIES

Student authorization cards

Class roster

Map of class location within assembly area

DAMAGE ASSESSMENT TEAM

PERSONNEL

RESPONSIBILITIES

Check areas of evacuation for safety; advise alternate routes if necessary

Check student assembly areas for gas and water pipe ruptures, downed power lines, etc.

Inspect all school buildings for damage; report structural problems, cracks in walls, damage to classrooms, science labs and other areas to EOC

Provide barriers for dangerous areas with barrier tape and danger signs

EQUIPMENT/SUPPLIES

Hard hat

School map

Damage-assessment forms

Barrier tape

Internet Resources for Terrorism/ Disaster Planning

For those agencies with access to the Internet there is a wealth of information available to assist you in the development and research of your plan. Listed below are a few links related to terrorism.

TERRORISM LINKS

Center for Disease Control and Prevention (CDC). Health related hoaxes and rumors.
http://www.cdc.gov/hoax_rumors.htm

U.S. State Department International Policy: Counterterrorism. Counterterrorism information site for the U.S. Department of State. Contains official statements, fact sheets, and speeches.
http://www.state.gov/www/global/terrorism/index.html

U.S. State Department "Patterns of Global Terrorism" Reports. The Hellenic Resources Institute offers the 1993, 1994 and 1995 "Patterns of Global Terrorism" reports on their site. These reports list the terrorist incidents during the year, as well as background information on the perpetrators.
http://www.hri.org/docs/USSD-Terror/

U.S. Department of State Travel Warnings: Consular Information Sheets. Official U.S. Department of State travel advisories for each country in the world. Also contains brief information on medical facilities, crime, infrastructure, legal issues, embassy locations, and other information useful for travelers.
http://travel.state.gov/travel_warnings.html

The Canadian Department of Foreign Affairs and International Trade's Travel Information & Advisory Reports. Travel information and advisories from the Canadian government.
http://www.voyage.gc.ca/destinations/menue_e.htm

The British Foreign and Commonwealth Office Travel Advice Notices. Material intended to advise and inform British citizens travelling abroad. Key official information from USIA concerning counterterrorism. Very similar to the U.S. Department of State Counterterrorism site.
http://www.fco.gov.uk/

U.S. Department of State HEROES
http://www.heroes.net/pub/heroes/index.html

FEMA Backgrounder and Fact Sheet on Terrorism. Basic information on terrorism from the FEMA. The fact sheet lists what to do before, during, and after a terrorist attack.
http://www.fema.gov/library/terror.htm

RCMP—Air India Bombing Disaster. The Royal Canadian Mounted Police's site on the 1985 Air India bombing disaster.
http://www.airindia.istar.ca

Canadian Security Intelligence Service 1996 Public Report. Canada's counterterrorism program.
http://www.csis-scrs.gc.ca

GOVERNMENT LINKS

Central Intelligence Agency
http://www.odci.gov/
http://www.cia.gov/

U.S. Department of State
http://www.state.gov/

U.S. Department of Justice
http://www.usdoj.gov/

Federal Bureau of Investigation
http://www.fbi.gov/

U.S. Marshals Service
http://www.usdoj.gov/marshals/

Immigration and Naturalization Service
http://www.usdoj.gov/ins/ins.html

U.S. Department of the Treasury
United States Secret Service
Bureau of Alcohol, Tobacco and Firearms
United States Customs Service
Office of Foreign Assets Control
http://www.treas.gov/sitemap.html

U.S. Department of Transportation
http://www.dot.gov

Federal Aviation Administration
http://www.faa.gov

National Security Agency
http://www.nsa.gov

U.S. Postal Inspection Service
http://www.usps.gov/websites/depart/inspect/

Canadian Security Intelligence Service
http://www.csis-scrs.gc.ca/

ACADEMIC AND INSTITUTIONS

Center for Democracy and Technology's Counterterrorism Issues. The CDT's page contains information and analysis relating to counterterrorism legislation.
http://www.cdt.org/policy/terrorism/

Electronic Privacy Information Center Counter-Terrorism Proposals. The Electronic Privacy Information Center is a resource for information on counterterrorism proposals and the restriction of free speech.
http://www.epic.org/privacy/terrorism/

Enough Terrorism. A project associated with the Terrorism Research Institute to educate the public, and help stop terrorism and gang/drug violence.
http://www.terrorism.org/

International Association of Counterterrorism and Security Professionals. The IACSP site has information concerning their organization as well as features from their magazine, *Counterterrorism and Security.*
http://www.worldonline.net

Censorship and Privacy: Terrorism Hysteria and Militia Fingerpointing Archive. The Electronic Frontier Foundation has information regarding legislative counterterrorism efforts that use "'terrorism' and 'militias' as excuses" for censorship.
http://www.eff.org/pub/Privacy/Terrorism_militias

Anti-Defamation League. The ADL is the leading organization against anti-Semitism and its site has information on terrorism, hate crimes, skinheads, militias, and extremists.
http://www.adl.org/

COUNTERTERRORISM LINKS

Terrorist Groups Profiles. Information about groups from the Naval Postgraduate School.
http://web.nps.navy.mil/~library/tgp/tgpndx.htm

The ERRI Counterterrorism Page. The Emergency Response & Research Institute has an excellent collection of news and summaries of world-wide terrorism events, groups, terrorist strategies and tactics.
http://www.emergency.com/cntrterr.htm

Package Bomb Indicators. Information concerning mail bomb indicators, courtesy of the ERRI and the U.S. Postal Inspector's Office.
http://www.emergency.com/pkgbomb.htm

Kim-spy: Paramilitary and Terrorism. The Kim-spy site contains a wide variety of links to paramilitary and terrorism information on Internet.
http://www.kimsoft.com/kim-spy2.htm

Vulnerability of the United States to Chemical and Biological Terrorism. EAI Corporation explains the vulnerability of the United States to terrorists' use of chemical and biological weapons.
http://eaicorp.com/cbvuln.html

Terror in Dhahran: Saudi Bomb Blast. CNN provides information on the blast, investigation, victims, and impact and analysis.
http://cnn.com/WORLD/1996/saudi.special/index.html

Olympic Park Bombing Special Section. CNN provides information on the blast, investigation, victims, and impact and analysis.
http://www.cnn.com/US/9607/27/olympic.bomb.main/index.html

The Police Officer's Internet Directory. A large collection of information relating to law enforcement.
http://www.officer.com/

BIOLOGICAL TERRORISM LINKS

Center for Disease Control
http://www.CDC.gov

Emergency First Responder Equipment Guides, October 2001. The National Institute of Justice (NIJ) is creating a series of guides for first-responder equipment that will provide agencies with information on the types and capabilities of available equipment. While only the first in this series has been published, NIJ is releasing working drafts of the remaining guides in response to the September 11th terrorist attacks as soon as they are available. These draft guides are only available electronically on NIJ's Web site and are subject to change prior to being posted in their final form.

Selection of Chemical Agent and Toxic Industrial Material Detection Equipment for Emergency First Responders, NIJ Guide 100-00

An Introduction to Biological Agent Detection Equipment for Emergency First Responders, NIJ Guide 101-11 (draft)

Guide for the Selection of Chemical and Biological Decontamination Equipment for Emergency First Responders, NIJ Guide 103-00 (draft)

Guide for the Selection of Communication Equipment for Emergency First Responders, NIJ Guide 104-00 (draft)

Guide for the Selection of Personal Protective Equipment for Emergency First Responders (working draft in progress)

http://www.ojp.usdoj.gov/nij/pubs-sum

American Red Cross—Terrorism: Preparing for the Unexpected
http://www.redcross.org/services/disaster/keepsafe/unexpected.html

Federal Response Plan April 1999 (includes revised Terrorism Incident Annex)
http://www.fema.gov/r-n-r/frp/

Texas Engineering Extension Service, Texas A & M University
http://www.tamu.edu/teex/

New Mexico Institute of Mining and Technology, Energetic Materials Research and Testing Center
http://www.emrtc.nmt.edu/

FEMA's Rapid Response Information System
http://www.rris.fema.gov/

CalPoly Chemical & Biological Warfare Course
http://projects.sipri.se/cbw/

FEMA/National Fire Academy Course "Emergency Response to Terrorism: Incident Management"
http://www.usfa.fema.gov/nfa/tr6m1.htm

FEMA/Emergency Management Institute Course "IS 195–Basic Incident Command System"
http://www.usfa.fema.gov/nfa/tr_ertss4.htm

CRITICAL INFRASTRUCTURE PROTECTION LINKS

The Infrastructure Protection Task Force (IPTF)
http://www.fbi.gov/contact/fo/norfolk/nipc.htm

MIPT—Oklahoma City National Memorial Institute for the Prevention of Terrorism
http://www.mipt.org

Department of Defense
http://www.defenselink.mil/

Office of Justice Programs, Office for State and Local Domestic Preparedness Support
http://www.ojp.usdoj.gov/osldps/

U.S. Army Soldiers and Biological Chemical Command
http://www.sbccom.apgea.army.mil/

State Health Departments
http://www.cdc.gov/search2.htm

State Emergency Management Agencies
http://www.ndpo.gov/stateema.htm

Alaska Department of Health BT Information
http://www.chems.alaska.gov/bioterrorism_home.htm

Health-care Association of Hawaii, Emergency Management Program
https://www.hah-emergency.net/

Los Angeles County Department of Health Services (See BT subheading)
http://lapublichealth.org/acd/

New Mexico Weapons of Mass Destruction Preparedness
http://www.wmd-nm.org/

New York City Department of Health
http://www.ci.nyc.ny.us/html/doh/home.html

Texas Department of Health BT Information
http://www.tdh.texas.gov/bioterrorism/default.htm

The Medical NBC Battlebook, US Army Center for Health Promotion and Preventive Medicine (USACHPPM)
http://chppm-www.apgea.army.mil/imo/ddb/dmd/DMD/TG/TECHGUID/batbooka.PDF

Metropolitan Medical Response System, Field Operating Guide
http://www.ndms.dhhs.gov/CT_Program/Response_Planning/response_planning.html

ANTHRAX ARTICLES ON THE WEB

Anthrax. Dixon, T. C. et al., New England Journal of Medicine, September 9, 1999
http://content.nejm.org/cgi/content/full/341/11/815

Anthrax as a biological weapon: Medical and pubic health management. Inglesby, T. V. et al., JAMA, May 12, 1999
http://jama.ama-assn.org/issues/v281n18/ffull/jst80027.html

Inhalational anthrax: Epidemiology, diagnosis, and management. Shafazand, S. et al., Chest, November, 1999
http://www.chestjournal.org/cgi/content/full/116/5/1369

Bioterrorism alleging use of anthrax and interim guidelines for management. MMWR, February 5, 1999
http://www.cdc.gov/mmwr/PDF/wk/mm4804.pdf

ANTHRAX WEB SITES

ASAP at GW Medical Center: Special Lectures and Press Conferences
http://www.gwumc.edu/asap/index2.htm

AMA Message to Physicians on Anthrax
http://www.ama-assn.org/ama/pub/category/6383.html

Basic Laboratory Protocols for the Presumptive Identification of *bacillus anthracis*
http://www.bt.cdc.gov/Agent/Anthrax/Anthracis20010417.pdf

CDC Health Advisory: How To Handle Anthrax and Other Biological Agent Threats, October 12, 2001
http://www.bt.cdc.gov/DocumentsApp/Anthrax/10122001Handle/10122001Handle.asp

CDC MMWR: Update—Investigation of Anthrax Associated with Intentional Exposure and Interim Public Health Guidelines, October 2001
http://www.cdc.gov/mmwr/preview/mmwrhtml/mm5041a1.htm

Centers for Disease Control and Prevention: Anthrax
http://www.cdc.gov/ncidod/dbmd/diseaseinfo/anthrax_g.htm

SMALLPOX ARTICLES ON THE WEB

Smallpox: Clinical and epidemiologic features. Henderson, D. A., Emerging Infectious Diseases, July–August 1999
http://www.cdc.gov/ncidod/EID/vol5no4/henderson.htm

Smallpox as a biological weapon: Medical and public health management. Henderson, D. A., et al., JAMA, 1999 Jun 9
http://jama.ama-assn.org/issues/v281n22/ffull/jst90000.html

SMALLPOX WEB SITES

Vaccinia (smallpox) Vaccine Recommendations of the Advisory Committee on Immunization Practices (ACIP)
http://www.cdc.gov/mmwr/preview/mmwrhtml/rr5010a1.htm
http://www.cdc.gov/mmwr/PDF/RR/RR5010.pdf

Centers for Disease Control and Prevention: Smallpox
http://www.bt.cdc.gov/Agent/Smallpox/Smallpox.asp

Medical Management of Biological Casualties Handbook, Chemical and Biological Defense Information Analysis Center
http://www.nbc-med.org/SiteContent/HomePage/WhatsNew/MedManual/

PLAGUE ARTICLES

Plague as a biological weapon: Medical and public health management. Inglesby, T. V. et al., JAMA, 2000
http://jama.ama-assn.org/issues/v283n17/ffull/jst90013.html

PLAGUE WEBSITES

Centers for Disease Control and Prevention: Plague
http://www.bt.cdc.gov/Agent/Plague/Plague.asp

Basic Laboratory Protocols for the Presumptive Identification of Yersinia Pestis (Plague)
http://www.bt.cdc.gov/Agent/Plague/Plague20010417.pdf

GENERAL INFORMATION

American Academy of Pediatrics. Chemical-biological terrorism and its impact on children: A subject review.
Pediatrics, March 2000
http://www.aap.org/policy/re9959.html

American College of Physicians—American Society of Internal Medicine Bioterrorism Resources
http://www.acponline.org/bioterro/index.html?hp

American Society of Microbiology: Resources Related to Biological Weapons Control and Bioterrorism Preparedness
http://www.asm.org/pcsrc/bioprep.htm

CDC MMWR: Recognition of Illness Associated with the Intentional Release of a Biologic Agent
http://www.cdc.gov/mmwr/preview/mmwrhtml/mm5041a2.htm

Centers for Disease Control & Prevention: Agent List—Biological Diseases/Chemical Agents
http://www.bt.cdc.gov/Agent/AgentList.asp

Centers for Disease Control and Prevention Health Advisory: How To Handle Anthrax

and Other Biological Agent Threats, October 12, 2001
http://www.bt.cdc.gov/DocumentsApp/Anthrax/10122001Handle/10122001Handle.asp

Medical Management of Biological Casualties Handbook, Chemical and Biological (Defense Information Analysis Center)
http://www.nbc-med.org/SiteContent/HomePage/WhatsNew/MedManual/Sep99/Current/sep99.htm

The Center for Civilian Bio-Defense Studies, John Hopkins University
http://www.hopkins-biodefense.org

The Centers for Disease Control and Prevention—Bioterrorism Preparedness and Response
http://www.bt.cdc.gov

The Center for Research on the Epidemiology of Disasters
http://www.md.ucl.ac.be/cred/front_uk.htm

The Defense Threat Reduction Agency—Chem-Bio Defense
http://www.dtra.mil

The Department of Health and Human Services, Office of Emergency Preparedness, National Disaster Medical System
http://ndms.dhhs.gov/NDMS/ndms.html

The Department of Health and Human Services, Office of Emergency Preparedness (OEP)
http://ndms.dhhs.gov

The Department of Defense, Nuclear, Biological, Chemical Medical References
http://www.nbc-med.org/others

Disaster Medicine and Medical Health
http://www.mentalhealth.org/cmhs/EmergencyServices

The Disaster Center
http://disastercenter.com

Center for Nonproliferation Studies
http://www.cns.miis.edu

Hospital Emergency Incident Command System (HSICS III)
http://emsa.ca.gov/dms2/history.htm

International Association of Emergency Managers
http://www.iaem.com

International Critical Incident Stress Foundation (ICISF)
http://icisf.org

United States Army Medical Research institute of Infectious Diseases (USAMRIID)
http://www.usamriid.army.mil

United States Army National Guard Bureau
http://www.ngb.dtic.mil

United States Army Soldier and Biological Chemical Command (SBCCOM)
http://www.sbccom.apgea.army.mil

Anthrax (Department of Defense)
http://www.anthrax.osd.mil.

STATE EMERGENCY OPERATION PLANS

Arizona
www.dem.state.az.us/serrp

Arkansas (State Plan)
www.adem.state.ar.us

California (Plans and Publications)
www.oes.ca.gov

Colorado
www.dlg.oem2.state.co.us/oem/Publications/BASPLAN.pdf

Delaware
www.state.de.us/dema/EMPLANS/deop/basic.pdf

Florida
www.dca.state.fl.us/bpr/Projects/CEMP%20Online/cemp2000.htm

Hawaii
scdwebmaster@scd.state.hi.us

Kansas (request via email)
Jleichem@AGTOP.STATE.KS.US

Kentucky
webserve.dma.state.ky.us/KY%20EOP/emergency_operations_plan_home.htm

Louisiana
199.188.3.91/Plans/LAEmergAssistDistAct.pdf

Michigan
www.msp.state.mi.us/division/emd/publst.htm

Minnesota
www.dps.state.mn.us/emermgt/EOP/MEOP/index.html

Nebraska
www.nebema.org/opns.html

New Jersey
www.state.nj.us/lps/njsp/ems/ems.html

New York (not available online)
ftp.nysemo.state.ny.us

North Carolina
www.dem.dcc.state.nc.us

South Carolina (SC Plans)
www.state.sc.us/epd

New Mexico
www.dps.nm.org/emergency/Plan/cover.htm

Texas
www.capitol.state.tx.us/statutes/go/go041800toc.html

Utah (request via email)
frontdsk@dps.state.ut.us

Virginia
www.vdem.state.va.us/library/eplan.cfm

Washington
www.wa.gov/wsem/2-ops/ops-plans/eop/eop-idx.htm

ACRONYMS

AAPCC	American Association of Poison Control Centers
AAR	After-Action Report
ABG	arterial blood gas
AC	hydrocyanic acid, chemical formula CHN
AC	hydrogen cyanide, (same compound) CHN
ACGIH	American Conference of Governmental Industrial Hygienists
ACH	acetylcholine
ACHE	acetylcholinesterase
ACI	Administrative/Communication/Information
ACLS	Advanced Cardiac Life Support
ACT	Area Command Teams
AEA	Atomic Energy Act
AFCEMA	Atlanta-Fulton County Emergency Management Agency
AFD	Atlanta Fire Department
AG	Attorney General
AHA	American Heart Association
AIHA	American Industrial Hygiene Association
ALS	advanced life support
ALT	alanine aminotransferase
AMA	Atlanta Metropolitan Area
AMC	Alternative Medical Centers
AMR	American Medical Response
ANG	Army National Guard
AO	Administrative Officer
AOEC	Association of Occupational and Environmental Clinics
AOR	Area of Responsibility
ARC	American Red Cross
ARDS	Adult Respiratory Distress Syndrome
ARES	Amateur Radio Emergency Services
ARS	Accute Radiation Syndrome
AS	arsine
ASAP	as soon as possible
ASH	Assistant Secretary for Health
AST	aspartate aminotransferase

ATFC	Assistant Task Force Commander
ATP	adenosine triphosphate
ATSDR	Agency for Toxic Substances and Disease Registry
AV	atrioventricular
BAL	British anti-Lewisite (dimercaprol)
BASEOPS	base of operations
BATF	Bureau of Alcohol, Tobacco, and Firearms
BCLS	Basic Cardiac Life Support
BLS	basic life support
B-NICE	Biological, Nuclear, Incendiary, Chemical, and Explosive
BP	blood pressure
BTLS	basic trauma life support
BUN	blood urea nitrogen
BW	biological warfare
BWAT	Biological Weapons Antiterrorism Act
C	Celsius
C/B	chemical/biological
CAM	Chemical Agent Monitor
CAS	Chemical Abstract Service
CAT	Crisis Advance Team
CAT	Crisis Advance Technology
CAT	Crisis Assessment Team
CBC	complete blood count
CBDCOM	Chemical and Biological Defense Command
CBIRF	Chemical/Biological Incident Response Force
CBR	Chemical, Biological, or Radiological
CBRIF	Chemical Biological Incident Response Force
CBRRF	Chemical Biological Rapid Response Force
CCC	Casualty Collection Center
CCP	Casualty Collection Points
CDC	Center(s) for Disease Control and Prevention
CDRG	Catastrophic Disaster Response Group
CERCLA	Comprehensive Environmental Response, Compensation, and Liability Act
CERT	Community Emergency Response Teams
CFR	Code of Federal Regulations
CG	Command Group
CHEMTREC	Chemical Transportation Emergency Center
CIDC	Crisis Intelligence Dissemination Center
CIS	Critical Incident Stress
CISD	Critical Incident Stress Debriefing

CISM	Critical Incident Stress Management
CK	cyanogen chloride, chemical formula CCIN
CMCHS	Civilian Military Contingency Hospital System
CNS	central nervous system
CO	Communications Officer
COG	Council of Governments
COPD	Chronic Obstructive Pulmonary Disease
CPAP	certified provider of airway products
CPC	chemical protective clothing
CPK	creatine phosphokinase
CPR	cardiopulmonary resuscitation
CX	phosgene oxime
DC	direct current
DCHAT	Disaster Community Health Assistance Teams
DCO	Defense Coordinating Officer
DDT	dichloro-diphenyl-trichloro-ethane
DEA	Drug Enforcement Administration
DEM	Department of Emergency Management
DES	Digital Encryption Service
DFO	Disaster Field Office
DHHS	Department of Health and Human Services
dL	deciliter
DMAT	Disaster Medical Assistance Team
DNA	deoxyribonucleic acid
DO	Duty Officer
DoD	Department of Defense
DOE	Department of Energy
DOJ	Department of Justice
DOS	Department of State
DOT	Department of Transportation
DP	diphosgene
EBR	Endogenous Biological Regulator
ECC	Emergency Communications Center (ATLANTA)
ECG	electrocardiogram
ED	emergency department
EEI	Essential Elements of Information
EICC	Emergency Information and Coordination Center
EIS	Emergency Information System
EM	Emergency Management
EMA	Emergency Management Agency
EMI	Emergency Management Institute

EMS	Emergency Medical Services
EMT	Emergency Medical Technician
EMT-I	Emergency Medical Technician-Intermediate
EMT-P	Emergency Medical Technician-Paramedic
EOC	Emergency Operations Center
EOD	Explosive Ordnance Disposal
EOP	Emergency Operations Plan
EPA	Environmental Protection Agency
EPD	Environmental Protection Division (Georgia Department of Natural Resources)
EPI	Emergency Public Information
ER	Emergency Room
ERDEC	Edgewood Research, Development, and Engineering Center
ERPG	Emergency Response Planning Guideline
ERT	Emergency Response Team
ERT-A	Advance Element of the Emergency Response Team
ESF	Emergency Support Function
EST	Emergency Support Team
F	Fahrenheit
FBI HQ	FBI Headquarters
FBI	Federal Bureau of Investigation
FCO	Federal Coordinating Officer
FECU	Federal Emergency Communications Coordinator
FEMA	Federal Emergency Management Agency
FMO	Field Medical Operations
FMOG	Field Medical Operations Group
FOG	Field Operations Guide
FRP	Federal Response Plan
g	gram
G6PD	glucose-6-phosphate dehydrogenase
GA	tabun
GAO	General Accounting Office
GB	sarin
GBI	Georgia Bureau of Investigation
GD	soman
GI	gastrointestinal
GMAG	Georgia Mutual Aid Group
GSA	General Services Administration
HAZMAT	hazardous material
HAZWOPER	Hazardous Waste Operations and Emergency Response Regulation

Hb	hemoglobin
HCL	hydrochloric acid
HDL	high-density lipoprotein
HEPA	High Efficiency Particulate Arresting
HIV	human immunodeficiency virus
HMO	Hazardous Materials Officer
HMO	Health Maintenance Office
HMOO	HAZMAT Operations Officer
HMTA	Hazardous Materials Transportation Act
HOG	Hospital Operations Group
hr	hour
HSO	Hospital Support Officer
HTH	calcium hypochlorite
IARC	International Agency for Research on Cancer
IC	Incident Command
IC	Incident Commander
ICP	Incident Command Post
ICS	Incident Command System
ID	identification
IDLH	immediately dangerous to life and health
ILI	influenza-like illness
IMS	Incident Management System
IND	Improvised Nuclear Device
IRR	immediate response resource
ISS	Incident Support System(s)
IV	intravenous
JIC	Joint Information Center
JOC	Joint Operations Center
KED	Kendrick Extrication Device
kg	kilogram
L	liter
LC_{50}	lethal concentration (fatal to 50% of a test population)
LD	lethal dose
LD_{50}	lethal dose (fatal to 50% of a test population)
LDH	lactate dehydrogenase
LDL	low-density lipoprotein
LDS	Lightweight Decon System
LEG	Law Enforcement Group
LFA	Lead Federal Agency
LOG	Logistics Operations Group
LP	Liver Profile

m	meter
MAC	Military Airlift Command or
MAG	Management Advisory Group
MARTA	Metropolitan Atlanta Rapid Transit Authority
MCI	Multiple Casualty Incident
MCIMS	Multi-Casualty Incident Management System
MD	Medical Doctor/Medical Director
MDI	methylene bisphengl isocyanete
MEEP	Medical Examiner Expansion Program
mEq	milliequivalent
Mg	magnesium
mg	milligram
MHz	megahertz
MIC	methyl isocyanate
MINICAD	Mini-Chemical Agent Detector
MIO	Medical Information Officer
MIRG	Medical Information and Research Group
MIRO	Medical Information Research Officer
mL	milliliter
mm Hg	millimeters of mercury
mm^3	cubic millimeters
MMR	measles, mumps, rubella
MMRS	Metropolitan Medical Response System
MMST	Metropolitan Medical Strike Team (Replaced with MMRS)
MOG	Medical Operations Group
MOO	Medical Operations Group Supervisor
MOP	Medical Operations Physician
MOU	Memorandum of Understanding
MRF-P	Medical Response Expansion Program
MS	Medical Specialist
MSDS	Material Safety Data Sheet
MSHA	Mine Safety and Health Administration
MSU	Management Support Unit
MSU	Medical Support Unit
MTM	Medical Team Manager
MW	molecular weight
NA	not applicable
NAC	National Agency Check
NBC	nuclear, biological, or chemical
NDMS	National Disaster Medical System
NE	norephinephrine

NECC	National Emergency Coordination Center
NEST	Nuclear Emergency Search Team
NFPA	National Fire Protection Association
NICP	Nuclear Incident Contingency Plan
NIH	National Institutes of Health
NIMMS	National Interagency Incident Management System
NIOSH	National Institute for Occupational Safety and Health
NMRT	National Medical Response Team
NOAEL	no-observed-adverse-effect level
NPSPAC	National Public Safety Planning Advisory Committee
NRC	Nuclear Regulatory Commission
NTP	National Toxicology Program
OCI	Office of Criminal Investigation (EPA)
OCME	The Office of the Chief Medical Examiner
OEP	Office of Emergency Preparedness
OPS	Operations
OPTC	Out Patient Treatment Centers
OSC	On-Scene Commander
OSD	Operational System Description
OSHA	Occupational Safety and Health Administration
PA	Physician Assistant
PAH	polyaromatic hydrocarbons
PAM	pralidoxime chloride
PAPR	Powered Air Purifying Respirators
PAS	Personnel Accountability System
PBB	polybrominated biphenyls
PC	personal computer
PCB	polychlorinated biphenyls
PCC	Poison Control Center
PCF	patient care forms
Pco_2	partial pressure of carbon dioxide
PCP	pentachloro phenol
PCS	Personal Communication System
PD	Police Department
PDD	Presidential Decision Directive
PEEP	positive end expiratory pressure
PEG	polythylene glycol
Pel	permissible exposure limit
PFT	pulmonary function test
PHN	Public Health Nurse
PHS	Public Health Service

PID	Photo Ionization Detector
PIO	Public Information Office(r)
PM	Program Management
PMT	Program Management Team
PNS	peripheral nervous system
Po_2	partial pressure of oxygen
POA	point of arrival
POC	point of contact
POD	point of distribution
PPC	personal protective clothing
PPE	personal protective equipment
ppm	parts per million
PPO	Preferred Provider Office
PSA	Patient Staging Areas
Pt	platinum
PT&E	Preparedness Training and Exercise
PVP	polyvinyl pyrrolidone
PWGH	Principal Working Group on Health
RADS	reactive airways dysfunction syndrome
RAM	Random Anti-terrorist Measures
RBC	red blood cell
RDD	Radiological Dispersal Device
REAC/TS	Radiation Emergency Assistance Center/Training Site
rem	roentgen man equivalent
RN	Registered Nurse
RNA	ribonucleic acid
SAIC	Special Agent in Charge
SARA	Superfund Amendments and Reauthorization Act
SAT	Standards and Training
sc	subcutaneous
SCBA	self-contained breathing apparatus
SCO	State Coordinating Officer
SEB	staphylococcal enterotoxin B
SIOC	Strategic Intelligence Operations Center
SLUDGE	salivation, lacrimation, urination, diarrhea, gastric distress, and emesis
SOG	Support Operations Group
SOP	standard operating procedures
SPG	Senior Policy Group
SRD	self-reading dosimeter
SSA	Supervisory Special Agent

SSN	Social Security Number
START	Simple Triage and Rapid Assessment Treatment
STEL	short-term exposure limit
T-2	trichothecene mycotoxin
TAC	Tactical
TERIS	Teratogen Information Service
TEU	Technical Escort Unit
TF	Task Force
TFC	Task Force Commander
TLD	Thermoluminescent Dosimeters
TMS	Tactical Medical Support
TNT	trinitrotoluene
TSG	Terrorism Stakeholders Group
TTF	Terrorism Task Force
TTT	Train-the-Trainer
TWA	time-weighted average
UHF	ultra-high frequency
ULC	Unified Local Command Teams
UN	United Nations
USAMRIC	U.S. Army Medical Research Institute of Chemical Defense
USAMRIID	U.S. Army Medical Research Institute of Infectious Diseases
USAR	U.S. Army Reserves
US&R	Urban Search and Rescue
USMC	U.S. Marine Corps
USMS	U.S. Marshal Service
USPHS MMST	United States Public Health Service Metropolitan Medical Strike Team
USPHS/OEP	United States Public Health Service/Office of Emergency Preparedness
USSR	United Soviet Socialist Republic
USSS	United States Secret Service
UV	ultraviolet
VA	Veterans Administration
VEE	Venezuelan equine encephalitis
VHF	very high frequency
VX	nerve agent persistent chemical
WDICP	Weapons of Mass Destruction Incident Contingency Plan
WMD	weapon(s) of mass destruction or weapon(s) of mass disruption

GLOSSARY

Absorption. The incorporation of liquids or gases into the body. Absorption is also the process by which liquid hazardous materials are soaked up by sand, sawdust, or other material to limit the spread of contamination.

Acute effect. A pathologic process caused by a single substantial exposure.

Acute exposure. A single encounter to toxic concentrations of a hazardous material or multiple encounters over a short period of time (usually 24 hours).

Adaptation. The tendency of certain receptors to become less responsive or cease to respond to repeated or continued stimuli.

Adsorption. The property of a substance to attract and hold to its surface a gas, liquid, or other substance.

Atlanta-Fulton County Emergency Management Agency. The emergency management agency that sponsors the Atlanta MMRS Response Program.

Air purification devices. Respirators or filtration devices that remove particulate matter, gases, or vapors from the atmosphere. These devices range from full-facepiece, dual-cartridge respirators with eye protection, to half-mask, facepiece-mounted cartridges with no eye protection.

Air-supplied respirators. A device that provides the user with compressed air for breathing.

Airways. Any parts of the respiratory tract through which air passes during breathing.

Albuminuria. Presence of protein (primarily albumin) in the urine; usually indicative of transient dysfunction or disease.

Alkali. A basic substance (pH greater than 7) that has the capacity to neutralize an acid and form a salt.

Alveolar ducts. The smallest of the lungs' airways that connect terminal bronchioles and alveolar sacs. Sometimes called bronchioles.

Alveoli (*singular* alveolus). Microscopic air sacs in which gas exchange between the blood and the lungs occurs.

Anemia. Any condition in which the number of red blood cells, the amount of hemoglobin, and the volume of packed red blood cells per 100 milliliters of blood are less than normal.

Anhydrous. Containing no water.

Anisocytosis. Considerable variation in the size of blood cells.

Anorexia. Lack of appetite; aversion to food.

Anoxia. Lack of oxygen in inspired air, blood, or tissues.

Anterior chamber of the eye. The fluid-filled front portion of the eye between the cornea and the lens.

Antidote. An agent that neutralizes a poison or counteracts its effects.

Anuria. Absence of urine production.

Aplastic anemia. Decrease in the amount of hemoglobin in the blood due to incomplete or defective development of red blood cells; usually accompanied by defective regeneration of white blood cells and platelets.

Apnea. Cessation of breathing.

Asphyxia. A condition in which the exchange of oxygen and carbon dioxide in the lungs is absent or impaired.

Aspiration pneumonia. Inflammation of the lungs due to inhalation of foreign material, usually food or vomitus, into the bronchi.

Asthma. A chronic condition in which constriction (spasm) of the bronchial tubes occurs in response to irritation, allergy, or other stimuli.

Ataxia. Incoordination of voluntary movement, especially affecting gait and speech.

Atelectasis. Lung collapse.

Atomic weight. The average weight (or mass) of all the isotopes of an element, as determined from the proportions in which they are present in a given element, compared with the mass of the 12 isotope of carbon (taken as precisely 12.000), which is the official international standard; measured in daltons.

Atopy. Tendency or predisposition to allergic reactions.

Autoignition temperature. The lowest temperature at which a gas or vapor-air mixture will ignite from its own heat source or a contacted heated surface without a spark or flame.

Axon. The part of a nerve cell that conducts nervous impulses away from the nerve cell body to the remainder of the cell (i.e., dendrites); large number of fibrils enveloped by a segmented myelin sheath.

Axonal. Pertaining to an axon.

Bilirubin. A red pigment that results from normal and abnormal destruction of red blood cells.

Blepharospasm. Involuntary spasmodic blinking or closing of the eyelids due to severe irritation.

Boiling point. The temperature at which the vapor pressure of a liquid equals the atmospheric pressure and the liquid becomes vapor.

Bradycardia. Slow heart rate, usually under sixty beats per minute.

Bronchi (*singular* bronchus). Large divisions of the trachea that convey air to and from the lungs.

Bronchiole. A small-diameter airway branching from a bronchus.

Bronchitis. Inflammation of the mucous membrane of the bronchial tubes, usually associated with a persistent cough and sputum production.

Bronchorrhea. Increased bronchial secretions.

Bronchospasm. Contraction of the smooth muscle of the bronchi, causing narrowing of the bronchi. This narrowing increases the resistance of air flow into the lungs and may cause a shortness of breath, typically associated with wheezing.

Bullae. Large fluid-filled blisters.

Carcinogenic. Having the ability to cause cancer.

Cardiac dysrhythmia. Abnormality in the rate, regularity, or sequence of the heart beat. Formerly referred to as cardiac arrhythmia.

CAT Team. Crisis Advance Technology (CAT) Team is a six-person MMRS advance component that will respond to a crisis management incident involving a potential or actual chemical, biological, or radiological threat.

Cataract. Loss of transparency (clouding) of the lens of the eye.

Catecholamines. Substances of a specific chemical nature (pyrocatechols with an alkylamine side chain). Catecholamines of biochemical interest are those produced by the nervous system (e.g., epinephrine [adrenaline] or dopamine) to increase heart rate and blood pressure, or medicines with the same general chemical structure and effect.

Caustic. Having the ability to strongly irritate, burn, corrode, or destroy living tissue.

Cerebellar abnormalities. Any irregularity in the cerebellum of the brain.

Cerebellum. The large brain mass located at the posterior base of the brain, responsible for balance and coordination of movement.

Cerebral infarctions. Death of tissue in the cerebrum due to lack of blood flow to the area.

Cerebrum. The largest portion of the brain; includes the cerebral hemispheres (cerebral cortex and basal ganglia).

Chemexfoliation. Chemical skin peeling; use of chemicals to remove scars or pigmentation defects.

Chemical formula. The collection of atomic symbols and numbers that indicates the chemical composition of a pure substance.

Chemical-protective clothing. Clothing specifically designed to protect the skin and eyes from direct chemical contact. Descriptions of chemical-protective apparel include nonencapsulating and encapsulating (referred to as liquid-splash protective clothing and vapor-protective clothing, respectively).

Chronic effect. A pathologic process caused by repeated exposures over a period of long duration.

Chronic exposure. Repeated encounters with a hazardous substance over a period of long duration.

Cognitive function. The ability to think.

Coma. State of profound unconsciousness from which the patient cannot be aroused.

Combustible liquid. Any liquid that has a flash point at or above 100 °F (37.7 °C) and below 200 °F (93.3 °C).

Compressed gas. A gas whose volume has been reduced by pressure.

Congenital anomalies. Birth defects.

Conjunctiva (*plural* conjunctivae). The delicate mucous membrane that covers the exposed surface of the eyeball and lines the eyelids.

Conjunctivitis. Inflammation of the conjunctiva; can result in redness, irritation, and tearing of the eye.

Contact dermatitis (allergic). Delayed-onset skin reaction caused by skin contact with a chemical to which the individual has been previously sensitized.

Contact dermatitis (irritant). Inflammatory skin reaction caused by a skin irritant.

Control zones. Areas at a hazardous materials incident whose boundaries are based on safety and the degree of hazard; generally includes the Hot Zone, Decontamination Zone, and Support Zone.

Cornea. The transparent membrane that covers the colored part of the eye.

Corneal opacification. Clouding of the cornea.

Corrosive. Having the ability to destroy the texture or substance of a tissue.

Consequence Management. Measures to protect public health and safety; restore essential government; and provide emergency relief to governments, businesses, and individuals from the consequences of terrorism. FEMA is the lead agency as stated in Presidential Decision Directive (PDD) 39.

Crisis Management. Measures to identify, acquire, and plan the use of resources needed to anticipate, prevent, and/or resolve a threat or act of terrorism. The FBI is the lead agency as in Presidential Decision Directive (PDD) 39.

Critical Care Area. The area in a hospital designated for the treatment of severely ill patients.

Cyanosis. Bluish discoloration of the skin and mucous membranes due to deficient oxygenation of the blood; usually evident when reduced hemoglobin (i.e., hemoglobin unable to carry oxygen) exceeds 5%.

Decon 1. A 28-foot MMRS Mass Decontamination Trailer System.

Decontamination. Removal of hazardous materials from exposed persons and equipment after a hazardous materials incident.

Decontamination Zone. The area surrounding a chemical hazard incident (between the Hot Zone and the Support Zone) in which contaminants are removed from exposed victims.

Defat. To remove natural oils from the skin.

Degradation. The process of decomposition. When applied to protective clothing, a molecular breakdown of material because of chemical contact; degradation is evidenced by visible signs such as charring, shrinking, or dissolving. Testing clothing material for weight changes, thickness changes, and loss of tensile strength will also reveal degradation.

Delirium. Extreme mental (and sometimes motor) excitement marked by defective perception, impaired memory, and a rapid succession of confused and unconnected ideas, often with illusions and hallucinations.

Dementia. General deterioration of mental abilities.

Demyelination. Removal (destruction) of the myelin sheath that surrounds and protects nerves.

Denervation atrophy. Shrinkage or wasting of muscles due to loss of nerve supply.

Dermal. Relating to the skin.

Dermatitis. Skin inflammation.

Dermis. The layer of the skin just below the epidermis or outer layer. The dermis has a rich supply of blood vessels, nerves, and skin structures.

Desiccation. Removal of moisture; drying.

Desiccant effect. Drying of the skin caused by removal of soluble oils.

Dilution. The use of water to lower the concentration or amount of a contaminant.

Diaphoresis. Excessive perspiration.

Diplopia. Double vision.

Dyscrasia. Blood disorder.

Dysphagia. Difficulty in swallowing.

Dyspnea. Shortness of breath; difficult or labored breathing.

Dysuria. Painful or difficult urination.

Edema. Accumulation of fluid in body cells or tissues; usually identified as swelling.

Embolization. Obstruction of a blood vessel by a transported clot or other mass.

Embryo. In humans, the developing conceptus up to eight weeks after fertilization of the egg. See also *fetus*.

Embryotoxicity. Ability to harm the embryo.

Emergency. A sudden and unexpected event requiring immediate remedial action.

Emesis. Vomiting.

Encephalopathy. Any disease of the brain.

Environmental hazard. A condition capable of posing an unreasonable risk to air, water, or soil quality; or plant or animal life.

Epidermis. The outermost layer of the skin.

Erythroderma. Intense, widespread reddening of the skin.

Erythema. Redness of the skin.

Esophageal strictures. Narrowing of the esophagus that causes difficulty in swallowing; often due to scar formation following extensive burns.

Esophagus. The portion of the digestive canal extending from the throat to the stomach. Also referred to as the gullet.

Euphoria. Intense and exaggerated feeling of well-being.

Exfoliative dermatitis. A skin condition that involves scaling or shedding of the superficial cells of the epidermis.

Exothermic reaction. Chemical reactions that produce heat.

Explosives. Compounds that are unstable and break down with the sudden release of large amounts of energy.

Explosivity. The characteristic of undergoing very rapid decomposition (or combustion) to release large amounts of energy.

Fasciculation. Muscle twitching.

Fetotoxicity. Ability to harm the fetus.

Fetus. In humans, the conceptus from eight weeks after fertilization until birth. See also *embryo*.

Flame-resistant. Slow or unable to burn.

Flammable. Able to ignite and burn.

Flammable (explosive) range. The range of gas or vapor concentration (percentage by volume in air) that will burn or explode if an ignition source is present. Limiting concentrations are commonly called the lower explosive limit and upper explosive limit. Below the lower explosive limit, the mixture is too lean to burn; above the upper explosive limit, the mixture is too rich to burn.

Flash point. The minimum temperature at which a liquid produces enough vapor to ignite.

Flashback. The movement of a flame to a fuel source; typically occurs via the vapor of a highly volatile liquid or by a flammable gas escaping from a cylinder.

Fluorosis. Accumulation of excessive fluoride in the body; characterized by increased bone density and mineral deposits in tendons, ligaments, and muscles.

Follow-up. Constant or intermittent contact with a patient after diagnosis or therapy.

Freezing point. Temperature at which crystals start to form as a liquid is slowly cooled; alternatively, the temperature at which a solid substance begins to melt as it is slowly heated.

Fume. Fine particles (typically of a metal oxide) dispersed in air that may be formed in various ways (e.g., condensation of vapors, chemical reaction).

Gangrene. Death of tissue due to lack of blood supply.

Gas. A physical state of matter that has low density and viscosity, can expand and contract greatly in response to changes in temperature and pressure, and readily and uniformly distributes itself throughout any container.

Glaucoma. A disease of the eye characterized by abnormal and damaging high pressure inside the eye; usually due to a blockage of the channel that normally allows the outflow of fluid from the eye.

Glomerulus (*plural* glomeruli). A tuft formed of capillary loops that filter blood in the kidney.

Hazard. A circumstance or condition that can cause harm.

Hazardous materials. Substances that, if not properly controlled, pose a risk to people, property, or the environment.

Hazardous materials incident. The uncontrolled release or potential release of a hazardous material from its container into the environment.

Hematuria. Condition in which the urine contains an abnormal amount of blood or red blood cells.

Hemodialysis. Removal of soluble substances from the blood by their diffusion through a semipermeable membrane.

Hemoglobinuria. Condition in which the urine contains an abnormal amount of hemoglobin.

Hemolysis. Destruction or dissolution of red blood cells in such a manner that hemoglobin is liberated into the medium in which the cells are suspended.

Hemolytic anemia. Any anemia resulting from destruction of red blood cells.

Hemoptysis. The spitting of blood derived from a hemorrhage in the lungs or bronchial tubes.

Hepatic. Pertaining to the liver.

Hepatomegaly. Enlargement of the liver.

Hot Zone. The area immediately surrounding a chemical hazard incident, such as a spill, in which contamination or other danger exists.

Hyperbilirubinemia. Condition in which an abnormally large amount of bilirubin is found in the blood. Jaundice becomes apparent when the level of bilirubin is double the normal level.

Hyperesthesia. Increased sensitivity to touch, pain, or other sensory stimuli.

Hyperpigmentation. An excess of pigment in a tissue or part of the body.

Hyperreflexia. Condition in which the deep tendon reflexes are exaggerated.

Hypersensitization. Increased sensitivity of the immune system; induced by initial exposure with subsequent exposures eliciting a greater than expected immunologic response.

Hypertension. High blood pressure.

Hypocalcemia. Condition in which an abnormally low concentration of calcium ions is present in the blood.

Hypokalemia. Condition in which an abnormally low concentration of potassium ions is present in the blood.

Hypomagnesemia. Condition in which the plasma concentration of magnesium is abnormally low; may cause convulsions and concurrent hypocalcemia.

Hypophosphatemia. Condition in which an abnormally low concentration of phosphate is found in the blood.

Hypotension. Low arterial blood pressure.

Hypotonia. Condition in which there is a loss of muscle tone.

Hypoxemia. Condition in which inadequate oxygen is present in arterial blood, short of anoxia.

Hypoxia. Condition in which below-normal levels of oxygen are present in the air, blood, or body tissues; short of anoxia.

Ignition (autoignition) temperature. The minimum temperature required to ignite gas or vapor without a spark or flame being present.

Immediately Dangerous to Life and Health (IDLH). The atmospheric concentration of a chemical that poses an immediate danger to the life or health of a person who is exposed but from which that person could escape without any escape-impairing symptoms or irreversible health effects. A companion measurement to the Permissible Exposure Limit (PEL), IDLH concentrations represent levels at which respiratory protection is required. IDLH is expressed in parts per million (ppm) or milligrams per cubic meter (mg/m^3).

Inadequate warning property. Characteristic (e.g., odor, irritation) of a substance that is not sufficient to cause a person to notice exposure.

Incident Commander. The person responsible for establishing and managing the overall operational plan at a hazardous material incident. The Incident Commander is responsible for developing an effective organizational structure, allocating resources, making appropriate assignments, managing information, and continually attempting to mitigate the incident.

Insecticide. An agent that has the ability to kill insects.

Intention tremor. Trembling of the extremities during movement.

Interstitial pneumonitis. Inflammation of the alveolar walls and the spaces between them.

Iritis. Inflammation of the colored part of the eye (iris).

Ischemia. Obstruction of blood flow (usually by arterial narrowing) that causes lack of oxygen and other bloodborne nutrients.

Ischemic necrosis. Death of cells as a result of decreased blood flow to affected tissues.

Jaundice. Yellowing of the skin and whites of the eyes due to an accumulation of bile pigments (e.g., bilirubin) in the circulating blood.

Keratitis. Inflammation of the cornea.

Lacrimation. Secretion of tears, especially in excess.

Laryngeal edema. Swelling of the voice box due to fluid accumulation.

Laryngitis. Inflammation of the mucous membrane of the larynx.

Laryngospasm. Spasmodic closure of the vocal apparatus.

Lead Agency. The FBI defines lead agency, as used in PDD-39, as the Federal department or agency assigned lead responsibility to manage and coordinate a specific function—either crisis management or consequence management. Lead agencies are designated on the basis of their having the most authorities, resources, capabilities, or expertise relative to accomplishment of the specific function. Lead agencies support the overall lead federal agency during all phases of the terrorism response.

Lethargy. State of extreme tiredness or fatigue.

Leukemia. Progressive proliferation of abnormal leukocytes found in blood and blood-forming tissues and organs; due to cancer of the bone marrow cells that form leukocytes.

Leukocyte. A white cell normally present in circulating blood.

Material Safety Data Sheet (MSDS). Documents prepared by the chemical industry to transmit information about the physical properties and health effects of chemicals and about emergency response plans.

Methemoglobin. A transformation product of hemoglobin in which Fe^{+2} is oxidized to Fe^{+3}. Methemoglobin contains oxygen that is firmly bound to the Fe^{+3} ion, which prevents the release of oxygen to the tissues.

Methemoglobinemia. Condition in which methemoglobin is present in the circulating blood.

Methemoglobinuria. Condition in which methemoglobin is present in the urine.

Miosis. Contraction of the pupil to a pinpoint.

Miscible. Having the ability to mix (but not chemically combine) in any ratio without separating into two phases (e.g., water and alcohol).

Mist. Liquid droplets dispersed in air.

Mitigation. Actions taken to prevent or reduce the severity of harm.

Monocytic leukemia. A form of bone marrow cancer characterized by an increase in the number of large, mononuclear white blood cells in tissues, organs, and the circulating blood.

Molecular weight. The sum of the atomic weights (q.v.) of the atoms in a molecule; measured in daltons.

Myalgia. Severe muscle pain.

Mydriasis. Dilation of the pupil.

Myelocytic leukemia. A form of bone marrow cancer characterized by the presence of large numbers of granular white blood cells in tissues, organs, and the circulating blood.

Myocardial ischemia. Insufficient oxygen supply to meet the metabolic demands of heart muscles.

Myocarditis. Inflammation of the muscles of the heart.

Myoclonus. Involuntary spasm or twitching of a muscle or group of muscles.

Myoglobin. The oxygen-transporting, pigmented protein of muscle; resembles blood hemoglobin in function.

Myoglobinuria. Presence of myoglobin in the urine.

Nasopharynx. Relating to the nasal cavity and that part of the throat that lies above the level of the soft palate.

Necrosis. Death of one or more cells or a portion of a tissue or organ.

Nephrotoxic. Capable of damaging the kidney.

Neuropathy. A disorder of the nervous system; in contemporary usage, a disease involving the cranial or spinal nerves.

Noncardiogenic pulmonary edema. Accumulation of an excessive amount of fluid in the lungs as a result of leakage from pulmonary capillaries; not due to heart failure.

Nystagmus. Involuntary rapid movements of the eyeballs, either rhythmical or jerky.

Ocular. Pertaining to the eye.

Odor threshold. The lowest concentration of a vapor or gas that can be detected by smell.

Off-gassing. Giving off a vapor or gas.

Olfactory fatigue. Temporary loss of the sense of smell due to repeated or continued stimulation.

Oliguria. Condition in which abnormally small amounts of urine are produced.

Opisthotonos. Tetanic spasm in which the spine and extremities are bent up and forward so that a reclining body rests on the head and the heels.

Optic atrophy. Shrinkage or wasting of the optic nerve that may lead to partial vision loss or blindness.

Optic neuritis. Inflammation of the optic nerve.

Osteosclerosis. Abnormal hardening or increase in density of the bone.

Paresthesias. Abnormal sensation such as burning, prickling, or tingling.

Percutaneous absorption. Passage of a substance through unbroken skin.

Peripheral neuropathy. A disorder of the peripheral nerves.

Permeation. The passage of chemicals, on a molecular level, through intact material, such as protective clothing.

Permissible Exposure Limit (PEL). The maximum time-weighted average concentration mandated by the Occupational Safety and Health Administration (OSHA) to which workers may be repeatedly exposed for eight hours per day, 40 hours per week without adverse health effects.

Photophobia. Abnormal sensitiveness to light, especially of the eyes.

Physical state. The state (solid, liquid, or gas) of a chemical under specific conditions of temperature and pressure.

Pneumonitis. Inflammation of the lungs.

Poikilocytosis. Presence of irregularly shaped red blood cells in the peripheral blood.

Posthypoxic encephalopathy. Condition in which the brain has been damaged as a result of insufficient oxygen.

Proteinuria. Condition in which an abnormal amount of protein is present in the urine. See also *albuminuria*.

Pruritic. Pertaining to itching.

Psychosis. A mental disorder characterized by derangement of personality and loss of touch with reality.

Pulmonary edema. Accumulation of extravascular fluid in the lungs that impairs gas exchange; usually due to either increased intravascular pressure or increased permeability of the pulmonary capillaries.

Pupil. The circular opening in the center of the iris through which light rays enter the eye.

Reactivity. Ability of a substance to chemically interact with other substances.

Rescuer protection equipment. Gear necessary to prevent injury to workers responding to chemical incidents.

Respiratory depression. Slowing or cessation of breathing due to suppression of the function of the respiratory center in the brain.

Response organization. An organization prepared to provide assistance in an emergency (e.g., fire department).

Response personnel. Staff attached to a response organization (e.g., HAZMAT Team).

Retrobulbar neuritis. Inflammation of the portion of the optic nerve behind the eyeball.

Rhinitis. Inflammation of the mucous membranes of the nasal passages.

Rhinorrhea. Discharge from the nasal mucous membrane.

Routes of exposure. The manner in which a chemical contaminant enters the body (e.g., inhalation, ingestion).

Sclera. The tough, white, supporting tunic of the eyeball.

Secondary contamination. Transfer of a harmful substance from one body (primary body) to another (secondary body), thus potentially permitting adverse effects to the secondary body.

Self-Contained Breathing Apparatus (SCBA). Protective equipment consisting of an enclosed facepiece and an independent, individual supply (tank) of air; used for breathing in atmospheres containing toxic substances or underwater.

Sensory neuropathy. Damage to the nerves that carry information about sensation (e.g., touch, pain, temperature) to the brain.

Sequela (*plural* sequelae). A condition that follows as a consequence of injury or disease.

Sloughing. Process by which necrotic cells separate from the tissues to which they have been attached.

Solubility. Ability of one material to dissolve in or blend uniformly with another.

Soluble. Capable of being dissolved.

Solution. A homogeneous mixture of two or more substances; usually a liquid.

Solvent. A substance that dissolves another substance.

Specific gravity. The ratio of the mass of a unit volume of a substance to the mass of the same volume of a standard substance (usually water) at a standard temperature.

Squad 4. MMRS Agent ID component response vehicle operated by the Atlanta Fire Department.

Status epilepticus. Severe seizures in which recovery does not occur between major episodes.

Stridor. A harsh, high-pitched respiratory sound often heard in acute respiratory obstruction.

SUPPORT 1. A Chevrolet Suburban unobtrusive response vehicle that provides tactical medical support and that also responds as part of the CAT Team for a Crisis Management Response. It also provides medical surveillance support for a full MMRS 43-member response deployment.

SUPPORT 2. A Chevrolet Stepvan MMRS Mobile Command Response unit that is equipped with VHF, UHF, 800, cellular, fax, satellite, and MMRS on-scene coordination equipment.

SUPPORT 3. A Ford E-150 four-door dual-wheel response vehicle that tows a 24-foot MMRS Incident Support Systems (ISS) response trailer that has MMRS decontamination, mass casualty, BLS, incident command, and control positions for the Medical Information and Research (MIR), Hospital Operations, Logistics, and Field Medical Operations Groups.

Support Zone. The area beyond the Decontamination Zone that surrounds a chemical hazard incident in which medical care can be freely administered to stabilize a victim.

Surfactant. An agent that reduces surface tension (e.g., wetting agents, detergents, dispersing agents).

TAC Unit. A two-person MMRS tactical medical team that provides tactical medical support to a federal, state, or local law-enforcement agency engaged in the interdiction of a potential or actual chemical, biological, radiological, or ordinance threat.

Tachycardia. Rapid heartbeat (typically greater than 100 beats per minute).

Tachypnea. Rapid breathing.

Teratogenic. Having the ability to cause congenital anomalies.

Tetany. Condition marked by involuntary muscle contractions or spasms.

Thrombocytopenia. Condition in which there is an abnormally small number of platelets in the blood.

Thrombosis. Blood vessel clotting.

Time-Weighted Average (TWA) air concentration. The concentration of a substance in the air that is measured by collecting it on a substrate at a known rate for a given period of time.

Tinnitus. Ringing in the ears.

Toxic. Having the ability to harm the body, especially by chemical means.

Toxic potential. The inherent ability of a substance to cause harm.

Tracheitis. Inflammation of the membrane lining the trachea.

Trismus. Lockjaw.

Tubular necrosis. Death of the cells lining the kidney tubules.

Uremia. Condition in which an abnormally high level of urea or other nitrogenous waste is found in the blood; due to kidney dysfunction.

Urticaria. Hives.

Vapor. The gaseous form of a substance that is normally a solid or a liquid at room temperature and pressure.

Vapor density. The weight of a given volume of a vapor or a gas compared to the weight of an equal volume of dry air, both measured at the same temperature and pressure.

Vapor pressure. A measure of the tendency of a liquid to become a gas at a given temperature.

Vascular. Pertaining to blood vessels.

Vasodilation. Increased diameter of the blood vessels.

Ventricular fibrillation. Rapid, tremulous movement of the ventricle that replaces normal contractions of the heart muscle; results in little or no blood being pumped from the heart.

Vertigo. Sensation of spinning or revolving.

Vesicant. An agent that produces blisters.

Vesiculation. Presence or formation of blisters.

Water-reactive material. A substance that readily reacts with water or decomposes in the presence of water, typically with substantial energy release.

Weapon of Mass Destruction (WMD). Title 18, U.S.C. 2332a, defines a weapon of mass destruction as (1) any destructive device as defined in section 921 of this title, [which reads] any explosive, incendiary, or poison gas, bomb, grenade, rocket having a propellant charge of more than four ounces, missile having an explosive or incendiary charge of more than one-quarter ounce, mine or device similar to the above; (2) poison gas; (3) any weapon involving a disease organism; or (4) any weapon that is designed to release radiation or radioactivity at a level dangerous to human life.

Wheezing. Breathing noisily and with difficulty; usually a sign of spasm or narrowing of the airways.

Numerics

1-800 number **224**
911 emergency centers **91**

A

Abu Nidal Organization **10**
acetyl coenzyme-A **152**
acetylcholine **146–147, 150**
acetylcholinesterase **146**
acetylcholinesterase crisis **150**
acidosis **152**
acids **159**
Action Direct **8**
activated charcoal **153**
active surveillance (epidemiological services) **204–206**
ACTs (Area Command Teams) **165**
acute delirium **208**
acute encephalopathy **208**
acute radiation syndrome (ARS) **155**
adamsite **162**
adenosine **151**
adenosine diphosphate (ADP) **151–152**
adenosine triphosphate (ATP) **151–152**
Administrative Officer **73**
ADP (adenosine diphosphate) **151–152**
adrenal glands **148, 149**
adrenalin **147, 198**
adrenergic sympathetic transmission **146**
advanced life support (ALS) intervention **141**
aerosol generators **190**
Afghanis mercenaries **9**
Afghanistan **9**
  drug-trafficking in **10**
  opium production in **11**
Africa **8**
agency operational centers **108**
agent treatment protocols **80**
agent-specific therapy **227**
air transport incidents **27**
airports **166, 247**
airway supplies **284**
albuterol inhalers **285**
alerting systems **41**
ALS (advanced life support) intervention **141**
alternative medical centers (AMCs) **177**
Amateur Radio Emergency Services (ARES) **178**
ambulatory patients **161**
AMCs (alternative medical centers) **177**
  in medical response expansion programs **225–227**
  staffing of **206**
American Red Cross (ARC) **103**
AMFs (alternate medical facilities) **237**
aminoglycosides **268**
ammonia **159**
ammunitions **190**
amoxicillin **278**
anarchists **3**
Ancillary Services Officer **87**
ANG (Army National Guard) **76**
anhydrous ammonia **159**
anhydrous calcium oxide **210**
aniline **156**
animal carcasses **178**
animal control agencies **204**
animal rights **10**

anoxia 154
anthrax 187, 197
  animal exposure and treatment 209
  decontamination of spores 210, 260–261, 264
  description of agent 263
  diagnosis of 263
  drug therapy 278
  human exposure and treatment 209
  incubation period 286
  mass prophylaxis for 218
  online resources 325
  outbreak control 264
  prophylaxis 264, 278
  sampling 209
  screening and confirmation times 216
  signs and symptoms of 263, 286
  spores 209–210
  threats of 259
    evacuation 260
    immediate actions 259–260
  treatment of 263–264, 278, 286
Anthrax Spore Vaccine® 209
anti-abortion groups 10
antibiotics 204, 283
anti-diarrhea medications 204
antidotes 79, 194
antimicrobial agents 192
AOR (area of responsibility) 164
ARC (American Red Cross) 103
Area Command Teams (ACTs) 165
area of responsibility (AOR) 164
ARES (Amateur Radio Emergency Services) 178
Army National Guard (ANG) 76
ARS (acute radiation syndrome) 155
arsine 156
arthralgias 265
assassinations 16
Assistant Task Force Leader 73
ATP (adenosine triphosphate) 151–152
atropine 79, 285
Attorney General 76
Aum Shinrikyo 8
autonomic nervous system 145–146
avalanche 27
axons 144–145
Azerbaijan 8

B

Bacillus anthracis 187, 210, 263
backdraft 25
back-up computers 66
bacteria 197
bandages 283
Bangladesh 7
basal metabolism 148, 149
base camps 230
basic life support (BLS) intervention 141
basic trauma life support (BTLS) 79
Basque separatists 7
bicarbonate ion 155
Bin-Laden, Usama 11
bio-bombs 171
bio-engineering facilities 171
biological agents 173
  actual release of 173, 190–194
    consequences 192
    device/release potential 191
    disease identification 192
    intelligence gathering 193
    notifications 191
    protective actions 192
    request of specialized resources on 193
    secondary attacks 193
    site security plans 193
  baseline agents 187–188
  characteristics of 197
  confirmed presence of 173, 186
  covert release of 172
  determining the use of 215
  dissemination of 198
  effects of weather on 199
  first responder concerns 198
  historical scenarios 187–188
  incident objectives 276
  incident types 172

indicators of **249, 275–276**
modes of delivery **171**
notification essentials **276**
online resources **326–327**
protection against **141**
recognition and evaluation of **175**
in suspected packages **172**
suspected release of **276**
threatened use of **172, 184–186**
training programs **297–304**
treatment vs. prophylaxis **188–190**
See also biological terrorism
See also biological warfare
biological emergencies **164**
biological poisoning **76**
biological response plan (BRP) **205**
biological terrorism **163**
active surveillance of **204–206**
alert on **183**
decision factors **180**
detection of **208**
emergencies **164**
end event **179**
event detection **176**
announced **176**
unannounced **176**
incident types **172**
indications of **205–206**
online resources **323, 326–327**
passive surveillance of **204**
preparedness programs **169**
preventive measures **169**
psychological aspects of **257**
recognition of **175**
responses to **164–165, 184**
command and control **168**
county level **165–167**
federal level **165–167**
integration of **165–167**
intelligence gathering **169**
local level **165–167**
state level **165–167**
secondary attacks **190, 193**
stages of severity of **179**
threatened use of **172**
threats of **171**
warning on **183**
See also biological warfare
See also bioterrorism events
See also terrorism
biological terrorism plan **59–63**
action planning **177–179**
internal notificationS **176**
maintenance of **175**
objectives of **163**
public notifications **179**
recognition and evaluation of agents in **175**
threat responses management **174–175**
training and exercising **175**
See also biological terrorism
biological terrorism stakeholders group (BTSC) **206–207**
biological warfare
alert on **183**
evaluation of **195**
first responder concerns **198, 249–251**
historical scenarios **187–188**
incident objectives **276**
indicators of **190, 275–276**
initial concerns
convergent casualties **194**
downwind potentials **194**
panic victims **194**
scarce supplies **194**
traffic congestion **194**
mass fatalities **200–201**
notification essentials **276**
recovery from **200–201**
responses to **184**
secondary attacks **190, 193**
threats of **184–186**
See also biological agents
See also biological terrorism
bioterrorism events
action planning **177–179**
announced **176**
consequences of **192**
decision factors **180**
detection of **176, 208**

bioterrorism events (continued)
end of **179**
indications of **205–206, 249**
initial concerns
convergent casualties **194**
downwind potential **194**
panic victims **194**
scarce supplies **194**
traffic congestion **194**
internal notificationS of **176**
public notifications of **179**
responses to **184**
stages of severity of **179**
surety measures **209**
treatment models in **181**
unannounced **176**
See also biological warfare
See also bioterrorism
bladder **148**
bleach **210**
blood **154**
as mediator of inflammation **157**
and respiratory system **157**
roles of **154**
blood agents **150, 151, 162**
blood vessels **148, 149**
BLS (basic life support) intervention **141**
B-NICE (Biological, Nuclear, Incendiary, Chemical. Explosive) **138**
body fluids **211**
bomb disposal teams **139**
bombings **16**
bone marrow **155**
Bosnia **9**
Branch Davidians **8**
bridges **113**
Brigate Rosse **8**
bronchioles **157**
bronchodilators **227**
BRP (biological response plan) **205**
Brucella **187, 264**
brucellosis **187**
animal exposure and treatment **211**
decontamination and isolation **265**
description of agent **264–265**
diagnosis of **265**
drug therapy **281**
exposure and sampling **211**
incubation period **286**
outbreak control **265–266**
prophylaxis **265**
screening and confirmation times **216**
signs and symptoms of **265, 286**
treatment of **265, 286**
BSTC (biological terrorism stakeholders group) **206–207**
BTLS (basic trauma life support) **79**
BTX **198**
bubo **212**
bubonic plague **187, 197, 266**
buildings **171**
burial **239**
burn injuries **141**
Burundi **8**

C

California **25, 95**
capability assessment **40**
carbamate insecticides **150**
carbaminohemoglobin **155**
carbon dioxide **154**
carbon monoxide **156**
carbon monoxide poisoning **156**
carboxyhemoglobin **156–157**
cardiac muscles **146**
Cardiopulmonary Services Manager **88**
Carlos **10**
cascade effect **38–40**
casualties **139, 194**
Catastrophic Disaster Response Group (CDRG) **107**
catheters **284**
Catholic churches **16**
Catholics **16**
Caucasus **11**
cave-ins **90**
CCPs (casualty collection points) **79, 142**
CDC (Centers for Disease Control and

Prevention) **76**
  Web site **323**
CDC stockpile **217**
CDRG (Catastrophic Disaster Response Group) **107**
cells **144–145**
cellular hydration **154**
cellular respiration **155**
Cellules Communistes Combatants **8**
CEM (Comprehensive Emergency Management) **30–32**
Center for Civilian Bio-Defense Studies **327**
Center for Research on the Epidemiology of Disasters **327**
Centers for Disease Control and Prevention (CDC) **76**
Central Asia
  as focal point for terrorism **10**
  human rights record in **11**
  regional security in **12**
  rise of Islamic fundamentalism in **10–11**
  role of Uzbekistan in **11**
central nervous system **144**
CEO (chief executive officer) **53**
CERT (community emergency response teams) **94–98**
  starting **95**
  training programs **96–98**
Chechen insurgents **171**
Chechnya **8**
chemical agents **275**
Chemical Biological Incident Response Force **75, 173, 189**
Chemical Biological Quick Response Force **173**
Chemical Biological Rapid Response Force **189**
chemical exposure **217**
chemical poisoning **76**
chemical warfare **275**
chemotaxis **157**
CHI (county health indicators) **205**
Chicago Tunnel Flood **25**
chief executive officer (CEO) **53**
China **9**
chloramphenicol **266, 279**
chlorine **158, 159, 162, 210**
cholinergic sympathetic transmission **146**
Christian fundamentalists **9**
churches **166**
ciliary muscles **148, 149**
ciprofloxacin **264, 278, 283**
CIS (critical incident stress) syndrome **77**
CIS Common Security Treaty **12**
civil defense **28**
civil disorders **27**
civil disturbances **25, 90**
Civil Preparedness Guide (CPG) **57**
civil war **14**
Claims Officer **86**
clean-team transfer **142–143**
clusters **235**
cold (support) zone **77**
Cold War, end of **8, 28**
Colombia **8, 10**
Colorado Serum **209**
commercial buildings **222**
commercial refrigerated warehouses **233**
communication centers **91**
Communication Specialists **309**
communications equipment **98**
Communications Group Supervisor **309**
Communications Officer **84**
communications specialists **132**
Communications Team **73**
community centers **226**
Comprehensive Emergency Management (CEM) **30–32**
comprehensive emergency management plan **90**
concert halls **171**
confined space emergencies **90**
consumer terrorism **10**
contagious diseases **141**
contaminated waste **255**
controlled substances **283**
convergent casualties **194**
coroners **166**

Corsican separatists **7**
Cost Officer **86**
County Health Department **245–246**
county health indicators (CHI) **205**
county public health clinics **273**
cowpox **269**
CPG (Civil Preparedness Guide) **57**
cranial nerves **146**
crisis management **170**
critical facilities **91**
critical incident stress (CIS) syndrome **77**
crop dusting aircraft **171**
crowd emergency decontamination **161**
crush syndrome **141**
cults **8**
cyanide poisoning **141**
cyanides **150–151, 154**
cyanogen chloride **162**
cyanosis **156, 266**
cytochromes **153**

D

dam failures **27, 90**
Damage Assessment and Control Manager **84**
DCHATs (disaster community health assistance teams) **98–99**
DCO (Defense Coordinating Officer) **114**
dead animals, disposal of **210–214**
deaths **200–201**
debris clearance **113**
deceased persons, disposition of **200–201**
decontamination **78, 140–141, 199–200, 252–253, 260–261**
  of antrhax spores **210**
  crowd emergency **161**
  factors in **160**
  mass patient **159–161**
  objectives in **160**
  of sites **201**
  tent showers **294**
  three-corridor **161**
  two-corridor **161**
decontamination systems **287**
  hoses **288–290**
  layouts **160**
  piping systems for **291**
  sprinklers **288**
  tent showers **294**
  three-corridor **161**
  two-corridor **161**
decontamination trailers **161**
Defense Civil Preparedness Agency **30**
Defense Coordinating Officer (DCO) **114**
Defense Threat Reduction Agency **327**
definitive decontamination **78**
Delayed Treatment Area Manager **87**
delirium **208**
dendrites **144–145**
Department of Defense **248**
  Web site **324**
Department of Energy (DOE) **76**
Department of Health (DOH) **246**
Department of Health and Human Services (DHHS) **327**
department stores **171**
Dependent Care Services Manager **88**
depolarization **145**
DHHS (Department of Health and Human Services) **327**
diagnosis **214–215**
diarrhea **150, 208**
diazepam **283**
diplopia **208**
Director of Emergency Management **47**
dirty bombs **171**
disaster community health assistance teams (DCHATs) **98–99**
disaster planning
  guidelines **55–56**
  IEMS approach **54–55**
  vs. improvisation **57**
  military approach **54**
  reasons for **55**
  writing the plan **56–57**
disaster plans **56–57**
  components of **57**
  good vs. bad plans **58**

improvisation in **57**
local emergency operations plans **58**
technical writing in **57**
disaster recovery **113**
disasters **102, 118**
information on **53**
notification of **114**
Discharge Area Manager **87**
discharge management **255**
disease outbreak **195–197**
laboratory detection of **215**
notification of **182**
diseases
anthrax **209–211**
brucellosis **211–212**
identification of **192, 196, 214–215**
online resources **325–326**
outbreak of **195–197**
plague **212**
disinfection **254**
dispatch personnel **251–252**
DOE (Department of Energy) **76**
DOH (Department of Health) **246**
domestic preparedness coodinator **296**
domestic preparedness specialists **296**
domestic preparedness training courses **296**
domestic terrorists **3**
dopamine **285**
downwind potentials **194**
doxycycline **264, 281, 283**
dressings **283**
drought **27**
drug cartels **10**
drug trafficking **10**
dysphagia **208**
dyspnea **208, 266**

E

early-warning capability **169**
earthquakes **25, 27, 66, 90**
Northridge **128**
Whittier Narrows **95**
Ebola virus **197**
EBRs (endogenous biological regulators) **197**
edema **158–159**
Edgewood Research, Development, and Engineering Center (ERDEC) **75**
effector organs **146**
EICC (Emergency Information and Coordination Center) **107**
elderly population **91**
elections **51**
emergencies **102**
biological **164**
elements of **37**
emergency communications **41**
emergency decontamination **78**
emergency generators **66**
Emergency Incident Commander **83**
Emergency Information and Coordination Center (EICC) **107**
emergency management **23**
basic tasks **23**
capability assessment **40–43**
effects of news media on **28**
exercises **46–47**
factors in **24–25**
fire services in **23–24**
goal-setting in **43**
hazard identification and planning **51**
information management **53**
and perception of threats **28–29**
resource database **52–53**
emergency management agencies **22**
Emergency Management Coordinator **47**
emergency management orqanizations
alerting and warning systems **41**
capability assessment of **40–43**
continuity of government **42**
direction and control **41**
emergency communications **41**
emergency operations planning **40**
emergency public information **41**
emergency reporting **43**
emergency support services **42**
evacuation **42**
protective measures **42**

emergency management orq. (continued)
resource management 40
shelter protection 42
staffing and training 41
emergency management systems 23–24
emergency medical services (EMS) 71
emergency medications 285
emergency operations plan. See EOP
emergency program managers 29, 52–53
hazard identification and planning 51
information management 53
job titles 51
leadership 53
non-emergency management-related duties 54
office administration duties 52
primary functions of 53
roles and responsibilities 29–30
roles of 50–51
staffing management 51
training 53
emergency public information 41
emergency reporting 43
Emergency Response Team. See ERT
emergency response training 301
emergency support functions. See ESFs
emergency support services 42
Emergency Support Team (EST) 107–108
emesis 150
EMI (Emergency Management Institute) 95, 98
EMS (emergency medical services) 18–19, 71
clusters 235
functional annexes 93
local hazards analysis 90–91
unusual high demand for 208
vulnerabilty analysis 91
encephalitis 208
endogenous biological regulators (EBRs) 197
environmental clean-up 240–241
environmental groups 10
environmental health specialists 296
Environmental Protection Agency (EPA) 76
EOC (Emergency Operations Center) 170
in bioterrorism response plan 167
duties and responsibilities of 245
EOP (emergency operations plan) 40, 43
local 58, 90
functional annexes 91–93
local hazards analysis 90–91
vulnerability analysis 91
online resources 327–328
planning 43
state 327–328
EPA (Environmental Protection Agency) 76
epidemics 90, 187
epinephrine 147
ERDEC (Edgewood Research, Development, and Engineering Center) 75
ERT (Emergency Response Team) 109
administration and logistics 111
Advance Element 109–110
information and planning 111
recovery operations 113
response operations 112–113
stucture of 110–111
erythromycin 283
ESFs (emergency support functions) 59, 236–237
annexes to 93
of EOC (Emergency Operation Center) 72
in ERT (Energency Response Team) 112–113
missions 104–106
EST (Emergency Support Team) 107–108
ET tubes 284
ETA (Euzkadi Ta Askatasuna) 7
ethnic groups 8, 91
Euro-terrorists 8
Euzkadi Ta Askatasuna (ETA) 7
evacuation 42, 260
exercises 46–47, 175
exotic diseases 204
Explorer Cadet Corps. 52

explosions 141
explosive devices 171
exposed animals 209
Exspor (sporicide) 210
extremist viewpoints 4
eyes 148, 149

F

FAC (family assistance center) 240
Facility Operations Officer 84
family assistance center (FAC) 240
far left groups 3
FARC (Fuerzas Armadas Revolutionaries de Colombia) 8
FBI (Federal Bureau of Investigation) 70
    JIC (Joint Information Center) 179
    response to bioterrorism 165–167
    SAIC (special agent in charge) 74
FCO (Federal Coordinating Officer) 103
    appointment of 104
fear as agent of change 2
Federal Bureau of Investigation. See FBI
Federal Coordinating Officer. See FCO
Federal Executive Agent 104
Federal Express 233
FEMA (Federal Emergency Management Agency) 30
    response to bioterrorism 165–167
    US&R (Urban Search & Rescue) Response System 129
    Web site 59, 324
Ferghana Valley 13
ferric iron 156
ferrous iron 156
fever 208
fibrotic tissues 157
field decontamination 160
Field Hazmat Group Supervisor 307
Field Hazmat Specialist 307
Field Medical Group Supervisor 306
Field Medical Operations Sector 73, 78
Finance Section Chief 85
Fire and Rescue 246–247
fire departments 44, 91
fire services 23–24
    in incident management systems 71
    role in bioterrorism response plans 167–168
firefighters 24
Firefighting Resources of Southern California Organized for Potential Emergencies (FIRESCOPE) 44–46
Fireground Command System 45
fires 24
FIRESCOPE (Firefighting Resources of Southern California Organized for Potential Emergencies) 44–46
first responders 251
flashover 25
fleas 197, 212
floods 27, 70, 90
    Midwest 127
Florida 25, 95
flu 176
flu season 204
fluoroacetate 153
FNLC (Front de la Liberation Nationale de la Corse) 7
food system 171
formaldehyde 210
Francisella tularensis 187, 267
freeways 194
Front de la Liberation Nationale de la Corse (FLNC) 7
FRP (Federal Response Plan)
    activation of 116
    assumptions of 102–103
    deployment 116
    ESFs (emergency support functions) in 59
    federal assistance in 74
    general operations 104–106
    integration of NDMS (National Disaster Medical System) 123
    notification 114
    regional actions 117
    response actions 116–117
    response structure

response structure (continued)
national-level 106–108
regional-level 108–117
Terrorism Incident Annex of 165–166
Web site 324
Fuerzas Armadas Revolutionaries de Colombia (FARC) 8
functional annexes 91–93

G

ganglion 146
garden spray devices 171
gastric distress 150
gastric glands 148, 149
gastrointestinal tract 146
General Services Administration (GSA) 110
genetically engineered pathogens 197
gentamicin 265, 268
Georgia 8
germs 173
glands 146, 148, 149
globins 155
glucose 152, 154
grants 113
GSA (General Services Administration) 110
Guam 26
guerrilla groups 8
guerrilla warfare 7
gut 148, 149

H

handicapped population 91
Hawaii 26
hazard analysis 37–38
capability assessment 40
cascade effect model 38–40
planning process 38
hazard identification 51
hazard mitigation assistance 114
hazard planning 51
hazard vulnerability analysis 56–57
hazardous materials 76, 90
training on 299–300
hazardous materials events 27, 70
hazardous materials specialists 132
hazards 27–28
analysis of 37–38
cascade effect 38–40
preventing exposure to 140
rating and scoring system 37
health laboratories 215
Healthcare Association of Hawaii 325
heart 148, 149
heavy equipment and rigging specialists 132
HEICS (Hospital Emergency Incident Command System)
Job Action Sheets 82–83
positions and missions 83–88
heme group 155
hemoglobin 155
hemolysis 156
hemoptysis 266
HEPA (high efficiency particulate arresting) filter masks 141
heparin lock flush kits 285
high winds 66
highly contagious agents 220
high-rise residential units 222
high-risk groups 15
Hindu Tamils 9
HIV (Human Immunodeficiency Virus) 197
home health nursing services 67
homeless population 223
hormones 154, 198
hospital cafeterias 226
Hospital Emergency Incident Command System (HSICS) 327
Hospital Liaison 306
Hospital Operations Sector 73, 78, 80
hospital specialists 296
hospitals 67, 235
admissions 208
decontamination system 287
disaster plans 143

listings of **271**
See also medical facilities
Hotel Association **248**
hotels **226**
household bleach **159**
HSICS (Hospital Emergency Incident Command System) **327**
humidity **199**
Hurricane Andrew **25, 125–127**
Hurricane Emily **128**
Hurricane Hugo **124–125**
Hurricane Iniki **26, 127**
hurricane season **51**
hurricanes **25, 27, 90, 124–128**
hydrogen sulfide **151, 154**
hydyrogen cyanide **162**
hyperbaric chamber **156**
hypochlorite solution **140**
hypochlorites **210**
hypothermia **208**
hypoxia **159, 208**

I

IAP (Incident Action Plan) **71**
ICISF (International Critical Incident Stress Foundation) **327**
ICS (Incident Command System) **70–71**
history of **44–46**
ideological groups **3**
IEMS (Integrated Emergency Management System) **32**
concept **33–34**
overview **32**
process **34–36**
ILS (influenza-like syndrome) **204**
Immediate Treatment Area Manager **86**
IMS (incident management systems) **70**
sections **70–71**
IMU (Islamic Movement of Uzbekistan) **11**
incapacitating agents **162**
incendiary devices **141**
Incident Action Plan (IAP) **71**
Incident Command Post (ICP) **76**
incident command system. See ICS
Incident Commander **47**
incident management systems. See IMS
incineration **210**
India **7, 9**
individual assistance **113**
Industrial Workers of the World (IWW) **16**
infected animals **209**
infection control management
discharge management **255**
isolation precautions **253–254**
patient placement **254**
patient transport **254**
post-mortem care **256**
sterilization of equipment and facilities **254**
infllammation **157**
influenza-like syndrome (ILS) **204**
Information Specialists **308**
infrastructure protection **324–325**
Infrastructure Protection Task Force (IP-TF) **324**
injuries **24**
insurgencies **7–8**
Integrated Emergency Management System. See IEMS
intelligence gathering **169, 193**
International Association of Emergency Managers **327**
International Critical Incident Stress Foundation (ICISF) **327**
international terrorism **6–7**
intervention **141**
intravenous catheters **285**
intravenous hydration **227**
intravenous sets **285**
involuntary nervous system **145–146**
ion transport chain **153–154**
ionizing radiation **155**
IPTF (Infrastructure Protection Task Force) **324**
Iran **9**
Islamic Caliphate **11**
Islamic fundamentalism **10–11**
Islamic Movement of Uzbekisan (IMU) **11**

isolation precautions 253–254
Israel 14, 287
IWW (Industrial Workers of the World) 16

J

Jammu and Kashmir Liberation Front (JKLF) 9
Japanese Red Army 10
JIC (Joint Information Center) 179
jihad 9
JKLF (Jammu and Kashmir Liberation Front) 9
Job Action Sheets 82–83
JOC (Joint Operations Center) 167
Joint Information Center (JIC) 179
Joint Operations Center (JOC) 167
jurisdictional boundaries 70

K

Kaczynski, Theodore 10
Kazakhstan 11
Kentucky 95
kidney 148, 149
Kosovo 8
Krebs cycle 152–153
Ku Klux Klan 9, 16
Kyrgyzstan 11

L

Labor Pool Officer 85
Laboratory Services Manager 87
lacrimal glands 148
lacrimation 150
lactic acidosis 152
LAFD (Los Angeles City Fire Department) 95
landslides 27
laryngoscopes 284
Law Enforcement Liaison 305
Law Enforcement Sector 73
Lebanon 9
left wing groups 8, 164
lesions 208
lewisite 162
Liaison Officer 83
liver 148, 149
loans 113
local EOP (emergency operations plan) 90
 functional annexes 91–93
 good vs. bad plans 58
 local hazards analysis 90–91
 vulnerability analysis 91
local government 77
local hazards analysis 90–91
local health departments 67, 176, 178
logisticians 308
Logistics Group Supervisor 308
Logistics Section Chief 83, 308
Logistics Sector 73
logistics specialists 132
lorazepam 283
Los Angeles City Fire Department 95
Los Angeles riots 25
Louisiana 25
low profile threats 28
low-risk groups 15
lungs 148, 149, 158
 tissue damage 158
lye 210
lymph nodes 212

M

Mace 162
Macedonia 8
malls 171
man-made threats 26
Marburg virus 197
Marx, Karl 4
Marxist left groups 4
mass burials 240
mass care 216, 229
mass care facilities 226
mass casualty incident emergency response

226, 237
mass casualty medical record **224**
mass fatalities **200–201**
mass immunization plan **225**
mass patient decontamination **159–161**
mass prevention/prophylaxis plan **216**
    augmentation activities in **224**
    distribution of medications **221–222**
    stage I (fewer than 100 patients) **217**
    stage II (101 to 1,000 patients) **217–219**
    stage III (101 to 1,000 patients) **219–221**
    treatment facilities and information **223–224**
mass prophylaxis centers **221**
Materials Supply Officer **84**
medical care **140**
Medical Care Officer **86**
Medical Director **306**
medical emergency personnel
    central processing of **230**
    duties of **229**
    prophylaxis **230**
    protective equipment for **230**
    regional call-outs for **228–229**
Medical Examiner Expansion Program (MEEP) **178**
medical examiners **204, 208**
Medical Examiners Office **200–201**
    duties and responsibilities of **246**
medical facilities **66**
    CCPs (casualty collection points) **79**
    clean-team transfer **142–143**
    interoperabiliity of **66–67**
    MMRS (Metropolitan Medical Response System) **79–80**
    non-traditional **142**
    receiving **142**
    responder casualties **67**
    roles and responsibilities of **67**
    security of **66**
    self-sufficiency of **66–67**
    traditional **142**
Medical Information/Research Sector **73**
Medical Operations Physician **73**
medical personnel **77**
medical prophylaxis **221–222**
    alternative methods **222–223**
    personnel for **223**
    treatment facilities and information **223–224**
medical response expansion program **225–227**
Medical Specialists **307**
Medical Staff Officer **85**
medical supplies **98**
    scarcity of **194**
    trransportation of **233**
medication distribution centers **222**
medulla oblongata **151**
MEEP (Medical Examiner Expansion Program) **178, 238–240**
Memorial Institute for the Prevention of Terrorism **324**
meningitis **208**
mental activities **148, 149**
mercenaries **10**
mesh stretchers **161**
messianic terrorism **8**
metabolic pathways **150–151**
methemoglobin formers **156**
methemoglobinemia **156, 157**
methylene blue **156**
methylprednisolone **285**
Metropolitan Medical Response System. See MMRS
Middle East **9**
Midwest Floods **127**
Miles Laboratories **209**
military tents **226**
militia groups **9**
minimal-risk groups **15**
Minor Treatment Area Manager **87**
minors **222**
miosis **150**
mission **227**
Mississippi River **127**
Missouri **95**
mitochondria **144, 152**

MMRS (Metropolitan Medical Response System) 73
- activation of 191
- components of 73
- elements of
  - command and control 168
  - intellligence gathering 169
- focus of 188
- integration with federal government 73–76
- integration with state and local government 77
- medical facilities 79–80
- medical management 77–79
- organizational chart 80
- Program Director 305
- response capabillity to bioterrorism 164
- role in bioterrorism response plans 167–168
- training
  - courses and course materials 302
  - documentation 302
  - emergency response 301
  - guidelines 298
  - hazadous materials 299–300
  - hospital provider training 303
  - NBC agents 301
  - non-team member 304
  - personnel 304
  - philosophy 298
  - proficiency training 303
  - programs 299
  - requirements 297, 305
  - schedules 309
  - sources for 309

mobile distribution vehicles 222
Mojahedin groups 9
monkeypox 269
morgue 227
Morgue Manager 87
morphine 159
Moscow, Russia 171
mosquito control trucks 171
MREP (medical response expansion program) 225–227
mucosal lesions 208
mudslides 90
multiple vectors 192
muscarine 150
muscarinic receptors 150
Muslim Kashmiris 9
myalgias 265
Myanmar 7

N

narcissistic terrorists 10
narco-terrorism 10
nasal glands 148, 149
National Disaster Medical System. See NDMS
National Earthquake Information Service (NEIS) 114
National Emergency Coordination Center (NECC) 115
National Fire Academy 95
National Fire Protection Association (NFPA) 45
National Governors Association 30
National Guard 248
National Hurricane Center 115
National Interagency Incident Management System (NIIMS) 129
National Medical Response Team for Weapons of Mass Destruction. See NMRT-WMD
National Threat Assessment Group 174
National Warning Center 115
nationalist groups 7–8
natural threats 26
NDMS (National Disaster Medical System) 75, 118
- history 121
  - integration into FRP (Federal Response Plan) 123
  - source and establishment 122–123
- missions of 118
- objectives of 118
- organizational resources 118–120

recent experiences **124–128**
NDMS member hospitals **67**
NECC (National Emergency Coordination Center) **115**
neighborhood canvassing **222**
NEIS (National Earthquake Information Service) **114**
neo-fascists **3**
nerve agents **146, 162**
nerves **144–145**
nervous system **144**
autonomic **145–146**
central **144**
involuntary **145–146**
parasympathetic **147–149**
peripheral **144**
sympathetic **146–147**
neurons **144–145**
neurotoxins **146**
neurotransmitters **145**
news media **115**
effects on emergency management **28**
news reporters **209**
NFPA (National Fire Protection Association) **45**
nicotine **150**
nicotinic receptors **150**
NIIMS (National Interagency Incident Management) **129**
nitrates **156, 159**
nitrites **156**
nitrobenzene **156**
nitrogen mustard **162**
nitroglycerine **159**
NMRT-WMD (National Medical Response Team for Weapons of Mass Destruction)
command and control **169**
response activities of **165**
nodern terrorism **18**
non-ambulatory patients **161**
non-cardiogenic pulmonary edema **159**
non-English speaking population **91**
non-invasive respiratory care **227**
non-traditional medical facilities **142**
non-water-soluble chemicals **158**
norephinephrine **146–147**
North Africa **9**
Northern Afghanistan **12**
Northridge earthquake (1994) **128**
NPS (National Pharmaceutical Stockpile) **227–228**
access to **231**
equipment **234**
local- and county-wide security **233**
obtaining and distributing medicines **231–233**
staffing **228–231**
staging Areas **234–235**
transportation **233**
nuclear attacks **27**
nuclear radiation **159**
Nuclear Regulatory Commission Operations Center **115**
nuclear terrorism **171**
nucleus **144**
nursing homes **67**
Nursing Service Officer **85**
nutrients **154**
Nutritional Supply Officer **84**

O

OCFD (Oklahoma City Fire Department) **24**
OCME (Office of the Chief Medical Examiner(**178**
office buildings **171**
Office of Territorial Affairs **115**
Office of the Chief Medical Examiner (OCME) **178**
ofloxacin **265, 281**
Oklahoma City bombing **2, 9, 17**
Oklahoma City Fire Department (OCFD) **24**
on-site treatment **79**
Operation Section Chief **86**
Operations Section Chief **306**
opium **11**

OPTCs (out patient training centers) 177
OPTCs (out patient treatment centers)
  in medical response expansion programs 225–227
  staffing of 206
Oregon 95
organelles 144
organophosphates 150
organs
  blood as 154
  parasympathetic effects on 149
  sympathetic effects on 148
osteomyelitis 265
out patient treatment centers (OPTCs), 177
outbreak 195–197
  laboratory detection of 215
  notification of 182
over-the-counter drugs 204
oxidative phosphorylation 141, 153–154
oxygen
  in the blood 155
  transport of 154–157
oxygen therapy 159, 227
oxyhemoglobin 155

P

pain management 227
pandemic 187
panic victims 194
paralysis 208
parasites 197
parasympathetic nervous system 147–149
paresis 208
parks 114
passive surveillance 204
pathogens 197
patient care 138–139
  mainstays of 140–143
  protective measures in 139
Patient Information Manager 85
Patient Tracking Coordinator 85
patients 139
  ambulatory 161
  decontamination of 140–141, 256
  discharge management 255
  infection control practices for 253–256
  isolation of 253–254
  medical receiving facilities for 142
  placement of 254
  post-mortem care 256
  priority classification of 78
  supportive medical care for 140
  tracking of 80
  transportation of 233, 254
PEEP (positive and expiratory pressure) 159
penicillin 278
pepper spray 162
peptides 198
peripheral nervous system 144
personal protective clothing (PPC) 200
personal protective equipment (PPE) 187, 230
personal protective measures 141
Peru 8
pH balance 154
phagocytosis 157
Pharmaceutical Push Package 283–284
pharmaceuticals
  local stockpiles of 224
  national stockpile 186
  redistribution of 80
  scarce supplies of 194
pharmacists 204
Pharmacy Services Manager 87
phenol 210
Philippines 9
Phoenix Fire Department 45
phogene oxime 162
phosgene 158, 162
phosphorylation 141, 153–154
phosphorylation uncouplers 154
PIRA (Provisional Irish Republican Army) 7
plague 187
  animal exposure 212
  decontamination of affected areas 213, 267

description of agent **266**
diagnosis of **266**
drug therapy **279**
exposure and sampling **212**
incubation period **286**
mass prophylaxis for **218**
online resources **326**
outbreak control **267**
prophylaxis **267, 279**
public information on **213**
screening and confirmation times **216**
signs and symptoms of **266, 286**
treatment of **266, 279, 286**
vector control **212**
plague meningitis **266**
plane crashes **90**
Planning Section Chief **84**
Plans Group Supervisor **308**
Plans Section Chief **307**
platelets **155**
pneumonic plague **187, 266**
PODs (points of distribution) **218–219, 231–233**
Poison Control Center **79**
poisonings **76**
political violence **10**
politicos **8**
port services **166**
positive end expiratory pressure (PEEP) **159**
postal workers **209**
post-exposure immunization **256**
post-exposure management **256–257**
postganglionic axions **146**
post-mortem care **256**
potassium ions **145**
potential hazards **27–28**
power failures **27, 90**
PPC (personal protective clothing) **200**
PPE (personal protective equipment) **187**
pralidoxime **285**
precipitation **199**
preganglionic axions **146**
pre-hospital specialists **296**
prescription drugs **204**
Presidential Decision Directive 39 **70**
preventive measures **169**
private hospitals **67**
Procurement Officer **85**
property loss **24**
prophylaxis **256**
protective measures **42, 139**
initiating **140**
personal **141**
Provisional Irish Republican Army (PIRA) **7**
PSAs (patient staging areas) **142**
Psychological Support Coordinator **88**
ptosis **208**
public assistance **113**
public buildings **114**
public health
catastrophic event in **219–221**
developing crisis **217**
disaster in **217–219**
public health clinics **273**
public hospitals **67**
public information **41**
Public Information Officer **83**
public notifications **179**
public transportation vehicles **79**
public utilities **114**
Puerto Rican Armed Resistance **17**
pulmonary agents **162**
pulmonary edema **158–159**
pulse-oximetery **157**
pumping **145**
pupils **148, 149**
Puritans **16**
pyruvic acid **152**

Q

Q fever **197**
Quakers **16**
quarantine **220**
quicklime **210**

R

racial supremacy groups **4**
radiation **155**
radiological incidents **27**
Radiology Services Manager **87**
rail transportation incidents **27**
RAM (random anti-terrorist measures) **169**
random anti-terrorist measures (RAM) **169**
Raven® stretcher **161**
RBCs (red blood cells) **154–156**
reactionary far right groups **4**
receptors **150**
record keeping **220**
recreational facilities **114**
red blood cells (RBCs) **154–156**
refrigerated trucks **233**
Regional Operations Center) **108**
religious groups **4, 8–10**
religious radicalism **9**
repolarization **145**
reporting **43**
resource database **52–53**
resource management **40**
respiratory irritants **141**
respiratory system **157**
responder casualties **67**
Revolutionary War **16**
ricin **198**
rickettsia **197**
rifampin **265, 281**
right wing groups **3, 164**
riot control agent **162**
riots **25**
risk analysis. See hazard analysis
River Forecast Center **115**
roads **113, 194**
ROC (Regional Operations Center) **108**
Rocky Mountain spotted fever **197**
Rote Armee Fraktion **8**
run-off **199**
Russia
    neo-imperial ambitions of **10**
    relationship with Uzbekistan **11–14**
    role in Central Asia **11**
    strategic partnerships with other countries **11**
Rwanda **8**

S

sacroiliitis **265**
Safety and Security Officer **83**
Safety Officer **73, 306**
salivary glands **148, 149**
salivation **150**
San Andreas Fault **25**
Sanitation Systems Manager **84**
SARA (Superfund Amendments and Reauthorization Act) **46**
sarin **143, 162**
saxitoxin **198**
SBCCOM (Soldier and Biological Chemical Command) **327**
scarification **269**
scarring **157**
SCBA (self-contained breathing apparatus) **200**
school emergency plan **311**
    damage assessment team **319**
    emergency operations center **312**
    fire/utility team **315**
    first aid team **314**
    search and rescue team **316**
    student/parent reunion team **318**
    student/staff supervision team **313**
    support/security team **317**
schools **166, 222, 226**
SCO (State Coordinating Officer) **104**
sea ports **233**
secondary attacks **193**
secondary contamination **200**
seizures **208**
self-contained breathing apparatus (SCBA) **200**
self-injectors **140**
self-referrals **80**
Sendero Luminoso **8**

separatist groups **4**
sepsis **208**
September 11 attack **171**
Severe Storms Forecast Center **115**
Shanghai Cooperation Organization **13**
Shays's Rebellion (1786) **16**
shelters **42**
sheriff's department **91**
    duties and responsibilities of **247**
Sikh Punjabis **9**
single-issue groups **10**
site decontamination **201**
skin lesions **208**
smallpox **187, 197, 213**
    decontamination **269**
    description of agent **269**
    diagnosis of **269**
    incubation period **286**
    online resources **326**
    outbreak control **270**
    prophylaxis **269, 281**
    screening and confirmation times **216**
    signs and symptoms of **269, 286**
    treatment of **269, 281, 286**
smooth muscles **146**
Social Services Officer **88**
social threats **25**
sodium hypochlorite **210**
sodium ions **145**
Soldier and Biological Chemical Command (SBCCOM) **327**
Somalia **9**
soman **162**
Southern California **44**
Soviet Union **9**
Soviet Union, dissolution of **28**
Spanish Influenza pandemic (1918) **187**
special agent in charge (SAIC) **74**
spinal column **146**
spinal nerves **146**
sports arenas **171**
sprayers **190**
Sri Lanka **7, 9**
St. Luke's Hospital **143**
Staff Support Manager **88**
staging areas **230**
State Coordinating Officer (SCO) **104**
State Emergency Management Agencies **324**
state government **77**
Status/Information Systems Officer **84**
sterile dressings **283**
sterilization **254**
steroid therapy **159**
streets **113**
streptomycin **265, 279**
stretcher basins **161**
stretchers **161**
stridor **266**
structural specialists **132**
stylets **284**
sub-conflict organizations **9**
subsidence **27**
Sudan **9**
sulfur dioxide **159**
sunlight **199**
Superfund Amendments and Reauthorization Act (SARA) **46**
support services **42**
surfactants **158**
surgical equipment **98**
surrogate terrorism **10**
sweat glands **148, 149**
sympathetic nervous system **146–147**
synapses **145**

T

tabun **162**
tachycardia **150**
tachypnea **150, 208**
TADMAT (Toledo Area DMAT) **99**
tag and release method **220**
Tajikistan **9**
Taliban **12**
Tamils **9**
tansportation of victims **79**
targeted population prophylaxis **218**
Task Force Leader **73, 80, 305**

TCA (Krebs) cycle 152–153
Team Medical Section Chief 309
Tech Escort Unit 173, 189
Technical Escort Unit (TEU) 75
technical information specialists 132
technical writing 57
temperature 199
temperature gradient 199
temporary housing 113
tent showers 294
tents 161
terminal cytochrome 153
terrorism 25, 90
- and Central Asia 10–14
- combatting 170
- commonly accepted variables of 2–3
- consumer 10
- definition of 2–3, 7
- early warning capability against 169
- EMS (emergency medical services) response to 18–19
- history of 15
  - eighteenth-century 16
  - nineteenth-century 16
  - twentieth-century 16, 17
- intended audience of 3
- international 6–7
- messianic 8
- narco-terrorism 10
- preparedness programs 169
- preventive measures 169
- recognition of 5–6
- responses to 170
  - command and control 168
  - intelligence gathering 169
- surrogate 10
- *See also* biological terrorism
- *See also* threats

terrorism consequence management plan 59–63
Terrorism Stakeholders Group. See TSG
terrorist events 5–6
- EMS (emergency medical services) response to 18–19

terrorist groups 166
- categories of 7–10
- categories of risks of 14–15
- left-wing 8
- mercenaries 10
- nationalists 7–8
- religiously-motivated 8, 8–10
- single-issue groups 10

terrorists
- beliefs of 5
- desired outcome of 3
- domestic 3
- left-wing 8
- narcissistic 10
- targets of 5
- types of 3–5

tetracycline 268, 280
TEU (Technical Escort Unit) 75
Texas Engineering Extension Service 324
thermonuclear weapons 28
Thraxol® 209
threats
- of biological terrorism 171, 184–186
- and emergency management 28–29
- evaluation of 172
- low-profile 28
- man-made 26
- natural 26
- ranking of 26
- social 25
- types of 26
- *See also* terrorism

three-corridor decontamination systems 161
ticks 197, 214
Time Officer 85
Tokyo sarin attack 143
Toledo Area DMAT (TADMAT) 99
tornadoes 27, 90
toxemia 266
toxic gases 139
toxins 197
traditional medical facilities 142
traffic congestion 194
train derailments 90
training programs 175

Train-the-Trainer (TTT) course **96–97**
transit systems **194**
Transporation Officer **84**
transportation terminals **171**
traumatic injuries **141**
treatment centers **223–224**
treatment priorities **77**
Treatnent Areas Officer **86**
triage **78, 257**
Triage Manager **86**
tropical storms **27**
Tsar Alexander II **16**
TSG (Terrorism Stakeholders Group)
  crisis management **170**
  early warning capability of **169**
tsunami **90**
tsunamis **28**
TTT (Train-the-Trainer) course **96–97**
tularemia **187, 197**
  animal exposure to **213**
  decontamination of affected areas **214, 268**
  description of agent **267**
  diagnosis of **268**
  drug therapy **280**
  exposure and sampling **213**
  forms of **214**
  human exposure to **213**
  incubation period **286**
  mass prophylaxis for **218**
  outbreak control **268**
  prophylaxis **268, 280**
  public information campaign on **214**
  screening and confirmation times **216**
  signs and symptoms of **267–268, 286**
  treatment of **268, 280, 286**
  vector control **214**
Turkestan **12**
Turkey **14**
Turkish Republic **11**
two-corridor decontamination systems **161**
Typhoon Brian **26**
Typhoon Omar **26**
typhoons **26**

U

U.S. Postal Service **233**
  duties and responsibilities of **248**
ulceroglandular tularemia **213**
ultraviolet light **199**
Unabomber **10**
unconscious victims **79**
unexplained animal deaths **204**
unexplained deaths **208**
United Parcel Service **233**
United States Geological Survey (USGS) **115**
United States Public Health Service (USPHS) **73**
United States Secret Service (USSS) **76**
UPSHS (United States Public Health Service) **73**
urban conflagrations **90**
urban fires **27**
Urban Search & Rescue (US&R) Response System **129**
urea **154**
urination **150**
US&R (Urban Search & Rescue) Response System **129**
US&R (Urban Search & Rescue) Task Force **129**
  Medical Team **131**
  Rescue Team **130**
  Search Team **130**
  Technical Team **131–132**
USGS (United States Geological Survey) **115**
USSS (United States Secret Service) **76**
Uzbekistan
  Jewish communities in **12**
  leadership role in Central Asia **11**
  relationship with Russia **11–14**
  relationship with the United States **13**
  threats to national sovereignty and independence of **10**

V

VA (Veteran's Administration) hospitals 67, 80
vaccinations 209
vaccines 192
Vaccinia Immune Globulin (VIG) 269
vapor pressure 158
Variola virus 187
vascular system 157
vectors 178, 192
ventilation system 171
vertebral osteomyelitis 265
very-high risk groups 14
vesicants 141, 162
veterinarians 204
veterinary diagnosis 215
victims 139
  panic 194
  self-transport of 194
  transportation of 79
  triage of 78
VIG (Vaccinia Immune Globulin) 269
viruses 197
vital signs 80
VMI (vendor management inventory) packages 231
volcanoes 28
vulnerability analysis 91
VX nerve agent 162

W

walk-in emergent medical practices 67
warning systems 41
Washington 95
water control facilities 113
water facilities 91
water system 171, 190
wave depolarization 145
weaponized contagious diseases 141
weapons of mass destruction (WMD) 164
weather 199
Web sites
  American Red Cross 324
  biological agents 326–327
  biological terrorism 326–327
  Center for Civilian Bio-Defense Studies 327
  Center for Research on the Epidemiology of Disasters 327
  Centers for Disease Control 129, 323
  critical infrastructure protection 324–325
  Defense Threat Reduction Agency 327
  Department of Defense 324
  DHHS (Department of Health and Human Services) 327
  diseases
    anthrax 325
    plague 326
    smallpox 326
  FEMA (Federal Emergency Management Agency) 59, 324
  FRP (Federal Response Plan) 324
  HSICS (Hospital Emergency Incident Command System) 327
  International Association of Emergency Managers 327
  Memorial Institute for the Prevention of Terrorism 324
  National Guard 327
  Public Health Services 129
  SBCCOM (Soldier and Biological Chemical Command) 327
  State Emergency Management Agencies 324
  state emergency operation plans 327–328
  Texas Engineering Exension Service 324
Whiskey Rebellion (1791) 16
white blood cells 155, 157
white supremacists 9
Whittier Narrows earthquake (1987) 95
wildland fires 25, 27, 44, 90
wind 199
winter storms 27
wire stokes baskets 161

WMD (weapons of mass destruction) **1, 164**
World Trade Center **171**
World Trade Center bombing **17**
World Trade Organization (WTO) **208**
WTO (World Trade Organization) **208**

X

x-ray equipment **98**

Y

yeast **197**
yellow fever **197**
Yemen **9**
Yersinia pestis **187, 266**
Yugoslavia **8**